Fractal-Based Methods in Analysis

T0138095

Herb Kunze • Davide La Torre • Franklin Mendivil
Edward R. Vrscay

Fractal-Based Methods
in Analysis

Springer

Herb Kunze
Department of Mathematics and Statistics
University of Guelph
Guelph Ontario
Canada
hkunze@uoguelph.ca

Franklin Mendivil
Department of Mathematics and Statistics
Acadia University
Wolfville Nova Scotia
Canada
franklin.mendivil@acadiau.ca

Davide La Torre
Department of Economics, Business
and Statistics
University of Milan
Milan
Italy
davide.latorre@unimi.it

Edward R. Vrscay
Department of Applied Mathematics
University of Waterloo
Waterloo Ontario
Canada
ervrscay@uwaterloo.ca

ISBN 978-1-4899-7374-0   ISBN 978-1-4614-1891-7 (eBook)
DOI 10.1007/978-1-4614-1891-7
Springer New York Dordrecht Heidelberg London

Mathematics Subject Classification (2010): 28A80, 28A33, 1AXX, 45Q05, 28CXX, 65L09, 65N21

Printed on acid-free paper

Springer is part of Springer Science+Business Media (www.springer.com)

In memory of Bruno Forte
mentor, teacher, collaborator, friend.

Preface

The idea of modeling the behaviour of phenomena at multiple scales has become a useful tool in both pure and applied mathematics. Fractal-based techniques lie at the heart of this area, as fractals are inherently multiscale objects. Fractals have increasingly become a useful tool in real applications; they very often describe such phenomena better than traditional mathematical models.

Fractal-Based Methods in Analysis draws together, for the first time in book form, methods and results from almost 20 years of research on this topic, including new viewpoints and results in many of the chapters. For each topic, the theoretical framework is carefully explained. Numerous examples and applications are presented.

The central themes are *self-similarity* across scales (exact or approximate) and *contractivity*. In applications, this involves introducing an appropriate space for contractive operators and approximating the "target" mathematical object by the fixed point of one of these contractions. Under fairly general conditions, this approximation can be extremely good. This idea emerged from *fractal image compression*, where an image is encoded by the parameters of a contractive transformation (see Sect. 3.1 and Figs. 3.3 and 3.4). The first step in extending this methodology is to construct interesting contractive operators on many different types of spaces. After this theoretical framework has been established, the next step is to apply the methodology in practical problems. In this book, we present extensive examples of both of these steps.

We originally conceived a document that we could give to our students to help them learn the background for their research. This document would contain an introduction to fractals via iterated function systems (IFSs) and some of the important subsequent developments, all from this IFS viewpoint. This document has since taken on a life of its own that has resulted in this book. The original goal is reflected in the second chapter, which is designed to serve as the basis for a course.

In the first chapter, we give a "bird's-eye" overview of the area, painting with broad brushstrokes to give the reader the feel and philosophy of the approach. We touch on many of the topics and applications, hoping to share our amazement at the breadth of interesting mathematics and applications and to entice the reader into learning more.

In Chapter 2, we present a brief course on the classical topics in the iterated function systems viewpoint on fractals. In order to help the reader who might be seeing the material for the first time, we have included many exercises in this chapter. This is in keeping with our own desire to use this particular chapter as the basis for a course on IFS fractals.

In Chapters 3–5 we carefully develop the IFS framework in a large variety of settings. In particular, in Chapter 3 we develop a theory of IFSs on various spaces of functions, including the interaction of IFSs with integral transforms and IFSs on wavelet spaces. IFSs on spaces of transforms have been used in mathematical imaging, with IFSs on Fourier transforms having applications in magnetic resonance imaging (MRI). Chapter 4 extends this to IFSs on multifunctions (set-valued functions) and measure-valued functions. Again, the primary motivation and application for this framework is in mathematical imaging. Chapter 5 proceeds to a careful construction of the framework in various spaces of measures, with many new results. We consider signed measures, vector measures, and multimeasures (set-valued measures). Furthermore, there is a discussion of "generalized" measures as dual objects to Lipschitz spaces. This is a very useful class of "measures" for many purposes.

In Chapter 6, we turn to another classical topic in IFS fractals, that of ergodic theory and the "chaos game." An IFS defines a dynamical system, which in turn generates an invariant measure. Chapter 6 extends the "chaos game" to IFSs on various function spaces, including random algorithms for generating wavelet analysis and wavelet synthesis.

Chapters 7 and 8 present an extensive range of applications of fractal-based methods to inverse problems. The models that we discuss span the range from ordinary differential equations (ODEs), partial differential equations (PDEs), and random differential equations to stochastic differential equations (SDEs) and more. The range of application topics presented is equally broad, from models in physics to biological, population, and economic models. These two chapters are just the start of the possible application areas and serve to illustrate the power of fractal-based methods.

The preface of a book is also the customary place for acknowledgments and expressions of thanks to appropriate people – collaborators, students, and individuals who have, in whatever way, helped the authors with their work or understanding of the subject material. In our case, the list of such people is quite long and the risk of omission quite large, so we shall keep our acknowledgments rather brief. First of all (and in a somewhat chronological order), Ed Vrscay would like to thank Michael F. Barnsley who, while at Georgia Tech, introduced him to the fields of fractal geometry and fractal image compression. He would also like to thank Jacques Lévy-Véhel, Dietmar Saupe and Claude Tricot for invaluable discussions, collaborations and assistance that began in the late-1980s and led to the formation of the "Waterloo Fractal Coding and Analysis Project." It was the attraction of Bruno Forte, former Chair of Applied Mathematics, University of Waterloo, to the "Waterloo Project" that contributed to its significant initial growth, in particular the mathematical formulation of generalized fractal transforms and associated inverse problems. Further growth of the project was made possible with the arrival of Franklin Mendivil. After retiring from Waterloo in 1995, Bruno would return to Italy to assume Emeritus Professorships, first at the University of Bari and then at the University of Verona. Here, he would eventually supervise Davide La Torre's Master's thesis on inverse problem for fractal transforms. Davide La Torre would like to thank Vincenzo Capasso and Bruno Forte for addressing him in these topics and for the suggestions, the inspiration, and the support that they have given him during his academic career. Bruno Forte was the first professor Herb Kunze met in the classroom while and undergraduate at the University of Waterloo. A few years later, Herb worked repeatedly as Bruno's teaching assistant for the same advanced Calculus course. Herb Kunze wishes to acknowledge the influential impact that Bruno had on his early academic life.

For these reasons we are dedicating this book to him.

Much of the research performed over the years by the authors has been made possible by support from the Natural Sciences and Engineering Research Council of Canada (NSERC) in various forms, all of which are very gratefully acknowledged here. These include the NSERC Discovery Grants of Ed Vrscay (1986–present), Franklin Mendivil (2000–present) and Herb Kunze (2001–present). Ed Vrscay also acknowledges the following NSERC support over the years: a Postdoctoral Fellowship (1984–86), a University Fellowship (1986–1996) and a Collaborative Research Grant (1995–1997) which made possible the birth of the "Waterloo Project."

We would also like to thank Achi Dosanjh and Donna Chernyk at Springer for all the help and guidance they have given us throughout this project. Finally, Karen, Cinzia, Emma, and Rosemary deserve a lifetime of thanks for so graciously enduring life with four "fractalators."

Guelph, Ontario   *Herb Kunze*
Milan, Italy   *Davide La Torre*
Wolfville, Nova Scotia   *Franklin Mendivil*
Waterloo, Ontario   *Edward R. Vrscay*

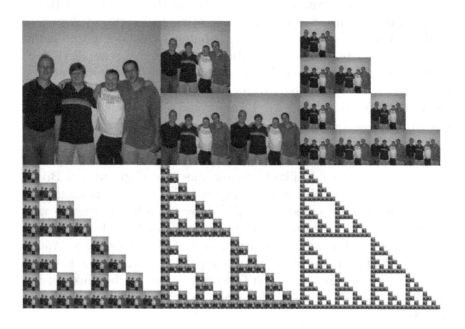

Contents

Chapter 1
What do we mean by "Fractal-Based Analysis"?

We consider "fractal-based analysis" to be the mathematics (and applications) associated with two fundamental ideas, namely, (1) *self-similarity* and (2) *contractivity*:

1. *Self-similarity.* This is the "fractals" part of "fractal-based analysis." Let $u \in X$ denote some mathematical object of interest, e.g., a set, a function, an image or a measure, in an appropriate space X. (For the moment, we may consider X to be a complete metric space.) The first step is to construct a number of "shrunken" and "distorted" copies of u, to be denoted as g_i, $1 \leq i \leq N$. Then combine these *fractal components* of u in some way to make a new object $v \in X$. We'll call the operation T that maps u to v a *fractal transform*. In summary, $T : X \to X$ and $v = Tu$.

2. *Contractivity.* Under suitable conditions on the parameters which characterize the shrinking and distortion mentioned above, the fractal transform T will be contractive in the metric space X. From Banach's Fixed Point Theorem [11] (which will clearly play a central role in this book), T has a unique and attractive fixed point $\bar{u} \in X$, i.e., $T\bar{u} = \bar{u}$.

Given the nature of fractal transforms T, their fixed points $\bar{u}$ will exhibit some kind of self-similarity, making them generally "fractal" in nature. In the particular case of *iterated function systems*(IFS), where fractal transforms T are defined by a set of contractive mappings in $\mathbb{R}^n$, their fixed points $\bar{u}$ can be made to resemble to a remarkable degree natural objects such as leaves or trees (see Fig. 1.1 for an example). Historically, this observation naturally led to the question of whether fixed points of fractal transforms could be used to approximate images in general. *Fractal image coding* is an example of such an inverse problem: Given a "target" image $u \in X$, can we find a contractive fractal transform T with fixed point $\bar{u}$ such that $\bar{u}$ approximates u to a desirable accuracy?

Fig. 1.1: Fern leaf as a fixed point of an IFS.

The above inverse problem may be generalized to "non-fractal" situations, namely, the problem of approximating elements of a metric space X by fixed points of a given class of contractive operators on X. This will be the subject of several chapters of this book.

The above discussion, which was intended to be a very brief overview, is admittedly incomplete and imprecise. It clearly raises a good number of questions. For example:

1. What types of spaces X are amenable to fractal transforms?
2. How do we "shrink" and "distort" copies of an element $u \in X$?
3. How do we combine its fractal components g_i to produce the element $v = Tu$?
4. What are the conditions for contractivity of T on X?
5. And finally, what are some applications of such fractal-based methods?

These are some of the questions that will be investigated and answered in this book.

For the remainder of this chapter, we provide a slightly deeper version of the above discussion, if only to whet the general reader's appetite and provide a slightly broader picture of fractal-based analysis. This is done with the understanding that a more complete and mathematically rigorous discussion will appear in later sections of this book.

1.1 Fractal transforms and self-similarity

Let $(\mathbb{Y}, d_\mathbb{Y})$ denote a complete metric space of mappings of a *base space* $\mathbb{X}$ (typically a subset of $\mathbb{R}^n$), to a *range space* $\mathcal{R}$ (typically a subset of $\mathbb{R}$). Consider an element $u \in \mathbb{Y}$. Then do the following:

1. Make N spatially contracted copies of u, i.e.,

$$f_i(x) = u(w_i^{-1}(x)), \quad x \in w_i(\mathbb{X}), \quad 1 \leq i \leq N,$$

where the $w_i : \mathbb{X} \to \mathbb{X}$ are contraction maps on $\mathbb{X}$. (It may be necessary to define $f_i(x) = 0$, or perhaps $f_i(x) = \emptyset$, the empty set, for $x \notin w_i(\mathbb{X})$.)

2. Modify the range values of these copies as follows,

$$g_i(x) = \phi_i(f_i(x)), \quad 1 \leq i \leq N,$$

where the "range maps" $\phi_i : \mathcal{R} \to \mathcal{R}$ are sufficiently regular, e.g., Lipschitz. We refer to the g_i, $1 \leq i \leq N$, as the *fractal components* of u.

3. Finally, combine the fractal components g_i in an appropriate manner (which may depend on the space $\mathcal{R}$), i.e.,

$$v(x) = \mathcal{O}(g_1(x), g_2(x), \cdots, g_N(x)), \tag{1.1}$$

where the operator $\mathcal{O} : \mathbb{Y}^N \to \mathbb{Y}$ may be required to satisfy some properties, the details of which will be omitted here. In many situations $\mathcal{O}$ is simply a pointwise sum.

At this point, we may wish to include an additional, "nonfractal" component, $h \in \mathbb{Y}$ in equation (1.1), i.e.,

$$v(x) = \mathcal{O}(g_1(x), g_2(x), \cdots, g_N(x), h(x)).$$

We shall refer to such a term, h, which may be independent of u, as a *condensation* term, as introduced by Barnsley and coworkers [16] (see also section 2.6.1). In the discussion that follows, we omit this term for notational simplicity, with the understanding that it may be included in our formalism if necessary.

The above procedure may be written compactly as follows,

$$v = Tu,$$

where $T : \mathbb{Y} \to \mathbb{Y}$ is the *fractal transform* operator associated with the set of contraction maps w_i, $1 \leq i \leq N$ and associated ϕ_i maps.

Under some conditions on the ϕ_i and the w_i (specifically their Lipschitz/contraction factors), the fractal transform T is contractive. Once again, from Banach's Fixed Point Theorem, there exists a unique $\bar{u} \in \mathbb{Y}$ which is the fixed point of T, i.e.,

$$T\bar{u} = \bar{u}. \tag{1.2}$$

Moreover, for any element $u_0 \in \mathbb{Y}$, the iteration sequence defined as follows,

$$u_{n+1} = Tu_n, \quad n = 0, 1, 2, \cdots,$$

converges to $\bar{u}$ (in metric $d_\mathbb{Y}$ on $\mathbb{Y}$). In other words, the unique fixed point $\bar{u}$ is globally attractive. This global attraction results from the global contractivity of T and is a very useful feature when present.

From equations (1.1) and (1.2), the fixed point $\bar{u}$ of a fractal transform T satisfies the equation,

$$\bar{u}(x) = \mathcal{O}(\phi_1(\bar{u}(w_1^{-1}(x))), \phi_2(\bar{u}(w_2^{-1}(x))), \cdots, \phi_N(\bar{u}(w_N^{-1}(x)))). \tag{1.3}$$

This functional equation be viewed as a kind of *self-similarity* property, i.e., that $\bar{u}$ may be expressed as a combination/union of spatially-contracted and distorted copies of itself.

Finally, we mention that Eq. (1.3) is a special form of the following functional equation,

$$f(x) = F(x, f(g_1(x)), \cdots, f(g_N(x))),$$

which was studied many years ago by Bajraktarevic [10], following the work of Read [143]. For this reason, various researchers in fractal imaging have referred to fractal transforms as *Read-Bajraktarevic operators*.

At this point, it is instructive to consider a couple of illustrative examples. The reader will probably have noticed that the above discussion was tailored for the treatment of functions. Indeed, much of this book is dedicated to the subject of fractal transforms on function and multifunction spaces. We often distinguish between a *geometric* fractal transform acting on sets from one acting on analytic objects in a vector space (such as functions or measures of various types).

1. **Scaling equation of multiresolution analysis.** $\mathbb{X} = \mathbb{R}$, the real line and $\mathbb{Y} = L^2(\mathbb{R})$. Let $u(x)$ be the scaling function associated with a given multiresolution analysis on $\mathbb{R}$ (or subset

thereof) [44, 123]. Define, as usual, $V_0 = \text{span}\{u(x-n)\}_{n \in \mathbb{Z}}$ and $V_1 = \text{span}\{u(2x-n)\}_{n \in \mathbb{Z}}$. Then $V_0 \subset V_1$, implying that u satisfies the *scaling equation* or *dilation equation*,

$$u(x) = \sum_k c_k \sqrt{2} u(2x - k).$$

In light of Eq. 1.3, u may be viewed as the fixed point of the fractal transform defined by the following contractive spatial maps w_i and associated range maps ϕ_k:

$$w_k(x) = \frac{1}{2}x + \frac{k}{2}, \quad \phi_k(t) = \sqrt{2}c_k t, \quad k \in \mathbb{Z}, \quad \text{such that} \quad c_k \neq 0.$$

Note that in this case the fractal transform is a linear operator, implying that $u(x) = 0$ is also a fixed point. In order to get a nontrivial fixed point function $u(x)$ we must place restrictions on the space of functions.

2. **Self-similar functions.** A standard definition of self-similarity of functions over sets satisfying the *open set condition* (see definition 2.35) is as follows [123, 82]: Let u be continuous and compactly supported and let Ω be the bounded open subset of $\mathbb{R}^n$ such that $\bar{\Omega} = \text{supp}\{u\}$. Further suppose that there exist disjoint open subsets $\Omega_k \subset \Omega$, $k = 1, 2, \cdots, N$, such that $\Omega_k = w_k(\Omega)$, where the w_k are contractive similitudes. Then u is self-similar on Ω if it satisfies the relation

$$u(x) = h(x) + \sum_k \alpha_k u(w_i^{-1}(x)),$$

where

$$h(x) = \begin{cases} h_k(x) & x \in \Omega_k, \\ f(x) & x \notin \bigcup_k \Omega_k. \end{cases}$$

and the functions f and h_k, $1 \leq k \leq k$, are assumed to be Lipschitz. Once again, the self-similar function $u(x)$ may be viewed as the fixed point of a fractal transform defined by the contraction maps $w_k(x)$, their associated range maps $\phi_k(t) = \alpha_k t$ and the condensation function $h(x)$. Alternatively, we could define the range map on each Ω_k as follows,

$$\phi_k(t) = \alpha_k t + h_k(x),$$

and redefine $h(x)$ appropriately. In many practical applications, $h_k(x)$ is constant, i.e., $\phi_k(t) = \alpha_k t + \beta_k$.

Example: Let $\Omega = (0,1)$, and

$$w_1(x) = \frac{1}{3}x, \quad w_2(x) = \frac{1}{3}x + \frac{1}{3}, \quad w_3(x) = \frac{1}{3}x + \frac{2}{3},$$

with associated ϕ_k maps,

$$\phi_1(t) = \frac{1}{2}t, \quad \phi_2(t) = \frac{1}{2}, \quad \phi_3(t) = \frac{1}{2}t + \frac{1}{2}.$$

Then $u(x)$ is the famous "Devil's staircase function," sketched below in Fig. 1.2. Clearly, $u(x)$ may be viewed as a union of three contracted copies of itself, with the middle copy being a "flattened" copy. The fixed point equation 1.3 satisfied by $u(x)$ becomes

$$u(x) = \begin{cases} \frac{1}{2}u(3x) & 0 \le x < \frac{1}{3}, \\ \frac{1}{2} & \frac{1}{3} < x < \frac{2}{3}, \\ \frac{1}{2}u(3x-2) + \frac{1}{2} & \frac{2}{3} < x \le 1. \end{cases}$$

In this and the previous example, the collection of w_k and associated ϕ_k maps is said to comprise a *iterated function system with greyscale maps* (IFSM), to be discussed in more detail in Chap. 2.

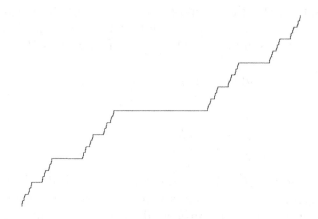

Fig. 1.2: "Devil's staircase function" on $[0,1]$

3. **Local self-similarity and "block fractal coding".** We now modify the previous definition to allow *subsets* of Ω to be mapped to *smaller subsets*. This is the essence of "block fractal coding" or "partitioned IFS" (see also the discussion of *local* or *partitioned IFSM* in chapter 3). For reasons that will become clear below (see

sections 1.4.1 and 3.1), we refer to the subsets Ω_k as *range* or *child blocks* and denote them as R_k. Now suppose that to each range block R_k there corresponds a set $D_{j(k)} \subset \Omega$ such that $\Omega_k = w_k(D_{j(k)})$, where w_k is a contractive similitude. $D_{j(k)}$ is the *domain* or *parent block* associated with R_k. Then u is affinely self-similar on Ω with respect to the range-domain associations $(R_k, D_{j(k)})$ if

$$u(x) = \alpha_k u(w_i^{-1}(x)) + \beta_k, \qquad x \in R_k. \qquad (1.4)$$

This relation, which forms the basis of most fractal signal/image coding methods, has its origins in the block fractal coding method of A. Jacquin [80]. For convenience, each range block R_k is often chosen to be an $n \times n$ block of pixels with associated $2n \times 2n$-pixel domain block $D_{j(k)}$. This subject will be discussed in greater detail in Chap. 2.

4. **"Multiparent" fractal coding:** We modify the above fractal coding scheme by allowing each range block R_j to be associated with *several* parent blocks. For simplicity, assume that a digital image is partitioned by $n \times n$-pixel range blocks R_k, $1 \leq k \leq N_R$. Furthermore assume the existence of a common *domain pool*, that is, a set of N_D $2n \times 2n$-pixel domain blocks selected from throughout the image. Let w_{jk} denote the contractive similitude that maps D_j to R_k. Given a function u, we define $v = Tu$, where T is a *multiparent fractal transform*, as follows: Over each range block, R_k, $1 \leq k \leq N_R$,

$$v(x) = (Tu)(x) = \sum_{j=1}^{N_D} c_{jk} \phi_{jk}(u(w_{jk}^{-1}(x))), \qquad x \in R_k, \qquad (1.5)$$

where $\phi_{jk}(t) = \alpha_{jk} t + \beta_{jk}$ and

$$\sum_{k=1}^{N_D} c_{jk} = 1, \qquad j = 1, 2, \cdots, N_R.$$

Multiparent fractal transforms may be used to perform image denoising. Consider the case of an image u that has been corrupted with additive Gaussian noise, i.e., $\tilde{u} = u + \mathcal{N}(0, \sigma)$. A well-known technique of noise variance reduction is to take repeated measurements/images and average over them. If we have several (noisy) domain blocks that provide good approximations to a given (noisy) range block $\tilde{u}(R_j)$, then averaging over them could produce a reasonable approximation to $u(R_j)$ the denoised image. This idea was exploited in [1].

In fact, multiparent fractal transforms may be viewed as a cross-scale, block-based version of the method of *nonlocal means denoising* [31] which, because of its simplicity and effectiveness, has received a great deal of interest in the context of *nonlocal image processing* [116]. Nonlocal means denoising and fractal image coding may be viewed as particular examples of a more general model of nonlocal affine self-similarity of images [2, 112].

1.2 Self-similarity: A brief historical review

The idea of self-similarity has been present in various forms in the mathematical and scientific literature for many years. Indeed, it provided the motivation for various set-valued mappings that perform the shrink-distort-combine procedure given above, including the methods of "iterated function systems" and fractal image coding. In what follows, we outline very briefly the historical development of the idea of self-similarity, once again with the disclaimer that no attempt at completeness has been made.

1.2.1 The construction of self-similar sets

Benoit Mandelbrot's classic work, *The Fractal Geometry of Nature* [124], presented the first description, along with an extensive catalog, of *self-similar sets*: sets that may be expressed as unions of contracted copies of themselves. Moreover, he called these sets "fractals," because their (fractional) Hausdorff dimensions exceeded their (integer-valued) topological dimension. The ternary Cantor set (Fig. 2.1) and the von Koch "snowflake curve" (Fig. 1.3) are perhaps the most celebrated examples of such fractal, self-similar sets. As many will recall, Mandelbrot introduced these ideas with the classical problem of finding the length of an irregular coastline.

Mandelbrot viewed fractal sets as limits of a recursive procedure that involved "generators." A generator G contains a set of rules – a kind of "grammar" – for operations on sets: Some generators act on line segments, others act on sets with area or volume. Starting with

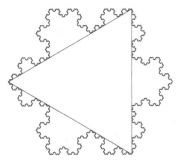

Fig. 1.3: The von Koch snowflake along with its defining equilateral triangle.

an appropriate "seed" set S_0 (in $\mathbb{R}^2$, for example), we then form the iteration sequence

$$S_{n+1} = G(S_n), \quad n = 0, 1, 2, \cdots.$$

In the limit $n \to \infty$, the sets S_n approach a limit set $\bar{S}$ which is typically a fractal. A well-known example of such a generator, relevant to the problem of fractal coastlines, is given in Fig. 1.4. This replacement rule will generate the von Koch curve (which forms one side of the boundary of the von Koch snowflake).

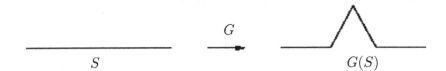

$$G$$

$$S \qquad\qquad\qquad\qquad\qquad G(S)$$

Fig. 1.4: Generator

Applying the above generator to a set (composed of line segments, e.g., a polygon in the plane) means replacing each line segment S in the set, with length l, by four segments of length $l/3$, as shown on the right. (We omit the technicalities regarding the orientation of the middle "tooth".) If we start with the segment $S_0 = S$ on the left above, the limiting set $\bar{S}$ is the celebrated von Koch curve presented in Fig. 1.5. This figure illustrates the first few iterations of this replacement rule.

Going back to the von Koch generator G, each one of the four line segments on the right can be viewed as a contracted (by one-third)

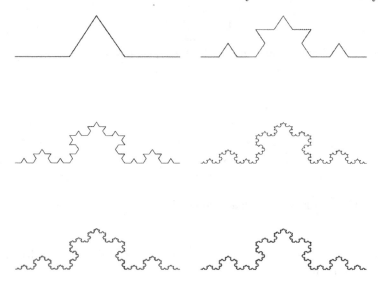

Fig. 1.5: First several stages in the construction of the von Koch curve

and translated copy of the original line segment. Consequently, each set S_{n+1} in the iteration sequence is a union of four contracted and translated copies of S_n. And in the limit, the von Koch curve S may be considered as a union of four contracted copies of itself, i.e.,

$$\bar{S} = \bigcup_{k=1}^{4} \bar{S}^{(k)} \qquad (1.6)$$

However, it was the landmark work of Hutchinson [75], followed shortly thereafter by that of Barnsley and Demko [17], that paved the way to being able to work with such fractal sets (and measures) analytically. Mandelbrot's generators were replaced by systems of geometric contraction maps in $\mathbb{R}^n$: Instead of focusing on the actual operation of replacing line segments or sets with contracted copies, one simply considered the geometric maps – we'll call them $w_i : \mathbb{R}^n \to \mathbb{R}^n$ – that performed these transformations.

For example, in the case of the von Koch curve, equation (1.6) would be rewritten as

$$\bar{S} = \mathbf{w}(\bar{S}) = \bigcup_{k=1}^{4} w_k(\bar{S}),$$

where the maps $w_i : \mathbb{R}^2 \to \mathbb{R}^2$ are given by

$$w_1 : \begin{pmatrix} \frac{1}{3} & 0 \\ 0 & \frac{1}{3} \end{pmatrix} \begin{pmatrix} x \\ y \end{pmatrix} \qquad w_2 : \begin{pmatrix} \frac{1}{6} & -\frac{\sqrt{3}}{6} \\ \frac{\sqrt{3}}{6} & \frac{1}{6} \end{pmatrix} \begin{pmatrix} x \\ y \end{pmatrix} + \begin{pmatrix} \frac{1}{3} \\ 0 \end{pmatrix}$$

$$w_3 : \begin{pmatrix} \frac{1}{6} & \frac{\sqrt{3}}{6} \\ -\frac{\sqrt{3}}{6} & \frac{1}{6} \end{pmatrix} \begin{pmatrix} x \\ y \end{pmatrix} + \begin{pmatrix} \frac{1}{3} \\ 0 \end{pmatrix} \qquad w_4 : \begin{pmatrix} \frac{1}{3} & 0 \\ 0 & \frac{1}{3} \end{pmatrix} \begin{pmatrix} x \\ y \end{pmatrix} + \begin{pmatrix} \frac{2}{3} \\ 0 \end{pmatrix}$$

As such, $\bar{S}$ is the fixed point of a *set-valued mapping* $\mathbf{w} : \mathbb{H}(\mathbb{X}) \to \mathbb{H}(\mathbb{X})$, where $\mathbb{H}(\mathbb{X})$ denotes the set of all nonempty compact subsets of $\mathbb{X} \subset \mathbb{R}^n$, $\mathbb{X}$ compact. Moreover, the mapping $\mathbf{w}$, which represents a kind of parallel action of the contractive maps w_k, is contractive in the complete metric space $(\mathbb{H}(\mathbb{X}), d_{\mathbb{H}})$, where $d_{\mathbb{H}}$ denotes the Hausdorff metric [75]. Thus the fixed point $\bar{S}$, often referred to as the *attractor*, is unique. Such a set of N contractive maps w_k is almost universally referred to as an N-map *iterated function system* (IFS), due to Barnsley and Demko [17]; we discuss this again in greater detail in section 2.2.

1.2.2 The construction of self-similar measures

If we associate a (nonnegative) probability p_k with each IFS map w_k, so that $\sum_{k=1}^{n} p_k = 1$, the resulting N-map *IFS with probabilities* (IFSP) defines an operator $\mathbb{M} : \mathcal{P}(\mathbb{X}) \to \mathcal{P}(\mathbb{X})$, where $\mathcal{P}(\mathbb{X})$ denotes the space of Borel probability measures on $\mathbb{X}$. Let $\mu, \nu \in \mathcal{P}(\mathbb{X})$ such that $\nu = \mathbb{M}\mu$. Then for any Borel set $S \subseteq \mathbb{X}$,

$$\nu(S) = (\mathbb{M}\mu)(S) = \sum_{k=1}^{N} p_k \mu(w_i^{-1}(S)). \qquad (1.7)$$

In light of our earlier discussion, $\mathbb{M}$ may be viewed as a fractal transform on measures: It produces N spatially-contracted copies $\mu(w_i^{-1})$ of the measure μ, alters their values by the p_i, and then combines them via the summation operator. As will be discussed in Sec. 2.5 and in greater generality in Chap 5, $\mathbb{M}$ is contractive in the complete metric space $(\mathcal{P}(\mathbb{X}), d_{MK})$ [75], where d_{MK} denotes the Monge-Kantorovich metric (often referred to as the "Hutchinson metric" in the IFS literature because of its use in [75]). This, of course, implies the existence of a fixed point $\bar{\mu} \in \mathcal{P}(\mathbb{X})$, the so-called *invariant measure* of the IFSP, which also possesses a self-similarity property. Moreover the support of $\bar{\mu}$ is the attractor $\bar{S}$ of the N-map IFS.

From a more practical perspective, the invariant measure provided by an IFSP can be used to provide black-and-white "shading" of the attractor $\bar{S}$ of its corresponding IFS. On a computer screen, the greyscale of a pixel representing a subset $S \in \bar{S}$ may be adjusted according to the invariant measure, $\bar{\mu}(S)$, supported on S – the darker the greyscale, the higher the measure. As such, IFSP invariant measures were viewed as a possible means of approximating natural images. Jacquin's original method of "block fractal coding" [81] was, in fact, formulated in terms of measures and IFSP operators. This approach naturally evolved to the consideration of images as functions, in particular L^2 functions, the usual scenario in signal and image processing.

At this point it is appropriate to mention that systems of contractive mappings had been previously studied by others, e.g., Williams [172], Nadler [137] and Karlin [83], to mention only a few. Interestingly, Karlin's studies of learning models involved random walks over Cantor-like sets, a precursor to the IFS "Chaos Game" of Barnsley and coworkers [16]

1.3 Induced fractal transforms

Suppose that there exists a $1 - 1$ mapping ψ between two complete metric spaces $\mathbb{Y}$ and $\mathbb{Z}$. Then a fractal transform operator $T : \mathbb{Y} \to \mathbb{Y}$ induces an equivalent operator $M : \mathbb{Z} \to \mathbb{Z}$, as sketched below. If T has a unique fixed point $\bar{u} \in \mathbb{Y}$, then M has a unique fixed point $\bar{z} = \psi(\bar{u}) \in \mathbb{Z}$. In what follows, we provide a brief overview of induced fractal transforms and their importance in applications. The subject is treated in greater detail in Sect. 3.4.

Induced fractal transforms have been useful in the treatment of inverse problems since it is sometimes easier – and possibly more fruitful – to work in the alternate space $\mathbb{Z}$. Some examples that have been useful include:

1. **Fourier transforms:** Let $\mathbb{Y} = L^2(\mathbb{X})$ where $\mathbb{X} = [-1, 1]$ and $\mathbb{Z} = L^2(\mathbb{R})$. Given a $u \in \mathbb{Y}$, we consider its Fourier transform $U \in \mathbb{Z}$ defined as follows,

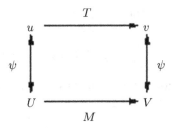

Fig. 1.6: Fractal transform $T : \mathbb{Y} \to \mathbb{Y}$ induces fractal transform $M : \mathbb{Z} \to \mathbb{Z}$.

$$U(k) = \int_{\mathbb{R}} u(x)\, e^{-i2\pi kx}\, dx, \quad k \in \mathbb{R}.$$

(Note that this is the signal processing definition, formulated in "k-space.")

Let T be the fractal transform defined by the following N-map *IFS with greyscale maps* on $\mathbb{R}$:

$$w_m(x) = s_m x + a_m, \quad \phi_m(t) = \alpha_m t + \beta_m, \quad 1 \le m \le N,$$

with $|s_m| < 1$. Then the associated fractal transform operator $M : \mathbb{Z} \to \mathbb{Z}$ is given by

$$V(k) = (MU)(k)$$
$$= \sum_{m=1}^{N} e^{-i2\pi a_m k} \left[\alpha_m s_m U(s_m k) + 2 s_m \beta_m \mathrm{sinc}(2 s_m k) \right].$$

The action of M on a Fourier transform $U(k)$ is to produce a set of *frequency-expanded* copies of $U(k)$ along with a condensation function composed of sinc functions. The former will come as no surprise to signal and image processors: the increase in resolution produced by the spatial contractions of T is equivalent to an increase in high-frequency content. The value of $U(k)$ for some $k \in \mathbb{R}$ is determined by values of U at *lower (magnitude) frequencies* $U(s_m k)$. The use of this operator to perform image enhancement by means of *frequency extrapolation* has been explored in [130, 131, 129].

2. **Fractal-wavelet transforms:** $\mathbb{Y} = L^2(\mathbb{X}))$ and $\mathbb{Z}$ is the space of wavelet representations of functions in $L^2(\mathbb{X})$ with l^2 metric. For simplicity, we consider $\mathbb{X} = [0,1]$. Let $\phi(x)$ (not to be confused with the greyscale maps $\phi_i(t)$) denote a scaling function that defines a standard multiresolution approximation of $L^2(\mathbb{X})$ and $\psi(x)$

the corresponding mother wavelet function. Then the functions $\{\phi, \psi_{ij}, i = 0, 1, 2, \cdots, j = 0, 1, \cdots, 2^i - 1\}$, where

$$\psi_{ij}(x) = 2^{1/2}\psi(2^i x - j),$$

form a complete basis of $L^2(\mathbb{X})$. Any $u \in L^2(\mathbb{X})$ admits an expansion in this basis of the form,

$$u(x) = b_{00}\phi(x) + \sum_{i=0}^{\infty} \sum_{j=0}^{2^i-1} c_{ij}\psi_{ij}(x).$$

The expansion coefficients are conveniently displayed in the following manner,

b_{00}							
c_{00}							
c_{10}				c_{11}			
c_{20}		c_{21}		c_{22}		c_{23}	
C_{30}	C_{31}	C_{32}	C_{33}	C_{34}	C_{35}	C_{36}	C_{37}

Here, C_{ij} represents a binary tree of infinite length with root c_{ij}. (The entire wavelet tree may be denoted as B_{00}, rooted at b_{00}.)

A *fractal-wavelet (FW) transform* M is defined by replacing blocks C_{kl} of the wavelet table with scaled copies of blocks C_{ij} that are rooted higher in the tree. The usual procedure is as follows: For a given $k^* > 0$ and $i^* = k^* - 1$, define

$$M\colon C_{k^*,l} \to \alpha_l C_{i^*,j(l)}, \text{where } j(l) \in \{0, 1, \cdots, 2^{i^*}-1\}, \ 0 \le l \le 2^{k^*}-1,$$

The α_l are the *fractal-wavelet scaling parameters*. (Note that constant β_l terms are not added to the coefficients.)

Example: For $k^* = 2$ and $i^* = 1$ consider the following FW transform:

$$C_{20} = \alpha_0 C_{10}, C_{21} = \alpha_1 C_{11}, C_{22} = \alpha_2 C_{11}, C_{23} = \alpha_3 C_{10}, |\alpha_i| < \frac{1}{\sqrt{2}}.$$

(The condition on the α_i guarantees contractivity.) Its action may be represented diagrammatically as follows,

$$M : B_{00} \to \begin{array}{|c|c|c|c|} \hline \multicolumn{4}{|c|}{b_{00}} \\ \hline \multicolumn{4}{|c|}{c_{00}} \\ \hline \multicolumn{2}{|c|}{c_{10}} & \multicolumn{2}{c|}{c_{11}} \\ \hline \alpha_0 C_{10} & \alpha_1 C_{11} & \alpha_2 C_{11} & \alpha_3 C_{10} \\ \hline \end{array} \qquad (1.8)$$

If we iterate this map, we obtain a sequence of wavelet trees $M^n(B_{00})$ that converge to a limit (in l^2 metric) which is a wavelet coefficient tree. The first few rows of this limit are given by

b_{00}							
c_{00}							
c_{10}				c_{11}			
$\alpha_0 c_{10}$		$\alpha_1 c_{11}$		$\alpha_2 c_{11}$		$\alpha_3 c_{10}$	
$\alpha_0^2 c_{10}$	$\alpha_0\alpha_1 c_{11}$	$\alpha_1\alpha_2 c_{11}$	$\alpha_1\alpha_3 c_{10}$	$\alpha_2^2 c_{11}$	$\alpha_2\alpha_3 c_{10}$	$\alpha_3\alpha_0 c_{10}$	$\alpha_3\alpha_1 c_{11}$

If we start with a wavelet tree in equation (1.8) with zeros below the entries c_{10} and c_{11}, then each application of M produces a new line of nonzero wavelet coefficients, representing an additional degree of spatial resolution. The production of wavelet coefficients c_{ij} with higher i-values from those of lower i-values is analogous to the frequency extrapolation method for Fourier transforms discussed earlier.

Note that our presentation of the fractal-wavelet transform has proceeded in the reverse direction: Instead of starting with an operator T in the spatial domain, we began with a transform M in the wavelet domain. It induces a block-based or partitioned fractal transform T with condensation function $h(x)$ in the spatial domain. More details may be found in [134]. Two-dimensional fractal-wavelet transforms, with particular application to image processing, are discussed in [166].

3. **Moments of probability measures:** $(\mathbb{Y}, d_{\mathbb{Y}}) = (\mathcal{P}, d_{MK})$, the space of probability measures on $\mathbb{X} = [0,1]$; $(\mathbb{Z}, d_{\mathbb{Z}}) = (D(\mathbb{X}), \bar{d}_2)$, the space of moment sequences of these measures,

$$D(\mathbb{X}) =$$

$$\{\mathbf{g} = (g_0 = 1, g_1, g_2, \cdots) \mid g_n = \int_{\mathbb{X}} x^n d\mu, \ n = 0, 1, 2, \cdots, \forall \mu \in \mathcal{P}(\mathbb{X})\}.$$

with weighted l^2 metric,

$$\bar{d}_2(\mathbf{g}, \mathbf{h}) = \sum_{k=1}^{\infty} \frac{1}{k^2}(g_k - h_k)^2, \quad \mathbf{g}, \mathbf{h} \in D(\mathbb{X}).$$

Here, we deviate from the notation used above and let $\mathbb{M} : \mathbb{Y} \to \mathbb{Y}$ denote the Markov operator associated with an N-map IFSP, as defined in equation (1.7). We consider the case of affine contractive IFS maps, $w_i(x) = s_i x + a_i$ with probabilities p_i.

As shown in [65], if we multiply both sides of equation (1.7) by x^n, $n = 0, 1, 2, \cdots$ and integrate over $\mathbb{X}$, we arrive at the following relation between the moments h_n of $\nu = \mathbb{M}\mu$ and the moments g_n of μ,

$$h_n = \sum_{i=1}^{N} A_{ni} p_i, \qquad n = 1, 2,$$

where

$$A_{ni} = \sum_{k=0}^{n} \binom{n}{k} s_i^k a_i^{n-k} g_k.$$

This shows that the induced operator $A : D(\mathbb{X}) \to D(\mathbb{X})$ is a linear operator, i.e., $\mathbf{h} = A\mathbf{g}$. Moreover, its (infinite) matrix representation is lower triangular. It is also contractive in $(D(\mathbb{X}), \bar{d}_2)$. Its unique fixed point $\bar{\mathbf{g}}$ is the vector of moments of the invariant/fixed point measure $\bar{\mu}$ of $\mathbb{M}$. [65]. The moments $\bar{g}_n$ of $\bar{\mu}$ may be computed recursively, as originally shown in [17].

1.4 Inverse problems for fractal transforms and "collage coding"

Barnsley and Demko were the first to see the potential of using IFS/fractal transforms for inverse problems of approximation [17]. In the context of their paper on IFS, they asked the following,

Given a set S, can we find an IFS $\mathbf{w} = \{w_1, \cdots, w_N\}$ such that S is the attractor of $\mathbf{w}$, i.e., $\mathbf{w}(S) = S$?

Realizing that this may be too much to ask,

Can we find an IFS $\mathbf{w}$ with fixed point attractor A such that $S \approx A$?

Mathematically, this may be understood as the *problem of approximation by fixed points of contractive maps*. Suppose that we have an element $u \in \mathbb{Y}$. Can we find a contractive fractal transform $T : \mathbb{Y} \to \mathbb{Y}$ with fixed point $\bar{u}$ such that $\bar{u} \approx u$, perhaps even to some desired degree of accuracy, i.e., $d_{\mathbb{Y}}(u, \bar{u}) < \epsilon$ for some specified $\epsilon > 0$? (Interestingly, Barnsley and Demko provided a simple, low-accuracy, solution to such an inverse problem – the "twin dragon" set in $\mathbb{R}^2$ – by attempting to match a couple of moments of this set by the moments of an IFS with probabilities.)

Even from its inception, the inverse problem of approximating sets/functions by fixed points of IFS was seen to be difficult. How does one vary the variable fractal parameters – call them π – that comprise a fractal transform $T(\pi)$ in an effort to steer the fixed point $\bar{u}(\pi)$ close to u? Such a method is complicated: One must compute the attractor/fixed point $u(\pi)$ at each step, then estimate the derivatives, etc.. A simpler, more automatic, method would be desirable.

Indeed, a great simplification results by means of a simple consequence of Banach's Fixed Point Theorem [155], which is referred to as the "Collage Theorem" [18, 16] in the IFS literature. Before stating this result, let us first motivate the idea of "collaging." Without loss of generality, we refer to the self-similarity relation for functions in equation (1.3), which has the form $u = Tu$.

In general, a function u will **not** be self-similar but with a proper selection of maps w_k and ϕ_k maps we may be able to arrive at an approximate self-similarity, i.e.,

$$u \approx Tu.$$

The error of this approximation, $d_{\mathbb{Y}}(u, Tu)$, is referred to as the *collage error* – it represents the error obtained by "tiling" u with copies of itself. Moreover, we expect that the smaller the collage error, the closer u is to being the fixed point $\bar{u}$ of T. Indeed, a simple application of the triangle inequality yields the mathematical relationship between the collage error $d_{\mathbb{Y}}(u, Tu)$ and the approximation error $d_{\mathbb{Y}}(u, \bar{u})$, known as the "Collage Theorem:"

$$d_{\mathbb{Y}}(u, \bar{u}) \leq \frac{1}{1-c} d_{\mathbb{Y}}(Tu, u),$$

where $c \in [0, 1)$ is the contraction factor of T. (The proof is presented in Chap. 2.)

The goal is now to make the approximation error $d_{\mathbb{Y}}(u, \bar{u})$ small by making the collage error $d_{\mathbb{Y}}(Tu, u)$ small (and perhaps keeping some control on c). This is the essence of *collage coding*, which is a central feature of fractal image coding and fractal-based methods in general: You try to "tile" a target $u \in \mathbb{Y}$ with copies of itself, as was shown nicely in the original paper of Barnsley and students [18] (and subsequently in the book by Barnsley [16]).

In general, whenever you have a family of contraction maps – fractal or non-fractal – over a complete metric space $(\mathbb{Y}, d_{\mathbb{Y}})$, you have the ingredients of an inverse problem: Given an element $u \in Y$, find a contraction map T with fixed point $\bar{u}$ such that the approximation error $d_{\mathbb{Y}}(u, \bar{u})$ is sufficiently small. Collage coding in a variety of spaces and applications – both fractal and non-fractal will be discussed extensively in Chaps. 7 and 8 and is the basis for the techniques developed in these chapters.

1.4.1 Fractal image coding

An overview of fractal transforms and self-similarity would not be complete without a brief description of the basics of fractal image coding. A more detailed discussion is to be found in Chap. 2.

A very simple prescription for the fractal coding of an $n \times n$-pixel image $u(x)$ is as follows. With reference to the earlier discussion on block fractal coding which included Eq. (1.4), we let R_k, $k = 1, 2, \cdots, N_R$, denote a set of $n_R \times n_R$-pixel nonoverlapping range blocks that form a partition of the image. Let D_j, $k = 1, 2, \cdots, N_D$ be a set of $2n_R \times 2n_R$-pixel domain blocks that are selected from throughout the image. (In order to keep the size down, we may consider the set of nonoverlapping $2n_R \times 2n_R$-pixel blocks that cover the image.

For each range block, R_k, compute the collage errors Δ_{kj} associated with all domain blocks, D_j, i.e.,

$$\Delta_{kj} = \min_{\alpha, \beta} \| u(R_k) - \alpha \tilde{u}(D_j) - \beta \|, \quad j = 1, 2, \cdots, N_D.$$

Here, $\tilde{u}(D_j)$ denotes the $n_R \times n_R$-pixel block image obtained by "decimating" the $2n_R \times 2n_R$-pixel domain block image $u(D_k)$. (For digital images, decimation is normally accomplished by replacing the image values over four neighbouring pixels that form a square in D_k by their average value placed on one pixel. Following the decimation, we may consider all eight possible isometries that map one block to another, i.e., four rotations and four reflections.) The block $D_{j(k)}$ yielding the lowest collage error Δ_{kj} is chosen to be the domain block associated with R_j.

The above procedure yields the fractal transform T which minimizes the total collage distance

$$\| u - Tu \| = \sum_{k=1}^{N_R} \Delta_{kj}.$$

over the nonoverlapping range blocks R_k. (Because the range blocks R_k were nonoverlapping, each approximation can be performed independently.) T is defined in terms of the range-domain assignments $(k, j(k))$ (along with isometries $i(k)$ if applicable) and ϕ-map parameters α_k, β_k: these parameters comprise the *fractal code* of the image u.

The fixed point $\bar{u}$ of T – the desired approximation to u – is then generated by iteration: Start with any $n \times n$-pixel "seed" image, u_0, for example the blank image $u_0 = 0$, and form the iteration sequence

$u_{n+1} = Tu_n$ until convergence is achieved (after roughly $\log_2 n$ iterations). At each step of the iteration procedure, each range block image $u(R_k)$ of u_n is replaced by $\alpha_k \tilde{u}(D_k) + \beta_k$. The result of this procedure, as applied to the 512×512-pixel test image *Boat*, is shown in the figure below.

For more detailed accounts of fractal coding, the reader is referred to [19, 60, 120].

Fig. 1.7: Clockwise, from upper left: The 512×512-pixel, 8 bit/pixel test image *Boat*. The images u_1, u_2 and fixed point images obtained from the iteration sequence $u_{n+1} = Tu_n$, where the seed u_0 is a blank (white) image and T is the fractal transform obtained by the collage coding method outlined in the text: 8×8-pixel range blocks, 16×16-pixel domain blocks.

Chapter 2
Basic IFS Theory

In this chapter, we give a short presentation of the classical theory of iterated function systems (IFSs). Our main purpose in doing this is to make this book as self-contained as possible. We also hope that this chapter might be used by a beginner to learn this theory. As such, the tone in this chapter is more expository than in some of the other chapters, and we give more attention to pedagogical motivation. Our choice of material includes all the standard results and constructions along with a selection of topics that are particularly useful in the rest of the book. There are several excellent books that provide alternative presentations of this material ([16, 56, 52] are three such books).

2.1 Contraction mappings and fixed points

The foundation of most of the constructions in IFS theory is the notion of contractive maps on some metric space and, in particular, the *contraction mapping theorem* (Theorem 2.2). Thus, we start our presentation with this classical material. Throughout this section, we assume that $(\mathbb{X}, d)$ is a complete metric space.

Definition 2.1. A function $f : \mathbb{X} \to \mathbb{X}$ is a *contraction* if there is some $c \in [0, 1)$ with $d(f(x), f(y)) \leq c\, d(x, y)$ for all $x, y \in \mathbb{X}$. The smallest such constant c is the *contraction factor* for the contraction f.

Notice that any contraction mapping is automatically continuous. The most important property of contraction mappings is the fact that all contractions have a unique fixed point. The following theorem is also called *Banach's fixed-point theorem*, as it appeared in his Ph.D. thesis (also see [11])!

Theorem 2.2. *(Contraction mapping theorem) Let $f : \mathbb{X} \to \mathbb{X}$ be a contraction with contraction factor $c < 1$. Then there is a unique $\bar{x} \in \mathbb{X}$ such that $f(\bar{x}) = \bar{x}$. Furthermore, for any x_0, the sequence defined by $x_{n+1} = f(x_n)$ converges to $\bar{x}$ with the estimate*

$$d(x_n, \bar{x}) \leq c^n \, d(x_0, \bar{x}).$$

Proof. We first prove the uniqueness. Suppose that there are two distinct points $x, y \in \mathbb{X}$ with $f(x) = x$ and $f(y) = y$. Then

$$d(x, y) = d(f(x), f(y)) \leq c \, d(x, y) < d(x, y),$$

which is a contradiction. Thus, if there is a fixed point, it must be unique.

To show existence, let $x_0 \in \mathbb{X}$ be arbitrary and define the sequence x_n by $x_{n+1} = f(x_n)$. We show that this sequence is a Cauchy sequence, and its limit is a fixed point of f.

First, we note that

$$\begin{aligned}
d(x_0, x_k) &\leq d(x_0, x_1) + d(x_1, x_2) + d(x_2, x_3) + \cdots + d(x_{k-1}, x_k) \\
&\leq d(x_0, x_1) + c \, d(x_0, x_1) + c^2 \, d(x_0, x_1) + \cdots + c^{k-1} d(x_0, x_1) \\
&\leq [1 + c + c^2 + \cdots + c^{k-1}] d(x_0, x_1) \\
&\leq \frac{1}{1 - c} d(x_0, x_1).
\end{aligned}$$

Furthermore, for $1 \leq n < m$, we have

$$\begin{aligned}
d(x_n, x_m) &= d(f(x_{n-1}), f(x_{m-1})) \leq c \, d(x_{n-1}, x_{m-1}) \\
&\leq c^2 \, d(x_{n-2}, x_{m-2}) \leq \cdots \leq c^n \, d(x_0, x_{m-n}) \leq \frac{c^n}{1 - c} d(x_0, x_1),
\end{aligned}$$

which immediately implies that the sequence x_n is a Cauchy sequence. Since $\mathbb{X}$ is complete, we know that $x_n \to \bar{x}$ for some $\bar{x} \in \mathbb{X}$. Since f is continuous, we have

$$\bar{x} = \lim x_n = \lim x_{n+1} = \lim f(x_n) = f(\bar{x}),$$

and so $\bar{x}$ is a fixed point of f. Finally, for the estimate, we see that

$$d(x_n, \bar{x}) = d(f(x_{n-1}), f(\bar{x})) \leq c \, d(x_{n-1}, \bar{x}) \leq \cdots \leq c^n d(x_0, \bar{x}).$$

$\square$

There are many other theorems that guarantee the existence of fixed points under various hypotheses. The particular strength of the contraction mapping theorem is that it not only guarantees a unique fixed point (some of the other fixed-point theorems simply assert there is at least one) but also gives an effective construction to approximate this fixed point to any degree of accuracy – we simply iterate the function sufficiently many times.

Exercise 2.3. Consider $X = [1, \infty)$ and $f(x) = x + 1/x$. Show that f satisfies $|f(x) - f(y)| < |x - y|$ for all $x, y \in X$ but that f has no fixed point in X. Notice that X is a complete metric space. Thus the condition that the contractivity of f be strictly bounded away from 1 is necessary.

Exercise 2.4. Suppose that X is compact and that $|f(x) - f(y)| < |x - y|$ for all $x, y \in X$. Show that f has a fixed point in X.

Exercise 2.5. We generalize the setting a little bit. Let X be complete and $f_n : X \to X$ satisfy $|f_n(x) - f_n(y)| \leq (1 - a_n)|x - y|$ for all n, where $\sum_n a_n = +\infty$ and $0 < a_n < 1$. Choose $x_0 \in X$, and define $x_{n+1} = f_n(x_n)$. Show that x_n converges.

The following proposition gives an often practical means of telling how close a given point in the space is to the fixed point. This estimate is often called the *collage theorem*.

Theorem 2.6. *(Collage theorem) Let $f : X \to X$ be contractive with contractivity $c < 1$ and $\bar{x}$ be its unique fixed point. Then, for any $y \in X$, we have*

$$d(y, \bar{x}) \leq \frac{d(y, f(y))}{1 - c}. \tag{2.1}$$

Proof. We see that

$$d(y, \bar{x}) \leq d(y, f(y)) + d(f(y), \bar{x}) = d(y, f(y)) + d(f(y), f(\bar{x}))$$
$$\leq d(y, f(y)) + c\, d(y, \bar{x}).$$

Rearranging this gives the desired result.   □

This simple result plays a major role in applications of the contraction mapping theorem, particularly in applications using IFSs. The meaning of the estimate is clear: the less the function f "moves" a given point $y \in X$, the closer this point y is to being the fixed point of f. Intuitively this makes perfect sense: points "close" to the fixed point $\bar{x}$ do not get "moved" very much by f, with this effect getting stronger the closer you are to $\bar{x}$.

We next turn to showing that the fixed point of a contraction is a continuous function of the contraction. That is, if you vary the contraction map, the resulting fixed point varies in a continuous fashion. First, we define a distance between two functions on $\mathbb{X}$.

Definition 2.7. Let $f, g : \mathbb{X} \to \mathbb{X}$ be two functions, and we define

$$d_\infty(f, g) = \sup_{x \in \mathbb{X}} d(f(x), g(x))$$

assuming this is finite.

Clearly this quantity is not always finite, even if both f and g are contractive. For instance, $f, g : \mathbb{R} \to \mathbb{R}$ given by $f(x) = x/2$ and $g(x) = x/3$ are both contractive but $d_\infty(f, g) = +\infty$. If we restrict our discussion to a suitable collection of functions, then d_∞ is in fact a metric. One instance of this is when $\mathbb{X}$ is compact, as then d_∞ is bounded (being a continuous real-valued function on a compact set).

The next result gives an estimate of the distance between the fixed points of two contractions based on the distance between the functions. The proof is extremely simple, but the result is important enough to record.

Proposition 2.8. *Let $f, g : \mathbb{X} \to \mathbb{X}$ be contractions with contractivities c_f, c_g and fixed points $\bar{x}_f, \bar{x}_g$. Then*

$$d(\bar{x}_f, \bar{x}_g) \le \frac{1}{1 - c_f} \, d_\infty(f, g).$$

Notice that in the estimate we could use either c_f or c_g, whichever one is more convenient or gives a better bound. Using this result, we can show that the fixed point is a continuous function of the contraction.

Theorem 2.9. *(Continuity of fixed points)* Let $f_n : \mathbb{X} \to \mathbb{X}$ be a *sequence of contractions with contractivities c_n and fixed points $\bar{x}_n$. Suppose that $c = \sup_n c_n < 1$ and that there is some $f : \mathbb{X} \to \mathbb{X}$ with $d_\infty(f_n, f) \to 0$. Then $\bar{x}_n \to \bar{x}$ and $\bar{x}$ is the fixed point of f.*

Proof. It is clear that f is a contraction with contractivity bounded from above by c since

$$d(f(x), f(y)) = \lim_n d(f_n(x), f_n(y)) \le \lim_n cd(x, y) = cd(x, y).$$

Thus f has a fixed point $\bar{x}$. We use the estimate in Proposition 2.8 with c and thus get

$$d(\bar{x}, \bar{x}_n) \leq \frac{1}{1-c} \, d_\infty(f, f_n) \to 0.$$

$\square$

This result will be very important in our later work. It establishes that continuous variations in the contraction maps (or parameters therein) will produce continuous variations in the fixed points. This will allow us to modify a contraction map in a controlled manner in order to produce a desired fixed point. The paper [103] contains a more general discussion of the issue of continuity of fixed points.

Exercise 2.10. There is another useful way to compare two Lipschitz functions. Fix a point $a \in X$, and suppose that $f, g : X \to X$ are Lipschitz with Lipschitz factors $c_x, c_y \geq 0$. Then it is easy to see that the quantity

$$d_L(f, g) = d(f(a), g(a)) + \inf\{\lambda > 0 : d(f(x), g(x)) \leq \lambda \, d(x, a)\}$$

is finite. Show that it is in fact a metric between Lipschitz functions. Prove a result on continuity of fixed points under this metric.

There is also another simple consequence of the proof of the contraction mapping theorem, which is sometimes called the *anti-collage theorem*.

Theorem 2.11. *(Anti-collage theorem) Let $f : X \to X$ be contractive with contractivity $c < 1$ and $\bar{x}$ be its unique fixed point. Then, for any $y \in X$, we have*

$$d(y, \bar{x}) \geq \frac{d(y, f(y))}{1+c}. \tag{2.2}$$

Exercise 2.12. Prove Theorem 2.11. The proof is similar to that of the collage theorem.

2.1.1 Inverse problem

These considerations lead naturally to an inverse problem, that is one of the central problems we investigate in this book. This inverse problem will arise in many different contexts. Our method of solution is at the heart of what we mean by "fractal-based methods in analysis."

Summarizing our results on contractions, if we are given a contraction mapping f, we can "find/construct" its fixed point $\bar{x}$ (or at least arbitrarily close approximations to it) by iterating the function f on any convenient starting point x_0.

Suppose instead we are given the following problem: "Given a $y \in \mathbb{X}$, can we find a 'suitably nice' contraction map f for which y is the fixed point; i.e., $f(y) = y$?" In general, the answer is *no*, especially as usually we are constrained to look in some family of contractions. The fruitful viewpoint is to relax the requirement for an exact match between the fixed point $\bar{x}$ of f and the given point y and instead ask for a small distance between them.

More formally, we have the following.

Definition 2.13. (Inverse problem of approximation by fixed points of contraction maps) Given a point $y \in \mathbb{Y}$ and an $\epsilon > 0$, can we find a contraction f whose fixed point $\bar{x}$ satisfies $d(\bar{x}, y) < \epsilon$?

One of the difficulties in trying to answer this question is estimating the distance $d(\bar{x}, y)$. The problem is that while it is relatively simple to generate $\bar{x}$ given the contraction f, it is not always simple to estimate $d(\bar{x}, y)$ without explicitly computing $\bar{x}$. In most of the situations we will discuss, we search for the contraction f out of some parameterized family, so we would like to be able to estimate the approximation error $d(\bar{x}, y)$ more directly from the parameters of the given f.

The collage theorem (Theorem 2.6) allows us to change the question into one that is more tractable; this is the main reason for the importance of this theorem. It allows us to control the distance $d(f(y), y)$ instead of $d(\bar{x}, y)$; the distance $d(f(y), y)$ involves only a single iteration of f, while, in principle, $d(\bar{x}, y)$ involves infinitely many iterations of f. It is the basis for the *collage coding* method, which is discussed in detail in Chapters 7 and 8 and in Sect. 3.1 on fractal imaging. The ideas around "collage coding" were first fully developed for application in fractal image compression and have since been found useful in many other contexts involving inverse problems.

There is another theorem that is also useful in this context. It was first noted in [91].

Theorem 2.14. *(Suboptimality theorem) Let $f_\lambda : \mathbb{X} \to \mathbb{X}$ be a family of contractions with the contractivity of f_λ being $c_\lambda < 1$. Denote the fixed point of f_λ by x_λ. Let $x \in \mathbb{X}$ be given and α be such that*

$$d(x, f_\alpha(x)) \leq d(x, f_\lambda(x)) \quad \text{for all } \lambda.$$

Finally, let β be such that

$$d(x, x_\beta) \leq d(x, x_\lambda) \quad \text{for all } \lambda.$$

Then we have that

$$d(x_\alpha, x_\beta) \leq \frac{2}{1 - c_\alpha} d(x, f_\alpha(x)).$$

This theorem is interesting because it indicates that the optimal "collage" approximation is not so far from the optimal approximation by a fixed point. It is clear that usually the "collage" approximation will be suboptimal, however.

Exercise 2.15. Prove Theorem 2.14. The proof is similar to that of the collage theorem.

2.2 Iterated Function System (IFS)

We now introduce the book's primary mathematical object – an *iterated function system*, or IFS. The seminal mathematical reference for the material in this section is the landmark paper by Hutchinson [75].

2.2.1 Motivating example: The Cantor set

We begin our discussion with the motivating example of the classical Cantor set $\mathcal{C}$ (sometimes also called the *middle 1/3-Cantor set*). In the classical construction, we start with the unit interval $I_0 = [0, 1]$ and remove an open interval of length $1/3$ from the center, leaving the two closed intervals $[0, 1/3]$ and $[2/3, 1]$. The union of these two intervals we will call I_1. See Fig. 2.1 for an illustration of this construction. This figure shows the first several stages of the construction.

For the next step, we remove an open interval of length $1/9$ from the center of each component of I_1, leaving the four closed intervals $[0, 1/9], [2/9, 1/3], [2/3, 7/9]$, and $[8/9, 1]$; we will call the union of these four closed intervals I_2. Continuing, at the nth stage we have the set I_n, which is composed of 2^n closed intervals, each of length 3^{-n}. We remove an open interval of length 3^{-n-1} from the center of each component of I_n. This will leave 2^{n+1} closed intervals, each of length 3^{-n-1}, whose union we call I_{n+1}. Notice that by this construction we have $I_{n+1} \subset I_n$, with each I_n a compact subset of $[0, 1]$.

The Cantor set $\mathcal{C}$ is defined to be the intersection of all the I_n; that is, $\mathcal{C} = \cap_n I_n$. By standard results, we know that $\mathcal{C} \neq \emptyset$ and is a compact subset of $[0, 1]$. Furthermore, it is not difficult to show that $\mathcal{C}$ is *totally disconnected*. What this means is that the largest connected pieces of $\mathcal{C}$ are single points. In addition, the cardinality of $\mathcal{C}$ is the same as the cardinality of $[0, 1]$, as there is a one-to-one correspondence between the points of $\mathcal{C}$ and the points of $[0, 1]$. To understand this, it is useful to have another description of $\mathcal{C}$. This alternate description will also be a foreshadowing of the more general idea of *addresses* for describing the points on the attractor of a generic IFS (see Sect. 2.3).

Fig. 2.1: Stages in the construction of the Cantor set.

The key idea in the alternate description is to notice that when we remove the interval $(1/3, 2/3)$ at the first stage, what we are doing is removing all the numbers in $[0, 1]$ whose base-3 expansion begins with the digit "1." This leaves all those numbers whose base-3 expansion begins with either one of the digits "0" or "2." In the second step, we now remove all those numbers whose base-3 expansion has a "1" as the *second* digit. In general, the set I_n contains all those numbers in $[0, 1]$ whose base-3 expansion contains no "1" digit in any of the first n places of the expansion. Thus the limiting set $\mathcal{C}$ contains all those numbers that have no "1" digit *anywhere* in their base-3 expansion. However, as there are still two other digits to choose from (the digits "0" and "2"), there are still infinitely many possible points left in $\mathcal{C}$. In fact, it is easy to define an injective function $f : [0, 1] \to \mathcal{C}$, which shows that $\mathcal{C}$ has cardinality at least as large as that of $[0, 1]$. To do this, let $\sum_i b_i 2^{-i}$ be a binary expansion for some $x \in [0, 1]$. We define

$$f(x) = \sum_i 2b_i 3^{-i}.$$

This defines some point of $\mathcal{C}$ since each base-3 digit of $f(x)$ will be either a "0" or a "2."

Another remarkable property of $\mathcal{C}$ is that it is composed of smaller pieces, each of which is an exactly scaled copy of $\mathcal{C}$. To be slightly more precise, if we let $\mathcal{C}_0 = \mathcal{C} \cap [0, 1/3]$ and $\mathcal{C}_1 = \mathcal{C} \cap [2/3, 1]$, we see that $\mathcal{C} = \mathcal{C}_0 \cup \mathcal{C}_1$ and that each $\mathcal{C}_0$ and $\mathcal{C}_1$ "look" just like smaller copies of $\mathcal{C}$. In particular, defining $w_0, w_1 : \mathbb{R} \to \mathbb{R}$ by

$$w_0(x) = x/3 \quad \text{and} \quad w_1(x) = x/3 + 2/3, \tag{2.3}$$

we see that $C_0 = w_0(\mathcal{C})$ and that $C_1 = w_1(\mathcal{C})$. This means that

$$\mathcal{C} = w_0(\mathcal{C}) \cup w_1(\mathcal{C}).$$ (2.4)

Equation (2.4) is a formal way to encode the *self-similarity* of the Cantor set. This equation also suggests another iterative construction of $\mathcal{C}$. Starting with $I_0 = [0, 1]$, we see that $w_0(I_0)$ and $w_1(I_0)$ are the two intervals in the second step of the construction above and $I_1 = w_0(I_0) \cup w_1(I_0)$. Continuing, we see that at each step $I_{n+1} = w_0(I_n) \cup w_1(I_n)$. So, instead of starting with some set and removing part of it at each stage, this time we iterate the combined action of the two functions w_0 and w_1 to produce the next stage.

The particular details of these "steps" in the construction are due to the initial set $I_0 = [0, 1]$. For instance, $I_{n+1} \subset I_n$ since $w_0(I_n) \subset I_n$ and $w_1(I_n) \subset I_n$. If we had started with some other initial set, there is no reason to believe that the intermediate steps would have been nested in this way. Figure 2.2 shows an alternate sequence of intermediate "steps" that starts with $I_0 = [1/3, 2/3]$. Defining $I_{n+1} = w_0(I_n) \cup w_1(I_n)$ as before, this time we do not have the same nested property of the sequence I_n. However, it still seems that the sets I_n "converge" to $\mathcal{C}$. The trick is finding the right metric in which to measure and show this convergence. The *Hausdorff metric* turns out to work perfectly for this.

Fig. 2.2: Stages in an alternate construction of the Cantor set.

Exercise 2.16. The Cantor set is also self-similar under other functions. Show that the functions $w_0(x) = 1/3 - x/3$ and $w_1(x) = 1 - x/3$ also work. Find other examples of functions that work, some involving more than two functions.

2.2.2 Space of compact subsets and the Hausdorff metric

Since we are interested in discussing convergence of sets, the proper setting for this is a space of sets. Again let $(\mathbb{X}, d)$ be a complete metric space.

Definition 2.17. Let

$$\mathbb{H}(\mathbb{X}) = \{A \subseteq \mathbb{X} : A \neq \emptyset \text{ and } A \text{ is compact}\}.$$

Note that $\mathcal{C} \in \mathbb{H}(\mathbb{R})$ and $I_n \in \mathbb{H}(\mathbb{R})$ for all n (these are the sets from the construction of the Cantor set and are from the previous section).

Now we need to define a metric on this space. The appropriate metric is the *Hausdorff metric*. Notice that the Hausdorff metric is a distance between compact sets and is based on an underlying metric that measures the distance between points. If you change the underlying metric, you also change the resulting Hausdorff metric.

Definition 2.18. (Hausdorff distance) For $A, B \in \mathbb{H}(\mathbb{X})$, we define

$$d_{\mathbb{H}}(A, B) = \max\left\{ \sup_{a \in A} \inf_{b \in B} d(a, b), \sup_{b \in B} \inf_{a \in A} d(a, b) \right\}.$$

It is not obvious that this defines a metric or what the meaning of the resulting distance is, so we will spend some time discussing it. The first observation is that

$$d(a, B) := \inf_{b \in B} d(a, b)$$

is the distance from the point a to the compact set B. Furthermore, the quantity

$$d(A, B) := \sup_{a \in A} \inf_{b \in B} d(a, b)$$

is like a one-sided distance from A to B. The Hausdorff distance is a maximum of two of these one-sided distances.

Exercise 2.19. Show that if $A, B \subseteq \mathbb{X}$ are such that $\mathrm{cl}(A), \mathrm{cl}(B) \in \mathbb{H}(\mathbb{X})$, then $d_{\mathbb{H}}(A, B) = d_{\mathbb{H}}(\mathrm{cl}(A), \mathrm{cl}(B))$.

Another way to start seeing the meaning of the Hausdorff distance is through the idea of an ϵ-dilation of a set.

Definition 2.20. Let A be a subset of a metric space $\mathbb{X}$ and $\epsilon > 0$. Then the ϵ-*dilation of A* is the set

$$A_\epsilon = \{x : d(x, a) < \epsilon \text{ for some } a \in A\}.$$

Clearly $A \subseteq A_\epsilon$ for any $\epsilon > 0$. The converse is not usually true.

Let $A, B \in \mathbb{H}(\mathbb{X})$ be given, and suppose that $A \subseteq B_\epsilon$. Then we know that for each $a \in A$ there is some $b \in B$ such that $d(a, b) < \epsilon$. However, this means that

$$d(A, B) = \sup_{b \in B} \inf_{a \in A} d(a, b) \leq \epsilon.$$

The converse is also equally simple: if $d(A, B) < \epsilon$, then $A \subseteq B_\epsilon$. See Fig. 2.3 for an illustration of this. This easily leads to the next result.

Proposition 2.21. *Let $A, B \in \mathbb{H}(\mathbb{X})$. Then*

$$d_{\mathbb{H}}(A, B) = \inf\{\epsilon > 0 : A \subseteq B_\epsilon \text{ and } B \subseteq A_\epsilon\}.$$

This result provides a nice "visual" characterization of the meaning of the Hausdorff distance.

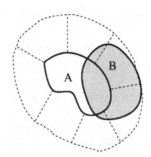

Fig. 2.3: An illustration for the Hausdorff distance.

Exercise 2.22. Show that $I_n \to \mathcal{C}$ in the Hausdorff metric, where I_n is the sequence of sets (the "steps") in either one of the constructions of the Cantor set (as illustrated in Figs. 2.1 and 2.2).

It is not difficult to show that Definition 2.18 defines a metric, so we do this now.

Theorem 2.23. *Let $(\mathbb{X}, d)$ be a metric. Then the Hausdorff metric based on d is also a metric.*

Proof. The symmetry in the definition easily shows that $d_{\mathbb{H}}(A, B) = d_{\mathbb{H}}(B, A)$.

It is easy to see that $d_{\mathbb{H}}(A, B) \geq 0$ for any $A, B \in \mathbb{H}(\mathbb{X})$ and that $d_{\mathbb{H}}(A, A) = 0$. On the other hand, if $d_{\mathbb{H}}(A, B) = 0$, then for each $a \in A$ we have $\inf_b d(a, b) = 0$. Since B is compact, this implies that $a \in B$ and thus $A \subseteq B$. In a similar fashion, we get that $B \subseteq A$ and thus $A = B$.

Finally, we show the triangle inequality, so let $A, B, C \in \mathbb{H}(\mathbb{X})$ be given. Then we have that for any $a \in A$ and $c \in C$

$$d(a, B) \leq \inf_{b \in B}(d(a, c) + d(c, b)) = d(a, c) + \inf_{b \in B} d(c, b) = d(a, c) + d(c, B).$$

This then gives

$$d(a, B) \leq \inf_{c \in C}(d(a, c) + d(c, B))$$
$$\leq \inf_{c \in C} d(a, c) + \sup_{c \in C} d(c, B) = d(a, C) + d(C, B)$$

and thus

$$d(A, B) = \sup_{a \in A} \inf_{b \in B} d(a, b) \leq d(A, C) + d(C, B).$$

The inequality $d(B, A) \leq d(B, C) + d(C, A)$ is similar, and thus we have shown that $d_{\mathbb{H}}(A, B) \leq d_{\mathbb{H}}(A, C) + d_{\mathbb{H}}(C, B)$.   □

Not only is the Hausdorff distance a metric, but if $\mathbb{X}$ is complete, then the space $(\mathbb{H}(\mathbb{X}), d_{\mathbb{H}})$ is a complete metric space, as we show next.

Theorem 2.24. *If $(\mathbb{X}, d)$ is complete, then so is $(\mathbb{H}(\mathbb{X}), d_{\mathbb{H}})$.*

Proof. Let $A_n \in \mathbb{H}(\mathbb{X})$ be a Cauchy sequence that we wish to show converges. Define

$$A = \bigcap_{m=1}^{\infty} \overline{\bigcup_{n \geq m} A_n},$$

where we use the bar to denote closure. We claim that $d_{\mathbb{H}}(A_n, A) \to 0$ and $A \in \mathbb{H}(\mathbb{X})$.

First we show that $A \in \mathbb{H}(\mathbb{X})$. Since $\mathbb{X}$ is a complete metric space, being compact is equivalent to being closed and totally bounded. We will show that each $B_m = \mathrm{cl}(\cup_{n \geq m} A_n)$ is compact and nonempty. This would imply that A is compact and nonempty as well. Now clearly each B_m is closed and $B_{m+1} \subseteq B_m$, so showing that B_1 is compact is sufficient. To show that B_1 is totally bounded, let $\epsilon > 0$ be given.

Then, since A_n is Cauchy, there is some m such that for any $n > m$ we have $d_{\mathbb{H}}(A_m, A_n) < \epsilon/2$, which implies that $A_n \subseteq (A_m)_{\epsilon/2}$, the $\epsilon/2$-dilation of A_m. Thus, $B_m \subseteq (A_m)_{\epsilon/2}$, so B_m is totally bounded (and thus compact) since A_m is totally bounded. Now notice that

$$B_1 = \mathrm{cl}(A_1 \cup A_2 \cup \cdots \cup A_{m-1} \cup B_m),$$

so it is the closure of a finite union of compact sets and thus is compact.

Next we show that $A_n \to A$ in the Hausdorff metric. Let $\epsilon > 0$ be given. Then there is some $m > 0$ such that for all $n \geq m$ we have $d_{\mathbb{H}}(A_n, A_m) < \epsilon/2$. However, this means that $A_n \subseteq (B_m)_{\epsilon/2}$ and $B_m \subseteq (A_n)_{\epsilon}$. Since $A \subseteq B_m$, this gives $A \subseteq (A_n)_{\epsilon}$. To show the reverse conclusion, let $x \in A_n$ be given. Then, since $d_{\mathbb{H}}(A_n, A_k) < \epsilon/2$ for all $k > n$, we have a sequence $x_k \in A_k$ with $d(x_k, x) < \epsilon/2$. However, then this means that $x_k \in B_n$, which is compact, and thus x_k has a cluster point $y \in B_n$ and so $d(y, x) \leq \epsilon/2$. Since the tail of the sequence x_k is in B_ℓ for all $\ell > n$, we have $y \in A$, which means that $A_n \subseteq A_\epsilon$. Thus $d_{\mathbb{H}}(A, A_n) < \epsilon$ as desired, so $A_n \to A$ in $\mathbb{H}(\mathbb{X})$. $\qquad\square$

Exercise 2.25. Does it make sense to extend $d_{\mathbb{H}}$ to closed and bounded (but not necessarily compact) sets? Investigate which properties remain and which (if any) change.

The Hausdorff distance has some interesting features. For instance, it is possible for a sequence of finite sets to converge to an uncountable set. As a simple example, for each $n \in \mathbb{N}$, let $A_n = \{i/n : 0 \leq i \leq n\} \subset [0, 1]$. Then $d_{\mathbb{H}}(A_n, [0, 1]) = 1/(2n)$, so $A_n \to [0, 1]$ in $\mathbb{H}(\mathbb{R})$. Another interesting property is that the addition or removal of a single point can drastically change the distance between two sets. This is also easy to see. Let $A = [0, 1]$ and $B = [0, 1] \cup \{x\}$, where $x \notin [0, 1]$. Then $d_{\mathbb{H}}(A, B) = \max\{-x, x-1\}$ and so can be arbitrarily large. This means that $d_{\mathbb{H}}$ might need to be modified or replaced for some applications, such as matching digital images.

Exercise 2.26. There are many other distances one could define between sets. For $A, B \subseteq \mathbb{R}^2$, investigate the properties of the "distance"

$$d(A, B) = \int_{\mathbb{X}} |\chi_A(t) - \chi_B(t)| \, d\lambda(t),$$

where χ_A is the characteristic function of A and λ is the Lebesgue measure. Does this define a metric? Why or why not? Is it necessary to restrict the sets to some class of sets in order for this to be a well-defined metric?

2.2.3 Definition of IFS

After setting the stage for the space in which our fractals will "live," we now define the notion of an iterated function system.

Definition 2.27. An *iterated function system*, or *IFS*, on a metric space $\mathbb{X}$ is a finite collection of mappings $w_i : \mathbb{X} \to \mathbb{X}$, $i = 1, 2, \ldots, N$, which are usually contractive. The *contractivity* of the IFS is the number $c := \max_i c_i$, where c_i is the contractivity of w_i.

Our example of the Cantor set $\mathcal{C}$ from Sect. 2.2.1 had the IFS $\{w_0, w_1\}$ on $\mathbb{X} = \mathbb{R}$, where $w_0(x) = x/3$ and $w_1(x) = x/3 + 2/3$. Thus $c_0 = c_1 = 1/3$, and so the contraction factor of the IFS is also $1/3$.

Exercise 2.28. Sometimes it is not necessary that all the maps be contractive, just that they be *eventually contractive*. Take the two maps $f, g : \mathbb{R}^2 \to \mathbb{R}^2$ given by

$$f \begin{pmatrix} x \\ y \end{pmatrix} = \begin{pmatrix} y \\ (x+1)/2 \end{pmatrix}, \quad g \begin{pmatrix} y \\ y \end{pmatrix} = \begin{pmatrix} y \\ x/2 \end{pmatrix}.$$

Show that the IFS $\{f \circ f, f \circ g, g \circ f, g \circ g\}$ is contractive even though $\{f, g\}$ is not.

One main idea of an IFS is that it formally encodes the idea of self-similarity. A *similarity* (or *similarity transform*) is a mapping $w : \mathbb{X} \to \mathbb{X}$ with the property that $d(w(x), w(y)) = r\, d(x, y)$ for all x, y, where $r > 0$ is the *similarity ratio*. This means that while distances are not preserved, all ratios of distances are. For the Cantor set $\mathcal{C}$, each part of $\mathcal{C}$ is clearly similar to the whole set, except scaled by the similarity ratio $r = 1/3$. Thus $\mathcal{C}$ is composed of two parts, where each is similar to the whole.

Exercise 2.29. Let $f : \mathbb{R}^n \to \mathbb{R}^n$ be distance preserving under the usual distance (that is, $\|x - y\| = \|f(x) - f(y)\|$ for all $x, y \in \mathbb{R}^n$). Show that f is linear. Use this to show that all similarities in $\mathbb{R}^n$ are linear.

As another example, consider the *Sierpinski triangle* $\mathcal{S}$, illustrated in Fig. 2.4. In this case, we can easily see that $\mathcal{S}$ is composed of three similar copies of itself, each with a similarity ratio $r = 1/2$. These three parts correspond to the three mappings

$$w_0\begin{pmatrix} x \\ y \end{pmatrix} = \frac{1}{2}\begin{pmatrix} x \\ y+1 \end{pmatrix}, \ w_1\begin{pmatrix} x \\ y \end{pmatrix} = \frac{1}{2}\begin{pmatrix} x \\ y \end{pmatrix}, \ w_2\begin{pmatrix} x \\ y \end{pmatrix} = \frac{1}{2}\begin{pmatrix} x+1 \\ y \end{pmatrix}.$$

The space $\mathbb{X}$ can be either $[0,1]^2$ or $\mathbb{R}^2$; it makes no difference. The former choice makes sense if we are thinking of the context of images on a computer screen.

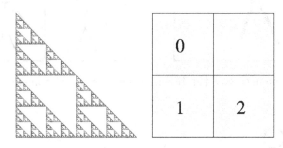

Fig. 2.4: The Sierpinski triangle and the three mappings in the IFS.

Now clearly any map $w : \mathbb{X} \to \mathbb{X}$ naturally induces a mapping $\hat{w} : \mathbb{H}(\mathbb{X}) \to \mathbb{H}(\mathbb{X})$ via $\hat{w}(A) = w(A) = \{w(a) : a \in A\}$. Often we will not distinguish between $w : \mathbb{X} \to \mathbb{X}$ and $\hat{w} : \mathbb{H}(\mathbb{X}) \to \mathbb{H}(\mathbb{X})$, but formally these two functions are different.

An IFS $\{w_1, w_2, \ldots, w_N\}$ induces a mapping on $\mathbb{H}(\mathbb{X})$ given by

$$W(S) = \bigcup_{i=1}^{N} \hat{w}_i(S). \tag{2.5}$$

The action of this mapping for the Sierpinski triangle IFS is illustrated in Fig. 2.5. We can see in this figure that the original set (the "stick man") is mapped by each of the three maps in the IFS and produces a new set that consists of three reduced copies of the original set.

Looking at Fig. 2.6, we see the repeated iteration of this process starting with the initial set S_0 of the "stick man." It seems clear that the limit should be the Sierpinski triangle and that the Sierpinski triangle will be the fixed point of the mapping W. This suggests that W is a contraction on $\mathbb{H}(\mathbb{R}^2)$. Our next task is to prove this in general for an IFS.

First we note that if $w : \mathbb{X} \to \mathbb{X}$ is a contraction with contractivity $c < 1$, then $\hat{w} : \mathbb{H}(\mathbb{X}) \to \mathbb{H}(\mathbb{X})$ is also a contraction with contractivity c. This is easy to see since $d(w(x), w(y)) \le c\,d(x,y)$ implies $\inf_{b \in B} d(w(a), w(b)) \le c \inf_{b \in B} d(a,b)$, and thus the one-sided distance satisfies this inequality as well; that is, $d(\hat{w}(A), \hat{w}(B)) \le c\,d(A, B)$.

Fig. 2.5: The action of the IFS mapping for the Sierpinski triangle.

Fig. 2.6: The iteration of the IFS mapping for the Sierpinski triangle.

Next we show that

$$d_{\mathbb{H}}(A_1 \cup A_2, B_1 \cup B_2) \leq \max\{d_{\mathbb{H}}(A_1, B_1), d_{\mathbb{H}}(A_2, B_2)\}. \qquad (2.6)$$

To see this, notice that for any $C \in \mathbb{H}(\mathbb{X})$ we have

$$d(A_1 \cup A_2, C) = \sup_{a \in A_1 \cup A_2} d(a, C)$$

$$= \max\left\{ \sup_{a \in A_1} d(a, C), \sup_{a \in A_2} d(a, C) \right\}$$

$$= \max\{d(A_1, C), d(A_2, C)\}$$

and

$$d(a, B_1 \cup B_2) = \min\{d(a, B_1), d(a, B_2)\}.$$

Thus,

$$d_{\mathbb{H}}(A_1 \cup A_2, B_1 \cup B_2)$$

$$= \max\left\{ \min\{d(A_1, B_1), d(A_1, B_2)\}, \min\{d(A_2, B_1), d(A_2, B_2)\}, \right.$$

$$\left. \min\{d(B_1, A_1), d(B_1, A_2)\}, \min\{d(B_2, A_1), d(B_2, A_2)\} \right\}$$

$$\leq \max\{d(A_1, B_1), d(B_1, A_1), d(A_2, B_2), d(B_2, A_2)\}$$

$$= \max\{d_{\mathbb{H}}(A_1, B_1), d_{\mathbb{H}}(A_2, B_2)\}.$$

This property of the Hausdorff distance leads to the next result.

Proposition 2.30. *Let $\{w_i\}$ be an IFS on $\mathbb{X}$ with contractivity $c < 1$. Then the induced map W on $\mathbb{H}(\mathbb{X})$ is also contractive with contractivity c.*

Corollary 2.31. *Every IFS induced mapping W has a unique fixed point $A \in \mathbb{H}(\mathbb{X})$, which satisfies the self-tiling equation*

$$A = \bigcup_{i=1}^{N} w_i(A). \tag{2.7}$$

The fixed point of the IFS is called the *attractor* of the IFS. In our two examples, the Cantor set was the attractor of the two-map IFS and the Sierpinski triangle was the attractor of the three-map IFS. Figure 2.6 shows the iterates of the IFS induced mapping W on a starting image. We clearly see the convergence of these iterates in these images.

In Fig. 2.7, we show several other attractors of different IFSs. It is instructive to try to determine the number of maps in the IFS from the image. It is even more instructive to try to find the IFS maps themselves. This is the *inverse problem*, and the collage theorem is very useful in this regard.

Exercise 2.32. For each attractor in Fig. 2.7, find an IFS.

Notice that not all these attractors are strictly self-similar, in that some of the maps in the IFS are not similarities – they distort the angles and distances. However, all of these attractors are *self-affine* in that all the IFS maps are affine (a linear function plus perhaps a translation).

Fig. 2.7: Some other attractors of various IFSs.

If $K \subseteq \mathbb{X}$ is closed and nonempty and $\cup_i w_i(K) \subseteq K$, then it is easy to see that the attractor $A \subseteq K$ (see [75], Sect. 3.1). Similarly, if $\cup_i w_i(K) \supseteq K$ and K is compact, then $K \subseteq A$.

Exercise 2.33. Show these two properties. That is, if $K \subseteq \mathbb{X}$ is closed and nonempty and $\cup_i w_i(K) \subseteq K$, then show that $A \subseteq K$. Also show that if K is compact and $\cup_i w_i(K) \supseteq K$, then $K \subseteq A$. Find an example of an IFS and a noncompact K such that $\cup_i w_i(K) \supseteq K$ but K is not contained in the attractor.

Topologically, the Cantor set is a model for a large class of attractors of IFSs. In fact, if the IFS is *nonoverlapping*, the attractor is always homeomorphic to the Cantor set. We say that the IFS is nonoverlapping if $w_i(A) \cap w_j(A) = \emptyset$ for $i \neq j$, so that the union in (2.7) is a disjoint union. Since A is compact, so are $w_i(A)$ for each i, and thus $w_i(A)$ is strictly separated from $w_j(A)$. That is, there is some $\epsilon > 0$ such that $\inf_{x \in A} \inf_{y \in A} d(w_i(x), w_i(y)) \geq \epsilon$. By the recursive nature of the IFS attractor A, this strict separation descends to all levels and thus A is totally disconnected (which means that the largest connected components are single points). Two of the attractors in Fig. 2.7 are totally disconnected and thus homeomorphic to the Cantor set.

The Cantor set is a topological model for any compact, totally disconnected and perfect metric space. A *perfect set* is a set that has no isolated points. This means that each point is a limit point of other points in the set. Any compact, totally disconnected, and perfect metric space S is homeomorphic to the Cantor set. It is not so hard to show that most attractors of IFSs are perfect sets. All the attractors we have given are perfect sets.

Exercise 2.34. Investigate sufficient conditions for an attractor of an IFS to be a perfect set.

There is another common "disjointness" condition, the *open set condition*.

Definition 2.35. (Open set condition) The IFS $\{w_1, w_2, \ldots, w_N\}$ satisfies the *open set condition* if there is some open set $U \subseteq \mathbb{X}$ such that $\cup_i w_i(U) \subseteq U$ and $w_i(U) \cap w_j(U) = \emptyset$ if $i \neq j$.

The open set condition does not mean that the images $w_i(A)$ are disjoint. An example is the Sierpinski triangle, which satisfies the open set condition with the open square $(0,1)^2$. In this case, this shows that the images are "almost" disjoint. The open set condition is very important in the measure-theoretic study of IFS attractors as it has implications for the dimension of the attractor (see [75, 55, 56, 57, 126]).

2.2.4 Collage theorem for IFS

The inspiration for the name "collage theorem" comes from its use in IFS inverse problems. Consider the attractor in Fig. 2.8, the "maple leaf" attractor. How would one design an IFS whose attractor resembled this set (or was exactly this set)? The idea is to "collage" the image with reduced copies of the entire image (reduced and perhaps distorted in a simple way). We see how to do this in an approximate way in Fig. 2.8. This image gives a good indication as to the reason for the name "collage theorem." Since we can approximately cover the maple leaf with four reduced and transformed copies of itself, we should be able to find an IFS with four maps whose attractor very closely resembles the maple leaf. The better the fit in the "collage," the better the attractor of the resulting IFS will resemble the maple leaf. This is the visual content of the collage theorem in the IFS setting.

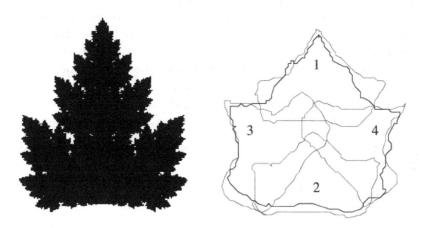

Fig. 2.8: (left) the maple leaf attractor and (right) its collage.

In the IFS context, the collage theorem takes the form

$$d_{\mathbb{H}}(B, A) \leq \frac{1}{1 - c}\, d_{\mathbb{H}}(B, w_1(B) \cup w_2(B) \cup \cdots \cup w_N(B)),$$

where A is the attractor and B is the "target" set.

Using this idea, it is easier to find an IFS that will generate a given image (at least in some cases). A nice example is the image in Fig. 2.9. We invite the reader to find a set of affine maps that will "collage" this attractor.

Exercise 2.36. Can you design an IFS whose attractor is a compact curve in $\mathbb{R}^2$? A curve in $\mathbb{R}^2$ is the continuous image of an interval. Is the Sierpinski triangle a curve?

2.2.5 Continuous dependence of the attractor

In this section, we prove that the attractor of an IFS is a continuous function of the parameters of the IFS. That is, if we vary the collection

Fig. 2.9: Can you find the IFS to generate this image?

of IFS maps in a continuous fashion, the attractor will change continuously. In the abstract, this is a direct consequence of Theorem 2.9; however, it is worth pointing out the aspects that make the IFS case different. In particular, the induced mapping on $\mathbb{H}(\mathbb{X})$ is composed of several individual mappings, and this makes the situation more complicated. Our presentation of this material borrows very heavily from [39]. Throughout this section, $(\mathbb{X}, d)$ is a complete metric space.

Definition 2.37. Let $0 \le s < 1$. Define

$$\text{Con}_s(\mathbb{X}) := \{f : \mathbb{X} \to \mathbb{X} : d(f(x), f(y)) \le s\, d(x, y) \text{ for all } x, y \in \mathbb{X}\}.$$

We place on $\text{Con}_s(\mathbb{X})$ the metric

$$d_\infty(f, g) = \max\left\{1, \sup_{x \in \mathbb{X}} d(f(x), g(x))\right\}.$$

We see that $\text{Con}_s(\mathbb{X})$ could also be named the Lipschitz functions of Lipschitz factor at most s. We use the name Con to emphasize that we are interested in contractions. It is straightforward to show that d_∞ is a metric. It is also easy to show that $(\text{Con}_s(\mathbb{X}), d_\infty)$ is a complete metric space; the idea is that if f_n is d_∞-Cauchy, then $f_n(x)$ is d-Cauchy for each x, so we obtain a limit in a pointwise fashion. The following theorem is a restatement of Theorem 2.9.

Theorem 2.38. *The fixed point of $f \in \text{Con}_s(\mathbb{X})$ is a continuous function of $f \in \text{Con}_s(\mathbb{X})$.*

Now, an IFS is a finite collection of contractions, not just a single contraction. While this collection does induce a single contraction on $\mathbb{H}(\mathbb{X})$, our development does not use this particular idea. Instead we think of an IFS as living in a finite product of copies of $\text{Con}_s(\mathbb{X})$.

Definition 2.39. Let $0 \le s < 1$ and $N \in \mathbb{N}$ be given. Then we define

$$\operatorname{Con}_s^N(\mathbb{X}) := \operatorname{Con}_s(\mathbb{X}) \times \operatorname{Con}_s(\mathbb{X}) \times \cdots \times \operatorname{Con}_s(\mathbb{X}),$$

where there are N factors in the product. We place the metric

$$d_\infty(F, G) = \max_i d_\infty(f_i, g_i)$$

on $\operatorname{Con}_s^N(\mathbb{X})$, where we have $F = (f_1, \ldots, f_N)$ and $G = (g_1, \ldots, g_N)$, so $F, G \in \operatorname{Con}_s^N(\mathbb{X})$.

It is also easy to see that $\operatorname{Con}_s^N(\mathbb{X})$ is a complete metric space, being the finite product of complete metric spaces.

Given $F \in \operatorname{Con}_s^N(\mathbb{X})$, we have the function $F \mapsto A(F)$, where $A(F)$ is the attractor of the IFS given by F. This is the function we are particularly interested in and that we show is continuous.

Theorem 2.40. *(Continuity of the attractor of an IFS) The function* $A : \operatorname{Con}_s^N(\mathbb{X}) \to \mathbb{H}(\mathbb{X})$ *that maps an IFS to its attractor is continuous.*

Proof. Let $0 < \epsilon < 1$ be given, and suppose that $F, G \in \operatorname{Con}_s^N(\mathbb{X})$ are such that $d_\infty(F, G) < \epsilon$. This means that $d(f_i(x), g_i(x)) < \epsilon$ for all $i = 1, 2, \ldots, N$ and all $x \in X$. Then, for any $B \in \mathbb{H}(\mathbb{X})$, we have

$$d_\mathbb{H}(F(B), G(B)) = d_\mathbb{H}\left(\bigcup_i f_i(B), \bigcup_i g_i(B)\right)$$
$$\le \max_i d_\mathbb{H}(f_i(B), g_i(B)) < \epsilon.$$

Thus, $\sup_{B \in \mathbb{H}(\mathbb{X})} d_\mathbb{H}(F(B), G(B)) \le \epsilon$ and so, by Proposition 2.8, we have $d_\mathbb{H}(A(F), A(G)) \le \epsilon/(1 - s)$. $\qquad\square$

As a simple corollary to this result, if the maps in the N-map IFS F depend on some parameters λ_j in a continuous way, then the attractor depends on the λ_j in a continuous way. Barnsley's picturesque description of this is that the attractor is "blowing in the wind" as it continuously reacts to the changes in the IFS parameters.

Corollary 2.41. *Suppose that* $F(\lambda_1, \lambda_2, \ldots, \lambda_k) \in \operatorname{Con}_s^N(\mathbb{X})$ *and* F *depends continuously on the parameters* $\lambda_j \in \mathbb{R}$. *Then the attractor of* F *depends continuously on the parameters* λ_j.

Exercise 2.42. Find a proof of Theorem 2.40 that is based on proving that the function $\operatorname{Con}_s^N(\mathbb{X}) \mapsto \operatorname{Con}_s(\mathbb{H}(\mathbb{X}))$ is continuous.

Exercise 2.43. Make an animation (movie) that illustrates the continuous dependence of an attractor of an IFS on the parameters in an IFS. For example, continuously interpolate between the fern leaf and the maple leaf.

2.3 Code space and the address map

There is a nice "addressing scheme" that labels all the points of the attractor of an IFS. This addressing scheme is driven by the recursive nature of the attractor, so it is not surprising that the structure of the addressing scheme is also recursive.

In order to explain this addressing scheme, we will concentrate on the example of the Sierpinski triangle $\mathcal{S}$, illustrated in Fig. 2.4. We notice that $\mathcal{S}$ satisfies the self-similarity identity

$$\mathcal{S} = w_0(\mathcal{S}) \cup w_1(\mathcal{S}) \cup w_2(\mathcal{S}),$$

and we have that the "parts," the various $w_i(\mathcal{S})$, are almost disjoint (they touch at a single point). These "first-level tiles" are illustrated in Fig. 2.10.

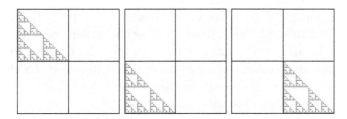

Fig. 2.10: The three "tiles" of the Sierpinski triangle.

We can perform another level in decomposing $\mathcal{S}$ into nine "second-level tiles" as

$$\begin{aligned}
\mathcal{S} = \; & w_0(w_0(\mathcal{S})) \; \cup w_0(w_1(\mathcal{S})) \cup w_0(w_2(\mathcal{S})) \\
& \cup \, w_1(w_0(\mathcal{S})) \cup w_1(w_1(\mathcal{S})) \cup w_1(w_2(\mathcal{S})) \\
& \cup \, w_2(w_0(\mathcal{S})) \cup w_2(w_1(\mathcal{S})) \cup w_2(w_2(\mathcal{S})).
\end{aligned}$$

Continuing, we get the "nth-level" decomposition

$$\mathcal{S} = \bigcup_{i_1=1}^{N} \bigcup_{i_2=1}^{N} \cdots \bigcup_{i_n=1}^{N} w_{i_1} \circ w_{i_2} \circ \cdots \circ w_{i_n}(\mathcal{S}).$$

At this point, we notice that the diameter of $w_{i_1} \circ w_{i_2} \circ \cdots \circ w_{i_n}(\mathcal{S})$ is equal to $2^{-n} \times \operatorname{diam}(\mathcal{S}) = 2^{-n}\sqrt{2}$, which goes to zero as $n \to \infty$. For the moment, fix an infinite sequence $i_n \in \{0, 1, 2\}$ and consider the sequence of sets

$$\mathcal{S}_{i_1 i_2 \cdots i_n} := w_{i_1} \circ w_{i_2} \circ \cdots \circ w_{i_{n-1}} \circ w_{i_n}(\mathcal{S}).$$

There is a nesting of these sets as

$$S_{i_1 i_2 \cdots i_{n-1} i_n} = w_{i_1} \circ w_{i_2} \circ \cdots \circ w_{i_{n-1}} \circ w_{i_n}(S)$$
$$\subseteq w_{i_1} \circ w_{i_2} \circ \cdots \circ w_{i_{n-1}}(S) = S_{i_1 i_2 \cdots i_{n-1}}$$

(notice that the difference is the extra map applied on the "inside" for the left side of this containment). These facts mean that

$$\bigcap_{n=1}^{\infty} S_{i_1 i_2 \cdots i_n} = \{x\},$$

where x is some point in $\mathbb{R}^2$. In fact, it is easy to see that $x \in S$ as $w_{i_1}(S) \subset S$ and thus each one of the sets in the sequence is a subset of S. Repeating with a different infinite sequence $i_n \in \{0, 1, 2\}$, we get another point of S; however, it is possible to start with two different sequences and obtain the same point of S.

Given a point $x \in S$, there must be some $i_1 \in \{0, 1, 2\}$ such that $x \in w_{i_1}(S)$. Similarly, because of the "second-order" decomposition, there must be an i_2 such that $x \in w_{i_1} \circ w_{i_2}(S)$. Continuing, to each x we can associate a sequence i_n such that $x \in w_{i_1} \circ \cdots \circ w_{i_n}(S)$ for each n; the more "digits" (terms in the sequence i_n) we know, the more precisely we know the "location" of the point x.

This situation for the Sierpinski triangle is illustrated in Fig. 2.11, where the first image shows the "first-level" addresses, the second image the "second-level" addresses, and the third image the "third-level" addresses. Every point in the region marked "021" is in the set $w_0 \circ w_2 \circ w_1(S)$, or has an address that starts with "021."

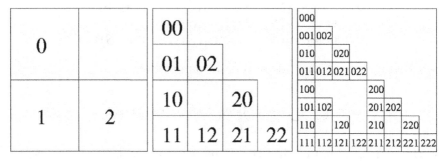

Fig. 2.11: Addresses on the Sierpinski triangle.

Another very common example is the example of base-10 representation of numbers in the interval $[0, 1]$. This corresponds to the 10-map

IFS $w_i(x) = x/10 + i/10$ for $i = 0, 1, 2, \ldots, 9$, whose attractor is the interval $[0, 1]$. The subset corresponding to $w_{i_0} \circ w_{i_1} \circ \cdots \circ w_{i_n}([0, 1])$ is the subset of $[0, 1]$ consisting of all those numbers whose base-10 expansion starts with $i_0 i_1 \ldots i_n$. In this special case, the "address" not only identifies (or labels) the point in $[0, 1]$ but also carries arithmetic information. This is completely atypical for a generic IFS, where the address is only a label (albeit a particularly nice, recursive label).

This mapping from the infinite sequence (the "address") to the point in $\mathcal{S}$ is the "address map," and we now move to a discussion of this in the general case.

The first step is to make a formal definition of "code space." For an IFS $\{w_1, w_2, \ldots, w_N\}$, our addresses will consist of infinite sequences where each term of the sequence can take any of the values $\{1, 2, \ldots, N\}$.

Definition 2.44. (Code space) The *code space* associated with the IFS $\{w_1, \ldots, w_N\}$ is the set

$$\Sigma = \{1, 2, \ldots, N\}^{\mathbb{N}} = \{\alpha : \alpha_n \in \{1, 2, \ldots, N\}, n \in \mathbb{N}\}.$$

We place the following metric on Σ:

$$d(\alpha, \beta) = \sum_{n=1}^{\infty} \frac{|\alpha_n - \beta_n|}{(N+1)^n}. \tag{2.8}$$

If $\alpha, \beta \in \Sigma$ agree in at least the first n positions (that is, $\alpha_i = \beta_i$ for $i = 1, 2, \ldots, n$), then we have

$$d(\alpha, \beta) \leq \sum_{i=n+1}^{\infty} \frac{N-1}{(N+1)^i} = \frac{N-1}{N(N+1)^n}. \tag{2.9}$$

This is because at worst α and β can disagree in every position after the nth and with the numerator in (2.8) being at worst $N - 1$.

Notice that (Σ, d) is a compact metric space (it is the countable product of the compact space $\{1, 2, \ldots, N\}$ endowed with the discrete topology). In fact, it is homeomorphic (but not isometric) to the standard Cantor set.

Exercise 2.45. Prove directly that (Σ, d) is a complete metric space.

We briefly explain why Σ is compact by showing that it is totally bounded. Let $\epsilon > 0$ be given. If we choose n such that $(N + 1)^{-n} < \epsilon$, then we know by (2.9) that any $\alpha, \beta \in \Sigma$ that agree in the first n places must satisfy $d(\alpha, \beta) < \epsilon$. There are N^n choices for these first n

places, so we can form an ϵ-net in Σ by taking each of these N^n choices for the first n places and then fixing all the rest of the positions to be equal to 1.

Definition 2.46. Fix $x \in \mathbb{X}$. The *address map* $\omega : \Sigma \to \mathbb{X}$ is defined as

$$\omega(\sigma) = \lim_{n \to \infty} w_{\sigma_1} \circ w_{\sigma_2} \circ \cdots \circ w_{\sigma_{n-1}} \circ w_{\sigma_n}(x). \qquad (2.10)$$

The address map is independent of the chosen point x. To see this, let $x \neq y$. Then, for a given $\sigma \in \Sigma$, we have that

$$d(w_{\sigma_1} \circ \cdots \circ w_{\sigma_n}(x), w_{\sigma_1} \circ \cdots \circ w_{\sigma_n}(y)) \leq c^n \, d(x, y),$$

where c is the contraction factor for the IFS. Thus, as $n \to \infty$, the distance goes to zero. We showed above that the limit does indeed exist. Thus the address map as given in (2.10) is well-defined and independent of the particular point that was chosen.

Just as in the specific case of the Sierpinski triangle above, we can label the images (or *subtiles*) as

$$A_{\sigma_1 \sigma_2 \cdots \sigma_n} = w_{\sigma_1} \circ w_{\sigma_2} \circ \cdots \circ w_{\sigma_n}(A). \qquad (2.11)$$

For notational simplicity, if $\sigma \in \Sigma$, we sometimes denote by $\sigma^n = \sigma_1 \sigma_2 \ldots \sigma_n$ the truncation of σ to the first n terms. This naturally leads to the notations

$$w_{\sigma^n} := w_{\sigma_1} \circ w_{\sigma_2} \circ \cdots \circ w_{\sigma_n}$$

and $A_{\sigma^n} = w_{\sigma^n}(A)$. Finally, it is also sometimes convenient to use $\Sigma^n = \{1, 2, \ldots, N\}^n$, the space of sequences of length n taken from the set $\{1, \ldots, N\}$.

Using this notation, the nth-level self-tiling equation becomes

$$A = \bigcup_{\alpha \in \Sigma^n} w_\alpha(A) = \bigcup_{\alpha \in \Sigma^n} A_\alpha. \qquad (2.12)$$

It is useful to notice that all the subtiles also satisfy a type of self-similarity condition,

$$A_\sigma = \bigcup_{i=1}^{n} w_\sigma \circ w_i(A) = w_\sigma \left(\bigcup_{i=1}^{N} w_i(A) \right), \qquad (2.13)$$

for any $\sigma \in \Sigma^n$.

We now show some simple properties of the address map. First we show that the range of ω is contained in the attractor A of the IFS.

To see this, we just note that ω is independent of the chosen point x, so we can start with some $x \in A$. But then $w_i(x) \in A$ as well for each $i = 1, 2, \ldots, N$ because of the tiling condition (2.7). Since A is closed, this means that the limit will also be a point in A. Thus $\omega(\Sigma) \subseteq A$. We will show that this is an equality.

Next we show that ω is continuous, in fact uniformly continuous. This is fairly easy to see. Let $\epsilon > 0$ be given, and recall that $c < 1$ is the contraction factor of the IFS, so each individual contraction factor c_i for w_i satisfies $c_i \leq c$. Then there is an n such that $c^n \operatorname{diam}(A) < \epsilon$. Let $x \in A$ be fixed (the "starting point" in the definition of ω). Choose δ such that

$$0 < \delta < \frac{N-1}{N(N+1)^n}.$$

Then, if $d(\alpha, \beta) < \delta$, we must have that α and β agree at least in the first n terms. But this means that

$$\omega(\alpha), \omega(\beta) \in w_{\alpha_1} \circ w_{\alpha_2} \circ \cdots \circ w_{\alpha_n}(A),$$

and since

$$\operatorname{diam}(w_{\alpha_1} \circ w_{\alpha_2} \circ \cdots \circ w_{\alpha_n}(A)) \leq c^n \operatorname{diam}(A) < \epsilon,$$

we have that $d(\omega(\alpha), \omega(\beta)) < \epsilon$, so ω is uniformly continuous.

We know that Σ is compact and ω is continuous, so $\omega(\Sigma) \subseteq A$ must also be compact. We now show that the address map ω is a surjection. We do this in a constructive way. So, let $x \in A$ be given. Then, from the nth-level self-tiling condition (2.12) for $n = 1$, we see that there is some $\sigma_1 \in \{1, 2, \ldots, N\}$ with $x \in A_{\sigma_1}$. However, since A satisfies (2.12) as well, there must be some σ_2 such that $x \in A_{\sigma_1 \sigma_2} \subset A_{\sigma_1}$. Continuing, we obtain a sequence $\sigma = \sigma_1 \sigma_2 \ldots \sigma_n \ldots$ such that $x \in A_{\sigma^n}$ for each n. Let $y = \omega(\sigma)$, the image of σ under the address map. Then, by construction of the address map, we know that $\omega(\sigma) \in A_{\sigma^n}$ for each n, which means that $d(x, y) \leq \operatorname{diam}(A_{\sigma^n}) \leq \operatorname{diam}(A) c^n \to 0$, and thus $x = y$. Therefore, the address map is surjective.

If $w_i(A) \cap w_j(A) \neq \emptyset$ for $i \neq j$, then any point in this intersection will have at least two different addresses. Thus, in general, ω is not injective. In fact, it is clear that it is injective if and only if the tiles are disjoint. Thus we have proved the following proposition.

Proposition 2.47. *The address map $\omega : \Sigma \to A$ is uniformly continuous and surjective; it is injective iff the IFS is disjoint.*

There are some points in A with particularly interesting addresses. First, the points in A with multiple addresses correspond to images of points in $w_i(A) \cap w_j(A)$ under some w_σ. Second, the fixed points of the maps w_i have particularly nice addresses. For instance, the fixed point of w_1 has the address $\sigma = 111\ldots$, and similarly for the other fixed points. Similarly, the point with address $21111\ldots$ is the image under w_2 of the fixed point of w_1. Those points $x \in A$ with periodic addresses of period length n are also interesting, as there is a finite sequence $\alpha \in \Sigma^n$ such that $x = w_\alpha(x)$. It is easy to show that periodic points are dense in the attractor.

The addresses give a nice encoding of the "dynamics" of the mappings w_i on the attractor A. For a good discussion of this, see [16].

Exercise 2.48. Consider the two IFSs $\mathcal{W}_1 = \{w_1, w_2, \ldots, w_N\}$ and $\mathcal{W}_2 = \{f_1, f_2, \ldots, f_M\}$ with $\mathcal{W}_2 \subseteq \mathcal{W}_1$. Use the idea of addresses and the address map to show that the attractor of $\mathcal{W}_2$ is a subset of the attractor of $\mathcal{W}_1$. Under what conditions could they be the same?

Exercise 2.49. A point $x \in A$ is a *periodic point* if some address of x is periodic. Show that periodic points are dense in the attractor A.

2.4 The chaos game

We have already seen one way to generate the attractor of an IFS – simply start with an initial set and iterate the mapping on $\mathbb{H}(\mathbb{X})$ that is induced by the IFS. This is often called the *deterministic method* of generating an attractor. There is another method, which is often called the *chaos game*. The first time most people see a description of the chaos game, they find it difficult to believe that it works. One very nice thing about the chaos game is that it quickly leads into ergodic theory, which is a very beautiful area of mathematics.

The chaos game is a surprising way to generate an image of an attractor of an IFS. The chaos game generates a random sequence of points that, when plotted, approximate the attractor to any desired degree of accuracy. The algorithm is very simple, and we describe it next. Again, we start with an IFS $\{w_1, w_2, \ldots, w_N\}$ and imagine that $\mathbb{X} = [0, 1]^2$, the computer screen.

1. Let $x_0 \in \mathbb{X}$ be chosen to start.
2. Choose $i_1 \in \{1, 2, \ldots, N\}$ randomly and equally likely. Compute $x_1 = w_{i_1}(x_0)$. Plot the point x_1.

3. For the general step, choose $i_{n+1} \in \{1, 2, \ldots, N\}$ randomly and equally likely, set $x_{n+1} = w_{n+1}(x_n)$, and then plot the point x_{n+1}.
4. Repeat the previous step as many times as desired until the image is close enough.

The first few points are unlikely to be points on the attractor A, so, if you wish, wait some number of iterations before plotting the points. An alternative is to start with $x_0 \in A$; choosing x_0 to be the fixed point of w_1 is a common choice.

The remarkable thing is that after some time the image of the attractor appears as if out of a cloud of smoke. The first few points are seemingly randomly distributed on the computer screen; in fact, they are somewhat randomly distributed on the attractor A. Figure 2.12 shows the image of the Sierpinski triangle after 400, 4000, and 10,000 points from the chaos game are plotted.

Fig. 2.12: Stages in the chaos game: 400, 4000, and 10,000 points plotted.

There is an elementary explanation as to why this algorithm works to produce an image of the attractor. What happens is that

$$A = \lim_{n \to \infty} \text{cl}(\{x_m : m \geq n\}),$$

where the limit is taken in the Hausdorff distance. Let $i_1 i_2 \ldots i_n \ldots$ be the infinite random sequence with $i_n \in \{1, 2, \ldots, N\}$ that is generated by the chaos game. Using this sequence of maps, the chaos game generates the sequence of points

$$x_n := w_{i_n} \circ w_{i_{n-1}} \circ \cdots w_{i_2} \circ w_{i_1}(x_0)$$

(notice that the composition of the maps is in the opposite order from the address map from (2.10)). Now, for any $a \in A_{i_n}$, we have $d(x_n, a) \leq \text{diam}(A)\, c$, while for any $a \in A_{i_n i_{n-1}}$ we have $d(x_n, a) \leq \text{diam}(A)\, c^2$. In general, if the first k positions in the address of $a \in A$ are $i_n i_{n-1} i_{n-2} \ldots i_{n-k+1}$, then $d(x_n, a) \leq \text{diam}(A)\, c^k$. If we have $a \in A$

and want to find some x_m such that $d(x_m, a) < \text{diam}(A)\, c^k$, all we need to do is find some n such that the first k positions of the address of a match $i_n i_{n-1} i_{n-2} \ldots i_{n-k+1}$. By the law of large numbers, with probability 1 this will eventually happen for any choice of a and any k. Thus, eventually the orbit generated by the chaos game will come arbitrarily close to any point of the attractor A.

The distribution of the points x_n on the attractor does not have to be uniform, however. To see this, one can imagine that we modify the IFS for the Sierpinski triangle by changing the first transformation to

$$w_0 \begin{pmatrix} x \\ y \end{pmatrix} = \begin{pmatrix} x/4 \\ (y+3)/4 \end{pmatrix}.$$

This new IFS produces the attractor shown in Fig. 2.13. The main difference on the "big" level is that the upper-left part is smaller than the two other parts, which corresponds to replacing the contraction factor of $1/2$ with the new contraction factor of $1/4$. If we run the chaos game as given with this new IFS, what will happen is that points in this "smaller" first-level tile will be plotted as frequently as in the other tiles, but since it is smaller in size this will result in a higher density of points in this region. As the IFS is recursive, this uneven density will be reflected on all smaller size scales. Thus this part of the image will appear darker than the rest, at least in the first part of the iterations.

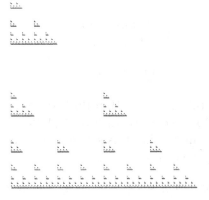

Fig. 2.13: A modified Sierpinski triangle.

The way to "correct" this (assuming we wish to correct it) is to use unequal weights in the chaos game. That is, at each step choose $i_{n+1} \in \{0, 1, 2\}$ with probabilities $p_0 = 1/9$ and $p_1 = p_2 = 4/9$. This

choice of probabilities will give a more uniform distribution of the
points plotted on the attractor.

This discussion leads naturally to the ideas of an IFS with probabilities and invariant measures of an IFS.

2.5 IFS with probabilities, self-similar measures, and the ergodic theorem

In this section, we define an IFS that acts on a new space, the space
of probability measures. The attractor of this IFS will no longer be
"geometric" (that is, a set) but an analytic object, a Borel probability
measure. This is the start of the program of defining appropriate IFS
operators on different spaces, all with the goal of introducing the idea
of self-similarity in these different contexts and with an eye toward
applying this self-similarity to modeling. In this book, we will carry
this program a significant distance, defining IFS operators on various
function spaces, spaces of measures, and also in the set-valued context.
In each case, the appropriate IFS framework is the basic requirement
for any further theory or application.

2.5.1 IFSP and invariant measures

Definition 2.50. (IFS with probabilities) An IFS with probabilities
is a finite collection of contractions $\{w_1, w_2, \ldots, w_N\}$ on the complete metric space $\mathbb{X}$ along with a finite collection of probabilities
$\{p_1, p_2, \ldots, p_N\}$ (that is, $p_i \geq 0$ and $\sum_i p_i = 1$).

An IFS with probabilities (IFSP) is the basic structure used in the
chaos game. Associated with each IFSP there is an *invariant measure*,
also called a *fractal measure* or sometimes *self-similar measure*. This
invariant measure is also given as a fixed point of a contraction, the
Markov operator associated with the IFS.

We use $\mathcal{P}(\mathbb{X})$ to denote the space of all Borel probability measures
on the metric space $\mathbb{X}$. For a brief discussion of measure theory, see
Appendix B.

Definition 2.51. (IFSP Markov operator) Let $\{w_i, p_i\}$ be an N-map
IFSP. Associated with this IFSP is the *Markov operator* $M : \mathcal{P}(\mathbb{X}) \to$
$\mathcal{P}(\mathbb{X})$ given as

$$(\mathbb{M}\mu)(B) = \sum_{i=1}^{N} p_i \mu(w_i^{-1}(B)) \tag{2.14}$$

for each Borel set $B \subseteq \mathbb{X}$.

The intuition behind the definition of $\mathbb{M}$ is that it uses the maps w_i to "move around" and "shrink" the support of μ and then the p_i scale the mass so that it remains a probability measure. As with the IFS operator on $\mathbb{H}(\mathbb{X})$, we have that $\mathbb{M}\mu$ is a combination of "smaller" and "distorted" copies of μ.

We notice that $\mathbb{M}$ is linear on $\mathcal{M}(\mathbb{X})$, the space of all signed measures on $\mathbb{X}$. Thus, if $\mathbb{M}$ is contractive on the whole space $\mathcal{M}(\mathbb{X})$, then, since it is linear, the only possible fixed point would be the zero measure, which would not be so interesting. Thus, to get a nontrivial theory, we need to restrict our attention to a subspace, in this case the probability measures. However, notice that if $\mu \in \mathcal{P}(\mathbb{X})$ is a fixed point for $\mathbb{M}$, then so is $\lambda\mu$ for any $\lambda \in \mathbb{R}$. This is completely analogous to the finite-dimensional case of Markovian transition matrices.

It is clear that $\mathbb{M}$ maps a probability measure to another probability measure, so this is a well-defined map. In order for $\mathbb{M}$ to be a contraction, we need a metric on $\mathcal{P}(\mathbb{X})$. We use the Monge-Kantorovich metric; see Sect. B.5.1 for a discussion of this metric. For the convenience of the reader, we will repeat some of the discussion here.

In order to guarantee a finite number from our formula for the Monge-Kantorovich metric, we will need to restrict the probability measures further. We define a subspace of the probability measures as follows.

Definition 2.52. Let $a \in \mathbb{X}$. Define

$$\mathcal{P}_1(\mathbb{X}) = \left\{ \mu \in \mathcal{P}(\mathbb{X}) : \int_{\mathbb{X}} d(a, x)\, d\mu(x) < \infty \right\}.$$

It is important to note that if the integral in this definition is finite for some given a, it is also finite for any other fixed b because of the triangle inequality and the fact that μ is a probability measure. We now define the Monge-Kantorovich metric on $\mathcal{P}_1(\mathbb{X})$.

Definition 2.53. (Monge-Kantorovich metric) For $\mu, \nu \in \mathcal{P}_1(\mathbb{X})$, we define

$$d_{MK}(\mu, \nu) = \sup \left\{ \int_{\Omega} f\, d(\mu - \nu) : f \in \text{Lip}_1(\Omega) \right\}. \tag{2.15}$$

Exercise 2.54. Show that if $\mu(\mathbb{X}) \neq \nu(\mathbb{X})$, then the right-hand side of (2.15) is infinite.

Exercise 2.55. Let $\mu(\mathbb{X}) = \nu(\mathbb{X})$. Show that

$$\int_{\mathbb{X}} f(x)d(\mu - \nu)(x) = \int_{\mathbb{X}} g(x)d(\mu - \nu)(x)$$

if $f(x) = g(x) + c$, where $c \in \mathbb{R}$.

A very important feature of the Monge-Kantorovich metric is that it incorporates the underlying distance on the metric space $\mathbb{X}$ into computing the distance between two measures. For instance, if the supports of μ and ν are strictly separated (that is, if $0 < \delta \leq \inf_x \inf_y d(x, y)$), where x is taken from the support of μ and y from the support of ν, then $d_{MK}(\mu, \nu) \geq \delta$. Thus $d_{MK}(\mu, \nu)$ combines information about both the supporting sets of μ and ν and how μ and ν are "distributed" on their respective supports.

Exercise 2.56. Let $x, y \in \mathbb{X}$ and δ_x, δ_y be the point masses at x and y. Show that $d_{MK}(\delta_x, \delta_y) = d(x, y)$. What if both μ and ν are convex combinations of point masses?

Exercise 2.57. Show that the set of probability measures that are convex combinations of point masses is dense in $(\mathcal{P}_1(\mathbb{X}), d_{MK})$.

Theorem 2.58. *(Completeness of space of probability measures) Let* $\mathbb{X}$ *be a complete and separable metric space. Then* $\mathcal{P}_1(\mathbb{X})$ *is a complete metric space under the Monge-Kantorovich metric. Furthermore, if* $\mathbb{X}$ *is compact, then* $\mathcal{P}(\mathbb{X}) = \mathcal{P}_1(\mathbb{X})$ *and both are also compact under the Monge-Kantorovich metric.*

Proof. We give a basic idea of the proof, leaving the technical details to the references [72, 73, 88, 171].

First, it is easy to see that $d_{MK}(\mu, \nu) \geq 0$. The fact that it is always finite follows from the estimate

$$|\int_{\mathbb{X}} f(x)d(\mu - \nu)(x)| \leq \int_{\mathbb{X}} |f(a) + d(x, a)|d(\mu + \nu)(x)$$

$$\leq 2f(a) + \int_{\mathbb{X}} d(x, a)d\mu(x) + \int_{\mathbb{X}} d(x, a)d\nu(x).$$

The fact that d_{MK} is symmetric is also obvious, as is the triangle inequality. The two more difficult things to show are the fact that $d_{MK}(\mu, \nu) = 0$ implies that $\mu = \nu$ and the completeness. Both of

these result from a fundamental duality, the duality of the space of
Lipschitz functions under a Lipschitz norm and the appropriate space
of measures. This duality was one of the fundamental contributions of
Kantorovich to this area.   □

Exercise 2.59. Give a complete proof of Theorem 2.58 in the special
case where $\mathbb{X}$ is compact. The Stone-Weierstrass, Banach-Alaoglu, and
Riesz representation theorems might be useful.

Now that we have the definition of the IFSP Markov operator $\mathbb{M}$
and a complete metric space in which this operator can be applied, we
show that $\mathbb{M}$ is contractive.

Theorem 2.60. *Let c_i be the contraction factor of w_i in the IFSP
$\{w_i, p_i\}$ and $c := \max_i c_i$. The IFSP Markov operator defined in (2.15)
satisfies*

$$d_{MK}(\mathbb{M}\mu, \mathbb{M}\nu) \leq \left(\sum_i p_i c_i\right) d_{MK}(\mu, \nu). \qquad (2.16)$$

*Thus if $\sum_i p_i c_i < 1$, $\mathbb{M}$ is a contraction on $(\mathcal{P}_1(\mathbb{X}), d_{MK})$ and there is
a unique Borel probability measure $\mu \in \mathcal{P}_1(\mathbb{X})$ such that $\mathbb{M}(\mu) = \mu$, or*

$$\mu(B) = \sum_i p_i \mu(w_i^{-1}(B)) \qquad (2.17)$$

for every Borel set $B \subseteq \mathbb{X}$. In particular, this happens if $c < 1$.

Proof. Take $\mu, \nu \in \mathcal{P}_1(\mathbb{X})$, and let $f : \mathbb{X} \to \mathbb{R}$ be a function with
Lipschitz constant equal to 1. Then we see that

$$\int_{\mathbb{X}} f(x) d(\mathbb{M}\mu - \mathbb{M}\nu)(x) = \int_{\mathbb{X}} f(x) \sum_i p_i d(\mu \circ w_i^{-1} - \nu \circ w_i^{-1})(x)$$

$$\leq \int_{\mathbb{X}} \left(\sum_i p_i f(w_i(y))\right) d(\mu - \nu)(y).$$

Let $\hat{f} = \sum_i p_i f \circ w_i$; we show that $\hat{f}$ is Lipschitz with factor $\sum_i p_i c_i$.
For $x, y \in \mathbb{X}$, we compute that

$$|\hat{f}(x) - \hat{f}(y)| = \left|\sum_i p_i [f(w_i(x)) - f(w_i(y))]\right|$$

$$\leq \sum_i p_i |f(w_i(x)) - f(w_i(y))|$$

$$\leq \sum_i p_i d(w_i(x), w_i(y))$$

$$\leq \left(\sum_i p_i c_i \right) d(x, y),$$

and thus $\hat{f}$ has a Lipschitz factor as desired. Since this is true for any Lipschitz f, the result follows.   □

We call the quantity $\sum_i p_i c_i$ the *average contractivity factor* of the IFSP.

In Fig. 2.14 we show that the invariant measure for the IFSP $w_0(x) = x/2$ and $w_1(x) = x/2 + 1/2$ and p_0, p_1 are as given in the figure. Each image is a histogram that has been normalized to have total mass 1024 (as there are 1024 entries in each histogram). For the first image, the vertical scale is $[0, 110]$, and in the second it is $[0, 2.5]$. This along with the image gives an indication of how unbalanced the first measure is. The scaling behaviour is also evident in the images.

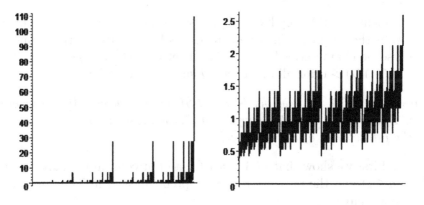

Fig. 2.14: Invariant measures for $p_0 = 0.2, p_1 = 0.8$ and $p_0 = 0.45, p_1 = 0.55$.

These two measures are part of a very interesting family of such measures that has a nice alternate description not involving IFSP. Take the two probabilities p_0 and p_1 and construct a number $x \in [0, 1]$ by randomly and independently choosing each binary digit of x. We choose the nth digit to be equal to 0 with probability p_0 and equal to 1 with probability p_1. The resulting distribution on $[0, 1]$ is also the invariant measure for the IFSP given above. If we take two such distributions μ and $\hat{\mu}$, with probabilities p_i and $\hat{p}_i$, respectively, then it turns out that unless $p_i = \hat{p}_i$, μ and $\hat{\mu}$ are mutually singular. To see this, let

$$B(p_0) = \left\{ x \in [0,1] : \lim_n \frac{1}{n} \#\{1 \le i \le n : x_i = 0\} = p_0 \right\},$$

where x_i is the ith binary digit of x. Then $\mu(B(p_0)) = 1$, while $\hat{\mu}(B(p_0)) = 0$ if $p_0 \ne \hat{p}_0$.

Exercise 2.61. Argue that $B(p_0)$ is an infinite set for any $p_0 \ne 0$. Prove that $\mu(B(p_0)) = 1$ while $\hat{\mu}(B(p_0)) = 0$ if $p_0 \ne \hat{p}_0$.

As one would expect, there is a close connection between the invariant measure μ of an IFSP and the geometric attractor of the related IFS (just the maps w_i without the probabilities p_i). To state this relationship, we recall the definition of the *support* of a measure.

Definition 2.62. (Support of a measure) The *support* of the measure μ on $\mathbb{X}$ is the closed set

$$\operatorname{supp}(\mu) = \mathbb{X} \setminus \bigcup \{U : U \text{ is an open set with } \mu(U) = 0\}.$$

As we mention in Appendix B, if $\mathbb{X}$ is separable and μ is Borel (which is always the case with invariant measures by definition), the support can also be characterized as the closed set C such that $\mu(\mathbb{X} \setminus C) = 0$ and $\mu(C \cap O) > 0$ for all open sets O such that $O \cap C \ne \emptyset$.

Theorem 2.63. Let $\{w_i, p_i\}$ be an IFSP with each $p_i > 0$ and where $c_i < 1$ for each i. Then the support of the invariant measure μ is equal to the attractor A of the IFS $\{w_i\}$.

Proof. First we show that $\mu(A_\epsilon) = 1$ for all but countably many $\epsilon > 0$. Take $a \in A$ and the measure $\nu = \delta_a$, a point mass at a. Then, for any $n \in \mathbb{N}$, we have

$$\mathbb{M}^n \nu = \sum_{\sigma \in \Sigma^n} p_\sigma \, \nu \circ w_\sigma^{-1}.$$

Now $\nu \circ w_\sigma^{-1}(A) = 1$ since $w_\sigma(a) \in A_\sigma \subset A$ for any $\sigma \in \Sigma^n$. This means that $\mathbb{M}^n \nu(A) = 1$ for all n as $\sum_\sigma p_\sigma = 1$ for $\sigma \in \Sigma^n$ for any n. Thus, for all $\epsilon > 0$, $\mathbb{M}^n \nu(A_\epsilon) = 1$ for all n. Now it is not necessarily the case that $\mathbb{M}^n \nu(B) \to \mu(B)$ for any Borel set B; this is only true for *continuity sets* of the limit measure μ. This means that if $\mu(\partial B) = 0$, then $\mathbb{M}^n \nu(B) \to \mu(B)$. Now it is not necessarily true that $\mu(\partial A_\epsilon) = 0$ for any given ϵ. However, there can be at most countably many $\epsilon > 0$ for which $\mu(\partial A_\epsilon) > 0$, and thus for all but countably many $\epsilon > 0$ we have $\mu(A_\epsilon) = 1$.

Suppose that $x \notin A$, so that there is some $\epsilon > 0$ with $B_\epsilon(x) \cap A = \emptyset$. By what we have just shown, $\mu(A_\delta) = 1$ for some $0 < \delta < \epsilon/2$. But this means that $\mu(B_\delta(x)) = 0$ and thus $x \notin \text{supp}(\mu)$ and so $\text{supp}(\mu) \subseteq A$.

For the other direction, let $a \in A$ and $\epsilon > 0$ be given. Take $\sigma \in \Sigma$ to be an address of a. Then, for large enough n, we have that $w_{\sigma^n}(A) \subseteq B_\epsilon(a)$. However, then $\mu(B_\epsilon(a)) \geq \mu(w_{\sigma^n}(A)) = p_{\sigma^n} > 0$. Since this is true for all ϵ, we know that $a \in \text{supp}(\mu)$ and so $A \subseteq \text{supp}(\mu)$. $\qquad \square$

Exercise 2.64. Prove Theorem 2.63 by showing that if B is the support of μ, then $B \in \mathbb{H}(\mathbb{X})$ and is invariant under the IFS.

Exercise 2.65. Let λ be the Lebesgue measure on $\mathbb{R}$ and $\{w_i, p_i\}$ be an IFSP on $\mathbb{R}$. If $\mu_0 \ll \lambda$, show that $\mathbb{M}^n \mu_0 \ll \lambda$ as well for any $n \in \mathbb{N}$. However, show that this does not mean that it is necessarily the case that $\mu \ll \lambda$. Construct an example where $\mu \ll \lambda$ and another example where $\mu \not\ll \lambda$.

Exercise 2.66. In the previous problem, it was shown that if $\mu_0 \ll \lambda$, then $\mathbb{M}\mu_0 \ll \lambda$. Under these situations, the IFSP Markov operator $\mathbb{M}$ induces an operator on the density functions of these measures. Is this well-defined? Investigate this operator and its convergence. Is there a reasonable space for this operator?

An even deeper relationship is the one between the invariant measure μ and an appropriate measure P on the code space Σ. First we define the measure P on Σ as the product measure (see Appendix B) given by the measure $\text{Pr}(x = i) = p_i$ on each factor $\{1, 2, \ldots, N\}$. As Σ is a countable product of copies of $\{1, 2, \ldots, N\}$, this will be a product measure with a countable number of factors. Next we recall that the address map ω is a continuous mapping from Σ to $\mathbb{X}$ (in fact, with range equal to the attractor A). Finally, we also recall that the push-forward measure of P via ω is the measure $\omega_\#(P)$ defined on $\mathbb{X}$ and given by

$$\omega_\#(P)(B) = P(\omega^{-1}(B))$$

for each Borel set $B \subseteq \mathbb{X}$. Since ω is continuous, the preimage of a Borel set is a Borel set and thus $\omega_\#(P)$ is a Borel measure.

Theorem 2.67. *The invariant measure of an IFSP $\{w_i, p_i\}$ is the push-forward measure $\omega_\#(P)$ of the natural product measure P defined on the code space Σ via the address map $\omega : \Sigma \to \mathbb{X}$.*

Proof. The proof simply shows that $\omega_\#(P)$ is invariant under the IFSP Markov operator and thus must be the invariant measure. We do this by introducing dynamics on Σ that match the dynamics generated by the maps w_i (this is also mentioned in Sect. 2.3). Then the fact that $\omega_\#(P)$ is invariant under $\mathbb{M}$ is equivalent to the fact that P on Σ is invariant under the action of dynamics on Σ.

We define the functions $\tau_i : \Sigma \to \Sigma$ by $\tau_i(\sigma_1, \sigma_2, \sigma_3, \ldots) = (i, \sigma_1, \sigma_2, \sigma_3, \ldots)$. By the definition of the product measure on Σ (see Appendix B), for any rectangle B we have $P(\tau_i(B)) = p_i P(B)$. What this means is that p_i times the restriction of P to $\tau_i(\Sigma)$ is equal to $(\tau_i)_\#(P)$. Putting this together for all i, we get for all P-measurable $B \subseteq \Sigma$

$$P(B) = \sum_i p_i (\tau_i)_\#(P)(B) = \sum_i P(\tau_i^{-1}(B)),$$

an invariance for P that is similar to that of μ under $\mathbb{M}$.

By our definition of the maps τ_i and the address map ω, it is clear that $\omega \circ \tau_i = w_i \circ \omega$ (so the diagram in Fig. 2.15 commutes) and thus $\tau_i^{-1} \circ \omega^{-1} = \omega^{-1} \circ w_i^{-1}$. But then

$$\mathbb{M}\omega_\#(P)(B) = \sum_i p_i \omega_\#(P)(w_i^{-1}(B))$$
$$= \sum_i p_i P(\omega^{-1}(w_i^{-1}(B)))$$
$$= \sum_i p_i P(\tau_i^{-1}(\omega^{-1}(B)))$$
$$= P(\omega^{-1}(B)) = \omega_\#(P)(B),$$

and thus $\omega_\#(P)$ is invariant under $\mathbb{M}$ and so must equal μ. $\square$

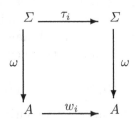

Fig. 2.15: Correspondence between dynamics on Σ and on A.

Exercise 2.68. Let $\{w_i, p_i\}$ be a disjoint IFSP and μ be its invariant distribution. Show that $\mu(w_i(A)) = p_i$ and, more generally, $\mu(A_\sigma) = p_\sigma$ for any $\sigma \in \Sigma^n$. Show that this uniquely defines the distribution μ.

2.5.1.1 Continuous dependence

We take a brief digression to show that the invariant measure of an IFSP is a continuous function of the operator $\mathbb{M}$. This is again a consequence of the general result, Proposition 2.8. The issue is to show that if the w_i and p_i depend continuously on some parameters λ_i then so does $\mathbb{M}$. Our standing assumption is that $(\mathbb{X}, d)$ is a compact metric space, and let $D := \sup_{x,y} d(x, y) < \infty$. Our discussion is taken from [39].

Let

$$\mathcal{H}^N = \left\{ (p_1, p_2, \ldots, p_N) : p_i \geq 0, \sum_i p_i = 1 \right\} \subset \mathbb{R}^N$$

be the space of probability vectors in $\mathbb{R}^N$ and we use the metric $d(p, q) = \max_i |p_i - q_i|$. Our space of N-map IFSP then will be

$$\mathcal{S}^N(\mathbb{X}) = \mathrm{Con}^N(\mathbb{X}) \times \mathcal{H}^N. \tag{2.18}$$

Notice that as we allow some $p_i = 0$, this will allow IFSs with different numbers of maps to be compared and so we can naturally think of $\mathcal{S}^N(\mathbb{X}) \subset \mathcal{S}^{N+1}(\mathbb{X})$. The metric we use on $\mathcal{S}^N$ is given by

$$d((\mathbf{w}_1, \mathbf{p}_1), (\mathbf{w}_2, \mathbf{p}_2)) =$$
$$\max\left\{ \max_i |p_{1,i} - p_{2,i}|, \max_i \sup_x d(w_{1,i}(x), w_{2,i}(x)) \right\}, \tag{2.19}$$

where we use a bold font to indicate a vector of contractions or probabilities. As $\mathbb{X}$ is bounded and each w_i is Lipschitz, this expression is bounded and thus d is also bounded. The metric in (2.19) gives the product topology on $\mathcal{S}^N(\mathbb{X})$ and is in fact a complete metric. This is easy to see as $\mathcal{H}^N$ and $\mathrm{Con}^N(\mathbb{X})$ are both complete under their respective metrics.

The following proposition is from [39] with only minor modifications.

Proposition 2.69. *Let* $(\mathbf{w}_1, \mathbf{p}_1) \in \mathcal{S}^N(\mathbb{X})$, *with contractivity factor* s_1, *associated Markov operator* $\mathbb{M}_1$, *and invariant measure* μ_1. *Then, for every* $\epsilon > 0$, *there is a* $\delta > 0$ *such that for all* $(\mathbf{w}_2, \mathbf{p}_2) \in \mathcal{S}^N(\mathbb{X})$ *that satisfy* $d((\mathbf{w}_1, \mathbf{p}_1), (\mathbf{w}_2, \mathbf{p}_2)) < \delta$ *it follows that* $d_{MK}(\mu_1, \mu_2) < \epsilon$, *where* μ_2 *is the invariant measure for* $(\mathbf{w}_2, \mathbf{p}_2)$.

Proof. Before we proceed, we note that the Markov operator $\mathbb{M}$ associated with an IFSP $(\mathbf{w}, \mathbf{p})$ is a contraction on $\mathcal{P}(\mathbb{X})$ and thus is an element of $\mathrm{Con}(\mathcal{P}(\mathbb{X}))$. As usual, we use the metric

$$d_\infty(\mathbb{M}_1, \mathbb{M}_2) = \sup_{\mu \in \mathcal{P}(\mathbb{X})} d_{MK}(\mathbb{M}_1(\mu), \mathbb{M}_2(\mu)).$$

As $\mathcal{P}(\mathbb{X})$ is compact under the Monge-Kantorovich metric (since $\mathbb{X}$ is compact), this is well-defined and is a metric. Our strategy is to show that the mapping $(\mathbf{w}, \mathbf{p}) \mapsto \mathbb{M}$ is continuous and use the basic result of Proposition 2.8 on $\mathrm{Con}(\mathcal{P}(\mathbb{X}))$.

Let $(\mathbf{w_2}, \mathbf{p_2}) \in \mathcal{S}^{\mathbf{N}}(\mathbb{X})$ be such that $d((\mathbf{w_1}, \mathbf{p_1}), (\mathbf{w_2}, \mathbf{p_2})) < \delta$, where

$$\delta < \frac{\epsilon(1 - s_1)}{1 + ND}. \tag{2.20}$$

The existence of such a $\mathbf{w_2}$ has been discussed in [16].

Now we have that

$$d_\infty(\mathbb{M}_1, \mathbb{M}_2) = \sup_\mu d_{MK}(\mathbb{M}_1\mu, \mathbb{M}_2\mu) \tag{2.21}$$

$$= \sup_\mu \sup_f \int_{\mathbb{X}} f \, d(\mathbb{M}_1\mu - \mathbb{M}_2\mu)$$

$$= \sup_\mu \sup_f \sum_i \int \left(p_{1,i} f(w_{1,i}(x)) - p_{2,i} f(w_{2,i}(x)) \right) d\mu$$

$$= \sup_\mu \sup_f \sum_i \int \Big(p_{1,i} f(w_{1,i}(x)) - p_{2,i} f(w_{1,i}(x))$$

$$+ p_{2,i} f(w_{1,i}(x)) - p_{2,i} f(w_{2,i}(x)) \Big) d\mu$$

$$= \sup_\mu \sup_f \left\{ \sum_i (p_{1,i} - p_{2,i}) \int f(w_{1,i}(x)) d\mu \right.$$

$$\left. + \sum_i p_{2,i} \int \left(f(w_{1,i}(x)) - f(w_{2,i}(x)) \right) d\mu \right\}$$

$$\leq \sup_\mu \sup_f \sum_i (p_{1,i} - p_{2,i}) \int f(w_{1,i}(x)) d\mu$$

$$+ \sup_\mu \sup_f \sum_i p_{2,i} \int \left(f(w_{1,i}(x)) - f(w_{2,i}(x)) \right) d\mu.$$

We now simplify the two expressions on the right-hand side of the inequality. Starting with the latter term,

$$\sup_f \sum_i p_{2,i} \int \left(f(w_{1,i}(x)) - f(w_{2,i}(x)) \right) d\mu$$

$$\leq \sup_f \sum_i p_{2,i} \int |f(w_{1,i}(x)) - f(w_{2,i}(x))|\, d\mu$$

$$\leq \sum_i p_{2,i} \int d(w_{1,i}(x), w_{2,i}(x)) d\mu \tag{2.22}$$

$$\leq \sum_i p_{2,i} \sup_x d(w_{1,i}(x), w_{2,i}(x))$$

$$\leq \max_i \sup_x d(w_{1,i}(x), w_{2,i}(x)) \sum_j p_{2,j} = \max_i d_\infty(w_{1,i}, w_{2,i}).$$

Now, for the first term of the right side of (2.21),

$$\sup_\mu \sup_f \sum_i (p_{1,i} - p_{2,i}) \int f(w_{1,i}(x)) d\mu$$

$$\leq \max_j |p_{1,j} - p_{2,j}| \sup_\mu \sup_f \sum_i \int f(w_{1,i}(x)) d\mu$$

$$\leq \max_j |p_{1,j} - p_{2,j}| \sup_\mu \sup_f \sum_i \int |f(w_{1,i}(x))| d\mu.$$

Now we use a nice property of the Monge-Kantorovich metric, namely that the functions $g(x)$ and $g(x)+c$ give the same value for the integral

$$\int_{\mathbb{X}} g(x) d(\mu - \nu),$$

where c is any constant. Thus we can assume that $f(y) = 0$ for some $y \in \mathbb{X}$ and so since $w_{1,i}$ is contractive and f has Lipschitz factor less than or equal to 1, we have $|f(w_{1,i}(x))| \leq \operatorname{diam}(\mathbb{X}) = D$. This means that

$$\sup_\mu \sup_f \sum_i (p_{1,i} - p_{2,i}) \int f(w_{1,i}(x)) d\mu \leq ND \max_i |p_{1,i} - p_{2,i}|. \tag{2.23}$$

Substituting (2.22) and (2.21) into (2.20), we get

$$d_{MK}(\mathbb{M}_1\nu, \mathbb{M}_2\nu) \leq d_\infty(\mathbb{M}_1, \mathbb{M}_2) \leq \delta + ND\delta < \epsilon(1 - s_1).$$

Using this inequality and Proposition 2.8, we obtain the desired result.

□

As a simple corollary, we obtain the following result.

Proposition 2.70. *Suppose that the IFSP* $(\mathbf{w}_\lambda, \mathbf{p}_\lambda)$ *depends continuously on the parameters* $\lambda = (\lambda_1, \lambda_2, \ldots, \lambda_n) \in \mathbb{R}^n$ *and that for all values of the parameters the average contractivity is bounded from above by* $c < 1$. *Then the invariant measure* μ_λ *depends continuously on the parameters* λ.

Notice that this is also a special case of Theorem 2.92 where the measures $p^{(n)}$ and p are all supported on subsets of size N in $\mathrm{Con}_s(\mathbb{X})$.

2.5.2 Moments of the invariant measure and $\mathbb{M}^*$

The invariance equation (2.17) of the fractal measure for an IFSP can help in calculating certain integrals. That is, if μ is the invariant measure, then for any measurable function we have

$$\int_{\mathbb{X}} f(x) \, d\mu(x) = \sum_i p_i \int_{\mathbb{X}} f(x) \, d\mu(w_i^{-1}(x))$$

$$= \int_{\mathbb{X}} \sum_i p_i f(w_i(y)) \, d\mu(y). \qquad (2.24)$$

This relation is especially useful in the case where $\mathbb{X} = \mathbb{R}^n$ and the w_i are affine maps. For simplicity, we restrict our discussion to the case $\mathbb{X} = \mathbb{R}$ and set $w_i(x) = s_i x + b_i$. Then (2.24) takes the form

$$\int_{\mathbb{X}} f(x) \, d\mu(x) = \sum_{i=1}^N p_i \int_{\mathbb{X}} f(s_i x + b_i) \, d\mu(x).$$

If we take $f(x) = x^n$ for $n \in \mathbb{N}$, then we can expand the integrand to obtain

$$\int_{\mathbb{X}} f(x) \, d\mu(x) = \sum_i p_i \int_{\mathbb{X}} \sum_{k=0}^n \binom{n}{k} s_i^k x^k b_i^{n-k} \, d\mu(x)$$

$$= \sum_{k=0}^n \left(\sum_{i=1}^N p_i s_i^k b_i^{n-k} \right) \int_{\mathbb{X}} x^k \, d\mu(x). \qquad (2.25)$$

Noting that the integral $g_n := \int_{\mathbb{X}} x^n \, d\mu(x)$ is the nth moment of μ, we for these moments the recursion

$$g_n \left[1 - \sum_{i=1}^N p_i s_i^n \right] = \sum_{k=0}^{n-1} \binom{n}{k} \left[\sum_{i=1}^N p_i s_i^k b_i^{n-k} \right] g_k, \quad n = 1, 2, \ldots \qquad (2.26)$$

with $g_0 = 1$. This recursion is very nice, as it gives an effective and efficient way to calculate moments of invariant measures of affine IFSPs. For example, for the Cantor set with $p_0 = p_1 = 1/2$, we get the moments

$$g_0 = 1, g_1 = \frac{2}{5}, g_2 = \frac{28}{85}, g_3 = \frac{1256}{4505}, g_4 = \frac{175408}{725305}, g_5 = \frac{15075232}{70354585}, \ldots$$

This recursion is the basis of the *moment matching* approach to the inverse problem for IFSP invariant measures (see [167]).

Exercise 2.71. What happens if the IFS maps are quadratic? Use the same idea to derive a recursion and comment on it. Do the same for the case where the IFS maps are general polynomials.

Equation (2.26) is also the basis of an induced IFS-type operation on the space of moment sequences (see [65]). Let $\nu = M\mu$, h_k be the moments of ν and g_k be the moments of μ. Then, using the same logic as above, we obtain the recursion relations

$$h_n = \sum_{k=0}^{n} \binom{n}{k} \left[\sum_{i=1}^{N} p_i s_i^k b_i^{n-k} \right] g_k, \quad i = 1, 2, \ldots,. \quad (2.27)$$

This equation defines a mapping from the moment sequence for ν (g_k) to the moment sequence for $\mu = M\nu$ (h_n). Since a compactly supported probability measure on $\mathbb{R}$ is completely defined by its moment sequence, this dual mapping is completely equivalent to the Markov operator on $\mathcal{P}([0,1])$. Using the natural basis for the space of moment sequences, this dual mapping is represented by a lower-triangular matrix. This is an example of the more general situation discussed in Sect. 3.4.

Something else we can obtain from (2.24) is a formula for the adjoint to M under the duality given by the Riesz representation theorem. This is the operator M^* defined by

$$M^*(f) = \sum_i p_i f \circ w_i, \quad (2.28)$$

which clearly maps any continuous $f : \mathbb{X} \to \mathbb{R}$ to another such continuous function. Iterating M^*, we get

$$(M^*)^2(f)(x) = \sum_{\sigma_1=1}^{N} \sum_{\sigma_2=1}^{N} p_{\sigma_1} p_{\sigma_2} f \circ w_{\sigma_1} \circ w_{\sigma_2}(x).$$

The important thing to notice is that the composition of the w_i is in the same fashion as in the address map. Thus,

$$(\mathbf{M}^*)^n f(x) = \sum_{\sigma \in \Sigma^n} p_\sigma f \circ w_\sigma(x).$$

Now, as $n \to \infty$, we have $w_{\sigma^n}(x) \to \omega(\sigma)$, the value of the address map at the point $\sigma \in \Sigma$, and it is independent of x. This means that each "term" of the summation converges to a function that is constant in x, and so the limit should also be a constant, independent of x. Roughly, $(\mathbf{M}^*)^n f$ should converge to the function on $\mathbb{X}$ with constant value equal to

$$\int_\Sigma f(\omega(\sigma)) \, dP(\sigma).$$

In fact, it is not difficult to make this precise. One of the key ideas is that the function

$$\phi_n(\sigma, x) = w_{\sigma^n}(x)$$

converges to the address map uniformly in σ and in x when x is taken in a bounded set (such as the attractor). This means that $f \circ \phi_n \to f \circ \omega$ uniformly. Another key idea is the convergence of the measures P_n to P on Σ, where P_n is the "restriction" of P to the first n factors in the product space Σ (more formally, P_n is a projection on the σ-algebra generated by these first n factors).

Using the duality, this means that we should have

$$\int_{\mathbb{X}} f \, d\mu \leftarrow \int_{\mathbb{X}} f \, d\mathbf{M}^n \mu = \int_{\mathbb{X}} (\mathbf{M}^*)^n f \, d\mu \to \int_{\mathbb{X}} \int_\Sigma f \circ \omega \, dP \, d\mu.$$

Notice that this is also consistent with the fact that the invariant measure is the push-forward of P under the address map, which gives

$$\int_{\mathbb{X}} f \, d\mu = \int_\Sigma f \circ \omega \, dP.$$

Exercise 2.72. Prove that $(\mathbf{M}^*)^n f$ converges to a constant function in the case where all the maps in the IFS are contractive.

Exercise 2.73. Take the IFS $w_i(x) = x/2 + i/2$ for $i = 0, 1$ along with the two probabilities p_0, p_1. Draw the graphs of $(\mathbf{M}^*)^n f$ for $n = 0, 1, 2, 3, 4, 5$ for $f(x) = c$, a constant. Do the same for the functions $f(x) = x$ and $f(x) = x^2$.

2.5.3 The ergodic theorem for IFSP

We have seen what happens if you compose the IFS maps in one particular direction,

$$w_{\sigma_1} \circ w_{\sigma_2} \circ \cdots \circ w_{\sigma_{n-1}} \circ w_{\sigma_n}(x) \to \omega(\sigma),$$

composing them with the "newest" map on the "inside," as in the address map. What happens if you compose them in the other direction? This is exactly the sort of thing that happens in the chaos game, where we have a sequence $\sigma \in \Sigma$ (which is randomly generated) and compute the sequence of points x_n given by

$$x_n = w_{\sigma_n} \circ w_{\sigma_{n-1}} \circ \cdots w_{\sigma_2} \circ w_{\sigma_1}(x). \tag{2.29}$$

It is clear that $x_n \in w_{\sigma_n}(A)$, so the points x_n and x_{n+1} could be quite distant from each other. This makes it clear that the sequence x_n given by (2.29) does not converge except in very exceptional cases. Can anything be said about this sequence?

This is exactly the sort of question that one encounters in ergodic theory. We will not go very far into this theory except to discuss the version of the ergodic theorem that one encounters in the chaos game and some small variations. We return to this subject in some more general situations in Chapter 6. There are many good introductions to ergodic theory, amongst which are the books [71, 141, 168].

The next result was first proved by J. Elton [53], who also proved a more general version [54]. Notice that the sequence x_n in the theorem is generated in the same way as in the chaos game. Thus, this theorem provides another justification for the chaos game. Moreover, it shows that not only can you generate an image of the attractor via the chaos game but you can also use it to generate approximations to the invariant measure. We will discuss both of these a bit more after the theorem. We give the proof from [61]. For a more elementary explanation and proof, see [162].

We introduce a bit of simplifying notation first. As the order of composition of the maps is different from that in the address map, we need different notation. Thus, for $\sigma \in \Sigma^n$, we define

$$w_{\widehat{\sigma}} = w_{\sigma_n} \circ w_{\sigma_{n-1}} \circ \cdots \circ w_{\sigma_2} \circ w_{\sigma_1}.$$

We only prove the case where $\mathbb{X}$ is compact. This is not usually a great restriction for us, as the attractor of an IFS is always compact and we can often restrict our attention to the attractor (or a compact neighbourhood of the attractor).

Theorem 2.74. *Let* $\mathbb{X}$ *be a compact metric space and* $\{w_i, p_i\}$ *be an IFSP that is contractive. Choose some* $x_0 \in \mathbb{X}$, *and define the sequence* x_n *by selecting* $\sigma \in \Sigma$ *according to the probability measure* P *and setting*

$$x_n = w_{\widehat{\sigma n}}(x) = w_{\sigma_n} \circ w_{\sigma_{n-1}} \circ \cdots \circ w_{\sigma_2} \circ w_{\sigma_1}(x_0).$$

Then, for any continuous $f : \mathbb{X} \to \mathbb{R}$ *and for* P *almost all* $\sigma \in \Sigma$, *we have*

$$\lim_n \frac{1}{n} \sum_{i \leq n} f(x_i) = \int_{\mathbb{X}} f(x) \, d\mu(x),$$

where μ *is the invariant measure for the IFSP.*

Proof. Let f be a continuous function on $\mathbb{X}$ and x be a fixed element of $\mathbb{X}$.

We wish to show that

$$1/n \sum_{i \leq n} f(w_{\widehat{\sigma i}}(x))$$

converges. Let ν-lim be a Banach limit on $l^\infty(\mathbb{N})$ (see [42], p. 82 for a nice discussion of Banach limits). Recall that a Banach limit is a "generalized limit" in the sense that it is a bounded linear functional on l^∞ that only depends on the "tail behaviour" of the sequence in l^∞. We will show that any two Banach limits will give the same value for P almost every $\sigma \in \Sigma$ so that the limit exists almost everywhere.

Define $\Psi_\nu : C(\mathbb{X}) \times \mathbb{X} \times \Sigma \to \mathbb{R}$ by

$$\Psi_\nu(f, x, \sigma) = \nu\text{-lim} \, 1/n \sum_{i \leq n} f(w_{\widehat{\sigma i}}(x).$$

The function $f \mapsto \Psi_\nu(f, x, \sigma)$ is a bounded linear functional on $C(\mathbb{X})$. Thus, by the Riesz representation theorem it corresponds to a (signed) measure μ_ν on $\mathbb{X}$. We know that μ_ν is a positive measure since for any positive f the value is positive. We know that it is a probability measure since if $f = 1$ the value is equal to 1.

We show that $\mu_\nu = \mu$ for P almost all σ. We do this by showing that μ_ν is invariant under $\mathbb{M}$.

Let $S : \Sigma \to \Sigma$ denote the shift map defined by $S(\sigma_1, \sigma_2, \ldots) = (\sigma_2, \sigma_3, \ldots)$. Since each w_i is contractive and $\mathbb{X}$ is compact, and f is uniformly continuous on $\mathbb{X}$, we know that

$$\left| 1/n \sum_{i \leq n} f\left(w_{\widehat{\sigma i}}(x)\right) - 1/n \sum_{i \leq n} f\left(w_{\widehat{S(\sigma)i}}(x)\right) \right| \to 0$$

as $n \to \infty$. Thus, since ν-lim depends only on the tail of the sequence in $\ell^\infty(\mathbb{N})$, we have that $\Psi_\nu(f, x, \sigma) = \Psi_\nu(f, x, S(\sigma))$, and so $\Psi_\nu(f, x, \sigma)$ is invariant with respect to S on Σ.

Since the shift map on Σ is ergodic [8], it follows that $\Psi_\nu(f, x, \sigma)$ is constant for P almost all σ.

To show that $\mu = \mu_\nu$, it suffices to show that

$$\int_X f(z) \, d\mu_\nu(z) = \int_X M^* f(z) \, d\mu_\nu(z)$$

which is the same as showing that

$$\nu\text{-}\lim_n 1/n \sum_{i \leq n} f(w_{\widehat{\sigma}i}(x)) = \nu\text{-}\lim_n 1/n \sum_{i \leq n} \sum_j p_j f(w_j \circ w_{\widehat{\sigma}i}(x)).$$

Computing we get

$$\nu\text{-}\lim_n 1/n \sum_{i \leq n} \sum_j p_j f(w_j \circ w_{\widehat{\sigma}i}(x))$$

$$= \int_{\sigma \in \Sigma} \nu\text{-}\lim_n 1/n \sum_{i \leq n} \sum_j p_j f(w_j \circ w_{\widehat{\sigma}i}(x)) \, dP(\sigma)$$

$$= \sum_j p_j \int_{\sigma \in \Sigma} \nu\text{-}\lim_n 1/n \sum_{i \leq n} f(w_j \circ w_{\widehat{\sigma}i}(x)) \, dP(\sigma).$$

Doing the change of variable $\sigma \equiv (\sigma_1, \sigma_2, \ldots) \to (j, \sigma_1, \sigma_2, \ldots)$, we get $dP \to dP/p_j$, so this integral becomes

$$\sum_j p_j/p_j \int_{\sigma_1 = j} \nu\text{-}\lim_n 1/n \sum_{i \leq n} f(w_{\widehat{\sigma i+1}}(x)) \, dP(\sigma)$$

$$= \int_{\sigma \in \Sigma} \nu\text{-}\lim_n 1/n \sum_{i \leq n} f(w_{\widehat{\sigma}i}(x)) \, dP(\sigma)$$

$$= \nu\text{-}\lim_n 1/n \sum_{i \leq n} f(w_{\widehat{\sigma}i}(x))$$

for P almost all σ.

Therefore, for P almost all σ, we know that μ_ν is invariant under M so $\mu_\nu = \mu$ for these σ. However, since ν was arbitrary, this shows that for P almost all σ

$$\lim_n 1/n \sum_{i \leq n} f(w_{\widehat{\sigma}i}(x)) = \int_X f(z) \, d\mu(z)$$

for all f and all $x \in X$. $\square$

Exercise 2.75. Let $x \in \mathbb{X}$ be fixed, and define the two sequences of random variables

$$Z_n = w_{\sigma_1} \circ w_{\sigma_2} \circ \cdots \circ w_{\sigma_n}(x) = w_{\sigma^n}(x)$$

and

$$\widehat{Z}_n = w_{\sigma_n} \circ w_{\sigma_{n-1}} \circ \cdots \circ w_{\sigma_1}(x) = w_{\widehat{\sigma^n}}(x).$$

Show that for each n the distribution of Z_n is the same as the distribution of $\widehat{Z}_n$.

The restriction that f be continuous is sometimes not convenient. The following is a version for functions f that are integrable with respect to μ. However, notice that the conclusion no longer works for all x but only for μ almost all x. The proof is taken from [162], for which we give only a sketch.

Theorem 2.76. *Let $\mathbb{X}$ be a complete metric space and $\{w_i, p_i\}$ be a contractive IFSP. Then, for every $f \in L^1(\mu)$, for μ almost every x in $\mathbb{X}$, and for P almost all σ in Σ, we have*

$$\lim_n 1/n \sum_{i \leq n} f\left(w_{\widehat{\sigma^i}}(x)\right) = \int_{\mathbb{X}} f(x) \, d\mu(x).$$

Proof. As mentioned, this elegant proof comes from [162] and we only give a sketch. The idea of the proof is to place the situation into the standard context of a measure preserving transformation and then use the classical pointwise ergodic theorem.

First notice that $\mu(A) = 1$, and thus we can assume that the point $x \in A$. Define the doubly infinite product space

$$\Omega := \{1, 2, \ldots, N\}^{\mathbb{Z}} = \{(\ldots, \sigma_{-1}, \sigma_0, \sigma_1, \sigma_2, \ldots) : \sigma_i \in \{1, 2, \ldots, N\}\}.$$

Each $\sigma \in \Omega$ is divided into two parts. The first part is $\sigma_L = (\ldots, \sigma_{-2}, \sigma_{-1}, \sigma_0)$. This part is used to determine a point in A by using the address map. The second part is $\sigma_R = (\sigma_1, \sigma_2, \ldots)$, which is thought of as coming from Σ (the usual code space). In this sense, $\Omega \cong A \times \Sigma$ is the space of interest. Define the function $U : \Omega \to \mathbb{X}$ by $U(\sigma) = \omega(\sigma_L)$; thus U maps the "left" part of σ to the corresponding point of A.

On Ω we define the transformation T given by $(T\sigma)_n = \sigma_{n+1}$ (that is, T "shifts" each bi-infinite sequence to the "left"). This gives $w_{\sigma_1} \circ U = U \circ T$. We also obtain that $\mu = U_\#(P)$, where P is the natural product measure on Ω. Now define $F = f \circ U : \Omega \to \mathbb{R}$. Then the standard pointwise ergodic theorem gives that

$$\int_\Omega F \, dP = \lim_n 1/n \sum_{i \le n} F(T^i \sigma)$$

for P almost all $\sigma \in \Omega$. However, by the definitions, we get

$$\int_\Omega F \, dP = \int_{\mathbb{X}} f \, d\mu$$

and

$$F(T^i \sigma) = f(w_{\widehat{(\sigma_L)^i}}(x)),$$

as we desired.   □

As a corollary to Theorem 2.76, we get the following result. This nice result says that not only can you generate an image of the attractor with the chaos game but you can also generate approximations to the invariant distribution. It says that the *empirical occupation distribution* will converge to the invariant distribution.

Corollary 2.77. *Let $\mathbb{X}$ be a complete metric space and $\{w_i, p_i\}$ be a contractive IFSP. Then, for μ almost all $x_0 \in \mathbb{X}$ and for P almost all $\sigma \in \Sigma$, the sequence generated by the chaos game, $x_n = w_{\widehat{\sigma^n}}(x_0)$, satisfies*

$$\mu(A) = \lim_n 1/n \#\{1 \le i \le n : x_i \in A\}$$

for all Borel sets A.

The ergodic theorem for IFSP (and the chaos game) is true in greater generality. Elton's result mentioned above will work for so-called *place-dependent* probabilities. In this case, the probabilities $p_i(x)$ are functions of $x \in \mathbb{X}$ and we require $\sum_i p_i(x) = 1$ for all x and that they be strictly bounded away from zero. The probabilities are also required to satisfy a Dini-type continuity condition, for details see [13]. In [54], the situation is more general in another direction, in which the sequence of maps does not have to be chosen independently and the collection can be infinite. All of these results are part of a general study of dynamics of the iteration of randomly chosen functions [47].

2.6 Some classical extensions

We now give a brief foray into a few extensions of the basic IFS framework for geometric IFS and IFSP. In later chapters, the extensions will focus on defining IFS-type operators on other mathematical objects; here we stay with the two basic objects (sets and measures) but widen the type of operator to enlarge the class of objects that may be considered.

2.6.1 IFS with condensation

The first type of extension we consider is adding a *condensation set* to a geometric IFS. The change from the standard IFS operator to one with condensation is similar to the change from a linear function to an affine one.

We take $\{w_i\}$ to be a (geometric) IFS on the complete metric space $\mathbb{X}$ and take $S \in \mathbb{H}(\mathbb{X})$ to be a fixed set. Then we define the *IFS operator with condensation* $W : \mathbb{H}(\mathbb{X}) \to \mathbb{H}(\mathbb{X})$ to be given by

$$W(B) = S \cup \bigcup_i w_i(B). \tag{2.30}$$

It is easy to see that

$$d_{\mathbb{H}}(W(B), W(C)) \leq \max\{d_{\mathbb{H}}(S, S), \max_i d_{\mathbb{H}}(w_i(B), w_i(C))\}$$
$$= \max_i d_{\mathbb{H}}(w_i(B), w_i(C)),$$

and so W is contractive on $\mathbb{H}(\mathbb{X})$. Thus W has a unique fixed point $A \in \mathbb{H}(\mathbb{X})$, which we again call the attractor. In this case, the set A satisfies the condition

$$A = S \cup \bigcup_i w_i(A).$$

If we start the deterministic iteration with the initial set S, then we see that

$$W^n(S) = S \cup \bigcup_i w_i(S) \cup \bigcup_{\sigma \in \Sigma^2} w_\sigma(S) \cup \cdots \cup \bigcup_{\sigma \in \Sigma^n} w_\sigma(S),$$

and so

$$A = S \cup \bigcup_{n=1}^{\infty} \bigcup_{\sigma \in \Sigma^n} w_\sigma(S). \tag{2.31}$$

Exercise 2.78. Devise an efficient algorithm to show the attractor of an IFS with condensation using deterministic iteration. Consider the special case where the condensation set S is also the attractor of an IFS (as in Fig. 2.16).

It is helpful to compare this with the situation of an affine map $\phi : \mathbb{R} \to \mathbb{R}$ given by $\phi(t) = at + b$. Assuming that $|a| < 1$ so that ϕ is a contraction, we get that the fixed point $\bar{x}$ of ϕ can be written as

$$\bar{x} = \frac{b}{1-a} = b + ab + a^2 b + a^3 b + \cdots + a^n b + \cdots = b + \sum_{n=1}^{\infty} a^n b. \quad (2.32)$$

Comparing (2.31) with (2.32), the analogy of an IFS with condensation to an affine contraction is clear. The "linear" part of the IFS with condensation is the same as a usual IFS mapping, while the "translation" part is the condensation set.

An example of an attractor of an IFS with condensation is shown in Fig. 2.16. Here the condensation set is the largest maple leaf in the center. We can clearly see all the "first-level" and "second-level" images of this condensation set. In this case, the condensation set is also the attractor of an IFS, so the image was generated by "mixing" two chaos games [12]. We will discuss this extension of the simple chaos game below in Sect. 2.6.3.

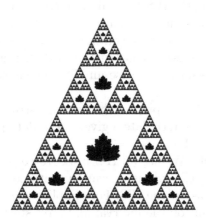

Fig. 2.16: Attractor for an IFS with condensation.

It is also possible to do the same type of extension for an IFSP, which we call an *IFSP with condensation*. Here we have a fixed probability measure $\nu_0 \in \mathcal{P}_1(\mathbb{X})$, IFS $\{w_i\}$, $i = 1, 2, \ldots, N$, and probabilities p_i for $i = 0, 1, 2, \ldots, N$. The Markov operator in this case is defined as

$$\mathbb{M}\nu(B) = p_0\nu_0(B) + \sum_i p_i\nu(w_i^{-1}(B)). \qquad (2.33)$$

Here $\mathbb{M}$ is affine, as the Markov operator without the condensation measure is linear. If we write this operator as

$$\mathbb{M}\nu = A\nu + \beta = \sum_i p_i\mu \circ w_i^{-1} + p_0\nu_0,$$

then the fixed point of $\mathbb{M}$ is

$$\beta + \sum_{n=1}^{\infty} A^n\beta = p_0\nu_0 + p_0 \sum_{n=1}^{\infty} \sum_{\sigma \in \Sigma^n} p_\sigma\nu_0 \circ w_\sigma^{-1}.$$

In particular, we notice that the fixed point is a linear function of the condensation measure ν_0.

Exercise 2.79. If μ is the invariant measure for an IFSP with condensation measure ν and $S \in \mathbb{H}(\mathbb{X})$ is the support of ν, show that the support of μ is the attractor of an IFS with condensation set S.

2.6.2 Fractal interpolation functions

The idea of applying the IFS formalism to construct self-similar functions is a very natural one. Ever since the construction by Weierstrass of a function that is everywhere continuous but nowhere differentiable there has been an interest in constructions of functions with pathological properties. It is natural to use an IFS to construct such "rough" functions.

In this short section, we present the basic construction of continuous fractal functions that interpolate a given set of data points in $\mathbb{R}^2$. This material is based on [15].

The idea is simple. We use an IFS in $\mathbb{R}^2$ to construct the graph of the desired function. Clearly, as we want a continuous function, we will need some conditions on the IFS. For simplicity, we assume that our function is to be built on the interval $[0, 1]$. We start with a collection of $N + 1$ data points (x_i, y_i) for $i = 0, 1, 2, \ldots, N$ with $x_0 = 0$ and $x_N = 1$. We further assume that $y_0 = y_N = 0$, again for simplicity.

The idea is that we have an N map IFS w_i, $i = 1, 2, \ldots, N$, consisting of affine functions of the form

$$w_i \begin{pmatrix} x \\ y \end{pmatrix} = \begin{pmatrix} x_i - x_{i-1} & 0 \\ y_i - y_{i-1} & d_i \end{pmatrix} \begin{pmatrix} x \\ y \end{pmatrix} + \begin{pmatrix} x_{i-1} \\ y_{i-1} \end{pmatrix}. \qquad (2.34)$$

Here d_i is a free parameter and we require $|d_i| < 1$ for contractivity. The idea is that the "horizontal" action of w_i should be to map $[0, 1]$ onto $[x_{i-1}, x_i]$, and the "vertical" action both forces the interpolation property and also contracts "vertically" by a factor of $|d_i|$. Figure 2.17 illustrates this idea, where the large rectangle is mapped onto the parallelogram. If $d_i > 0$, then the bottom edge of the parallelogram is the line segment connecting (x_{i-1}, y_{i-1}) to (x_i, y_i). If $d_i < 0$, then it is the top edge that is this line segment.

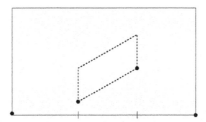

Fig. 2.17: Illustration of the affine maps for a fractal interpolation function.

Exercise 2.80. Show that the attractor of the IFS given in (2.34) is the graph of a continuous function $f : [0, 1] \to \mathbb{R}$. Is it possible for it to be differentiable? Under what conditions will it be differentiable?

Exercise 2.81. Suppose we remove the restriction that $y_0 = y_N = 0$. Find the formula for the map w_i in an IFS for a fractal interpolation function for this situation. What conditions on the parameters of the w_i are necessary for the attractor to be a continuous function? If we no longer require the attractor to be continuous, do we still obtain a well-defined function? And what are the conditions?

In Fig. 2.18, we show the first few iterations of an IFS for a fractal interpolation function with the data points $(0, 0)$, $(0.25, 0.5)$, $(0.5, 0.25)$, $(1, 0)$ and $d_1 = 0.5, d_2 = -0.25, d_3 = 0.5$. Notice that the first image is the piecewise linear approximation through these points. From this figure, it is not obvious that the limiting function will interpolate the data. To see this more clearly, it is useful to overlay the iterations as in Fig. 2.19. In Fig. 2.19, we see that each successive iteration is a refinement of the previous iteration. This viewpoint is worth commenting on, as it has links to affine IFSMs (to be discussed in Sect. 3.2.3).

Continuing with our example fractal interpolation function, let f_1 be the piecewise linear function that interpolates the data. Then we see from Fig. 2.19 that the next iteration, f_2, is equal to f_1 plus the

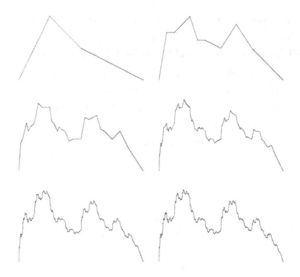

Fig. 2.18: First few iterations for a fractal interpolation function.

image of f_1 under the IFS; call this image $g_1 = T(f_1)$, so that $f_2 = f_1 + T(f_1) = f_1 + g_1$. Continuing, we see that

$$f_{n+1} = f_1 + T(f_n) = f_1 + g_1 + g_2 + \cdots + g_{n-1}$$
$$= f_1 + T(f_1) + T^2(f_1) + T^3(f_1) + \cdots + T^{n-1}(f_1).$$

In this way, the IFS for the fractal interpolation function is very much like an IFS with condensation. In fact, this fits exactly into the context of an affine IFSM (IFS on functions) as discussed in Sect. 3.2.3. In Fig. 2.20, we show f_1 and the first few summands g_n for $n = 1, 2, 3$.

2.6.3 Graph-directed IFS

There is a nice generalization of an IFS that can be formalized in different ways, but the general idea is that we replace a single fixed-point equation with a system of equations. As long as some simple consistency condition is met, this results in an attractor of a more complicated type of IFS. These types of IFS are also called *recurrent IFSs*. Our basic references for this section are [12, 14, 128]. We do not give anywhere near a complete discussion of this topic, as our goal is simply to introduce the basic ideas.

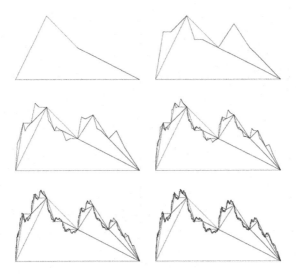

Fig. 2.19: Overlaying the first few iterations for a fractal interpolation function.

Let V be a finite set of vertices and E be a multiset of directed edges (by multiset we mean that we allow there to be multiple directed edges between two vertices). For each $v \in V$, let $\mathbb{X}_v$ be a complete metric space and for each $e = (a, b) \in E_{ab} \subset E$ let $f_e : \mathbb{X}_b \to \mathbb{X}_a$ be a contraction (this "backwards" direction is that chosen in [128]). The "fractal" set constructed in this manner is the vector of sets A_v, one for each $v \in V$, which satisfies

$$A_v = \bigcup_{\substack{u \in V \\ e \in E_{vu}}} f_e(A_u). \qquad (2.35)$$

Clearly we require some properties of the graph. The most basic is that for every $v \in V$ there is at least one $u \in V$ such that (v, u) is a directed edge.

If each $\mathbb{X}_i \subseteq \mathbb{X}$, then it may be convenient to consider the attractor to be $A = \cup_v A_v$. This is particularly the case where one wants to consider the boundary of a self-similar set to be self-similar in this more general sense. For instance, the boundary of the twin dragon fractal, Fig. 2.21, is itself self-similar in this more general graph-directed sense. That is, parts of the boundary are made up of contracted versions of other parts of the boundary. In this particular case, the maps in the graph-directed IFS that defines the boundary of the twin dragon are

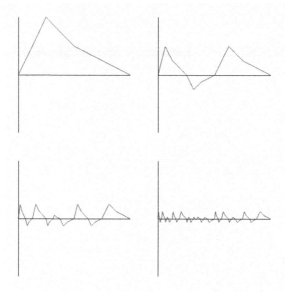

Fig. 2.20: The individual summands in the fractal interpolation function.

modifications of restrictions of the two maps in the IFS for the entire twin dragon (for a nice explanation of this, see [122, 159]).

Fig. 2.21: The boundary of the twin dragon tile is graph-directed self-similar.

Exercise 2.82. Find a graph-directed IFS whose attractor is a circle.

Exercise 2.83. Find a graph-directed IFS whose attractor is the boundary of the twin dragon fractal.

The existence of an attractor for a graph-directed IFS is also based on a contraction map. One formulation is to set $\mathbb{Y} = \prod_v \mathbb{H}(\mathbb{X}_v)$ with the metric given by $\mathbf{d}_{\mathbb{H}}(\mathbf{A}, \mathbf{B}) = \max_v d_{\mathbb{H}}(A_v, B_v)$. For a generic $\mathbf{A} \in \mathbb{Y}$, we define $F : \mathbb{Y} \to \mathbb{Y}$ by

$$F(\mathbf{A})_v = \bigcup_{\substack{u \in V \\ e \in E_{vu}}} f_e(A_u).$$

It is straightforward to show that F is a contraction with contractivity equal to the maximum of the contractivities of the f_e.

Exercise 2.84. Prove the claim above that F is contractive with contractivity equal to the maximum of the contractivities of the f_e.

The usual IFS is a special case when there is only one vertex and one loop for each contraction. A slightly more general case is also worth mentioning. Let $\{f_i\}$ be an IFS on $\mathbb{X}$ where $i = 1, 2, \ldots, N$. We take as our vertex set $V = \{1, 2, \ldots, N\}$ and let $G = (V, E)$ be any directed graph on V that is *strongly connected* (there is a directed path from any vertex to any other vertex). For each $e = (i, j) \in E$, let $f_e = f_i$. This defines a graph-directed IFS where G specifies the allowable compositions, so on the address space side this specifies the allowable addresses. In fact, the space of allowable addresses forms a *subshift of finite type*. In this situation, the attractor of the graph-directed IFS given by G is a subset of the attractor of the original IFS $\{f_i\}$. Using the same IFS as for the Sierpinski triangle as an example, Fig. 2.22 shows the graph and the corresponding attractor.

Fig. 2.22: The graph and attractor for a modified (recurrent) Sierpinski triangle.

A fractal such as that in Fig. 2.16 is also the attractor of a graph-directed IFS. In this case, the graph is as given in Fig. 2.23. Notice that this graph is not strongly connected but has two strongly connected components, corresponding to the two different fractals (the maple leaf and the Sierpinski triangle). We have the IFS for the Sierpinski triangle, $\{w_1, w_2, w_3\}$, and the IFS for the maple leaf, $\{f_1, f_2, f_3, f_4\}$ (as in Fig. 2.8), along with g as the similitude that scales the maple leaf attractor and places it in the center of the triangle.

The invariance equation for this graph-directed IFS is

$$A_1 = \bigcup_i w_i(A_1) \cup g(A_2),$$
$$A_2 = \bigcup_i f_i(A_2),$$

where A_2 is the maple leaf and A_1 is the combination of the Sierpinski triangle and maple leaf.

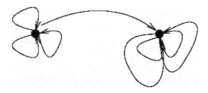

Fig. 2.23: The graph for the Sierpinski/maple leaf attractor of Fig. 2.16.

There is also a similar type of construction for an IFSP where the "invariant measure" is a vector solution to a system of linear equations of IFS type [12, 14]. As an example, for the mixture of the Sierpinski triangle and maple leaf in Fig. 2.16, we can choose any $p_1, p_2, p_3, q_1, q_2, q_3, q_4, r$ such that each is strictly positive and their sum is equal to 1. Then the invariance equations are

$$\mu_1(B) = \sum_{i=1}^{3} p_i \mu_1(w_i^{-1}(B)) + r\mu_2(g^{-1}(B)),$$
$$\mu_2(B) = \sum_{j=1}^{4} q_j \mu_2(f_j^{-1}(B)).$$

In this equation, μ_1 is a measure supported on the triangle and μ_2 is a measure supported on the maple leaf. As these two equations are uncoupled, they are relatively easy to solve. First we find the fixed point for the second equation and then use this as a "condensation measure" for the first equation. Much more general situations are possible. For a nice discussion of this, see [12, 23, 22].

Before leaving this topic, we also should mention that it is possible to construct a chaos game (and prove a corresponding ergodic theorem) for these mixed models. This works by running parallel chaos games for each vertex and then periodically mixing. We will again use the mixture of the Sierpinski triangle and maple leaf as our example. Let $\{w_1, w_2, w_3\}$ be an IFS for the Sierpinski triangle and p_1, p_2, p_3 be corresponding probabilities. Furthermore, let $\{f_1, f_2, f_3, f_4\}$ be an IFS for the maple leaf and q_1, q_2, q_3, q_4 be corresponding probabilities. Finally, let $\rho \in (0, 1)$ be chosen and let $g : \mathbb{R}^2 \to \mathbb{R}^2$ be the function that first shrinks the maple leaf and then places it in the center of the triangle. The algorithm is taken from [12] and is as follows:

1. Initialize X_1, Y_1 and set $n = 1$.
2. Choose $I = 0$ or $I = 1$ with probability ρ or $1 - \rho$, respectively.
3. Choose f_k randomly according to the probabilities q_1, q_2, q_3, q_4.
4. If $I = 0$, set $Y_{n+1} = g(X_n)$ and $X_{n+1} = f_k(X_n)$ and go to step 7.
5. If $I = 1$, choose w_ℓ randomly according to the probabilities p_1, p_2, p_3.
6. Set $Y_{n+1} = w_\ell(Y_n)$ and $X_{n+1} = f_k(X_n)$.
7. Plot the point Y_n if n is large enough (often, $n > 1000$ is good).
8. Increment n and go to step 2.

The image that is plotted will be as given in Fig. 2.16. Furthermore, the empirical occupancy distribution will converge to the attractor of the corresponding IFSP with condensation. For details and proofs of these results, see [12]. This reference also gives an ergodic theorem for these mixed chaos games. We return to this topic in Sect. 6.2. Chapter 6 gives many extensions of the chaos game and the ergodic theorem for more general objects.

Exercise 2.85. Suppose we are given L different IFSPs and an $L \times L$ probability transition matrix P. If $p_{i,j} > 0$, then we set a directed edge from vertex j (which corresponds to the jth IFSP) to vertex i, thus defining a graph-directed IFSP. Devise a chaos game algorithm to generate an approximation to the vector of invariant probability distributions for this graph-directed IFSP.

2.6.4 IFS with infinitely many maps

The idea of allowing an IFS to contain infinitely many maps is very natural. There have been several different approaches to this, both for geometric a IFS [114, 127] and for an IFSP [54, 23, 132]. We will discuss two related and quite general frameworks; however, this is only a sampling.

2.6.4.1 IFS indexed by a compact set

Our first framework will be one where there are possibly uncountably many maps in the IFS but the maps have to depend continuously on the parameter. The primary reference for this material is [114]. Let $(\mathbb{X}, d_{\mathbb{X}})$ be a complete metric space and (Λ, d_Λ) be a compact metric space. The IFS will be a continuous function $w : \Lambda \times \mathbb{X} \to \mathbb{X}$ with $w_\lambda : \mathbb{X} \to \mathbb{X}$ contractive with contractivity c_λ for each $\lambda \in \Lambda$. We assume that $c = \sup_\lambda c_\lambda < 1$. The induced IFS map on the space $\mathbb{H}(\mathbb{X})$ is given by

$$W(B) = \bigcup_{\lambda \in \Lambda} w_\lambda(B). \tag{2.36}$$

This can also be thought of as the image of the compact set $\Lambda \times B \subset \Lambda \times \mathbb{X}$, and thus we know that $W(B) \in \mathbb{H}(\mathbb{X})$ for any $B \in \mathbb{H}(\mathbb{X})$. We now show that W is contractive with contractivity c by showing that

$$W(L) \subset W(K)_{c\,d_{\mathbb{H}}(K,L)+c\epsilon} \text{ and } W(K) \subset W(L)_{c\,d_{\mathbb{H}}(K,L)+c\epsilon}$$

for all $\epsilon > 0$. For each $x \in K$ and $\lambda \in \Lambda$, there is a $y \in L$ so that $d(x,y) < d_{\mathbb{H}}(K,L) + \epsilon$ and thus

$$d(w_\lambda(x), w_\lambda(y)) \le cd(x,y) + c\,d_{\mathbb{H}}(K,L) + c\epsilon$$

which implies that $W(K) \subset W(K)_{cd_{\mathbb{H}}(K,L)+c\epsilon}$. The reverse inclusion is similar, and thus $d_{\mathbb{H}}(W(K), W(L)) \le cd_{\mathbb{H}}(K,L) + c\epsilon$ for any ϵ and so $d_{\mathbb{H}}(W(K), W(L)) \le cd_{\mathbb{H}}(K,L)$.

Proposition 2.86. *Suppose that $w : \Lambda \times \mathbb{X} \to \mathbb{X}$ is continuous and w_λ is contractive with contractivity bounded by $c < 1$. Then the map W as defined in (2.36) is contractive and thus has a unique fixed point $A \in \mathbb{H}(\mathbb{X})$ that satisfies*

$$A = \bigcup_\lambda w_\lambda(A).$$

As in the case of an IFS with finitely many maps, there is a corresponding code space and address map. In this case, the code space is

$$\Sigma = \Lambda^{\mathbb{N}} = \{\sigma = (\sigma_1, \sigma_2, \ldots) : \sigma_n \in \Lambda\}$$

and the address map $\omega : \Sigma \to X$ is defined in the same way as

$$\omega(\sigma) = \lim_n w_{\sigma^n}(x).$$

The address map is again surjective onto the attractor A, and there is again a correspondence between the shift dynamics on Σ and the dynamics on A induced by the IFS w_λ. The metric on Σ is slightly different but still quite simple, as

$$d(\sigma, \alpha) = \sum_{i=1}^{\infty} \frac{d_\Lambda(\sigma_i, \lambda_i)}{2^i}.$$

Exercise 2.87. Prove that the collection of all finite IFSs on Λ is dense in the set of all possible IFSs on Λ.

There is also a very nice general framework in which we can say that the attractor of an infinite IFS is a continuous function of the IFS. Now we let Λ be any metric space, not necessarily compact. We continue to assume that $w : \Lambda \times X \to X$ is continuous and w_λ is contractive with contractivity bounded by $c < 1$. To each $\Psi \in \mathbb{H}(\Lambda)$ there is an associated IFS $\{w_\lambda : \lambda \in \Psi\}$ and an associated attractor $A(\Psi)$.

For a subset $\Psi \subseteq \Lambda$, we denote by Σ_Ψ the product space $\Psi^{\mathbb{N}} \subseteq \Lambda^{\mathbb{N}}$, the code space for the sub-IFS $\{w_\lambda : \lambda \in \Psi\}$.

Theorem 2.88. *[114]* *The function A that associates to each $\Psi \in \mathbb{H}(X)$ its attractor $A(\Psi) \in \mathbb{H}(X)$ is uniformly continuous. Thus, if $K_n \in \mathbb{H}(\Lambda)$ converge to K, then $A(K_n) \to A(K)$.*

Proof. The proof works by going through code space and using the fact that the address map is uniformly continuous.

Let $\epsilon > 0$ be given. Then there is a $\delta > 0$ such that if $\alpha, \sigma \in \Sigma$ satisfy $d_\Sigma(\alpha, \sigma) < \delta$, then $d_X(\omega(\alpha), \omega(\sigma)) < \epsilon$. Take $\Psi, \Upsilon \in \mathbb{H}(\Lambda)$ with $d_\mathbb{H}(\Psi, \Upsilon) < \delta$. Then $\Upsilon \subseteq \Psi_\delta$, so for $\alpha \in \Sigma_\Upsilon$ we have $\alpha_i \in \Upsilon$ for all i and thus there is $\beta_i \in \Psi$ with $d_\Lambda(\alpha_i, \beta_i) < \delta$. But then $\beta \in \Sigma_\Psi$ and $d_\Sigma(\alpha, \beta) = \sum_i 2^{-i} d_\Lambda(\alpha_i, \beta_i) < \delta$, and so $d_X(\omega(\alpha), \omega(\beta)) < \epsilon$. Since this is true for any $\alpha \in \Sigma_\Upsilon$, we have $\omega(\Sigma_\Upsilon) \subset \omega(\Sigma_\Psi)_\epsilon$. The reverse containment is similar, and thus $d_\mathbb{H}(\omega(\Sigma_\Psi), \omega(\Sigma_\Upsilon)) < \epsilon$ and so $d_\mathbb{H}(A_\Psi, A_\Upsilon) < \epsilon$. $\square$

We point out that this notion of self-similarity is perhaps too broad, as any compact subset $K \subseteq X$ is the attractor of an IFS of this type. The simplest way to do this is to let $\Lambda = K$ and $w_\lambda(x) = \lambda$. Then clearly $K = \cup_\lambda w_\lambda(K)$.

There is an intermediate notion that restricts the collection to be countable, $\{w_i\}$. Without any structure on the indexing set, it is necessary to take a closure in the induced mapping

$$W(B) = \overline{\bigcup_n w_n(B)}$$

if one wishes to obtain a closed set; guaranteeing a compact set is yet another task. The direction taken in [127] is somewhat different in that the invariant set is defined as the image of the natural addressing map. This set may be extremely complicated in a descriptive set-theoretic sense, as it could be a general analytic set. However, the paper [127] is mainly concerned with infinite conformal iterated function systems, as its author's motivation comes from questions in dynamical systems.

2.6.4.2 IFSP indexed by a compact set

Now we turn our attention to IFSs with probabilities, again with an indexing set that is a compact metric space Λ but this time where the space $\mathbb{X}$ is also compact. The material in this part is from [132]. Let $w : \Lambda \times \mathbb{X} \to \mathbb{X}$ be continuous and $\in \mathcal{P}(\Lambda)$, so that p is a Borel probability measure on Λ. We define the Markov operator in this case to be

$$\mathbb{M}_p\mu(B) = \int_\Lambda \mu(w_\lambda^{-1}(B)) \, dp(\lambda) \qquad (2.37)$$

for all Borel sets $B \subseteq \mathbb{X}$. We will usually leave out the subscript on $\mathbb{M}$ when the measure p is fixed.

By the Riesz representation theorem, an equivalent way to define $\mathbb{M}$ is as

$$\int_\mathbb{X} f(x) \, d\mathbb{M}(\mu)(x) = \int_\Lambda \int_\mathbb{X} f(w_\lambda(x)) \, d\mu(x) \, dp(\lambda), \qquad (2.38)$$

where f is a continuous bounded real-valued function on $\mathbb{X}$.

It is fairly straightforward to show that $\mathbb{M}$ maps a probability measure on $\mathbb{X}$ to another probability measure on $\mathbb{X}$, so we leave out the details.

Exercise 2.89. Show that $\mathbb{M}\mu \in \mathcal{P}(\mathbb{X})$ if $\mu \in \mathcal{P}(\mathbb{X})$.

Definition 2.90. We say that w is *contractive on average* with contractivity $0 \le c < 1$ if for all $x, y \in \mathbb{X}$ we have

$$\int_\Lambda d(w_\lambda(x), w_\lambda(y)) \, dp(\lambda) \le cd(x, y)$$

Theorem 2.91. *If w is contractive on average, then $\mathbb{M}$ is contractive in the Monge-Kantorovich metric on $\mathcal{P}(\mathbb{X})$.*

Proof. Let $f \in \mathrm{Lip}_1(\mathbb{X})$. We calculate for μ and ν in $\mathcal{P}(\mathbb{X})$

$$
\begin{aligned}
\int_{\mathbb{X}} f \, d(\mathbb{M}\mu - \mathbb{M}\nu) &= \int_{\mathbb{X}} f(x) \, d\left(\int_{\Lambda} (\mu(w_\lambda^{-1}(x)) - \nu(w_\lambda^{-1}(x))) \, dp(\lambda) \right) \\
&= \int_{\Lambda} \int_{\mathbb{X}} f(x) \, d\left(\mu(w_\lambda^{-1}(x)) - \nu(w_\lambda^{-1}(x)) \right) \, dp(\lambda) \\
&= \int_{\Lambda} \int_{\mathbb{X}} f(w_\lambda(y)) \, d(\mu(y) - \nu(y)) \, dp(\lambda) \\
&= \int_{\mathbb{X}} \left(\int_{\Lambda} f(w_\lambda(y)) \, dp(\lambda) \right) d(\mu(x) - \nu(x)) \\
&= c \int_{\mathbb{X}} \phi(y) \, d(\mu(y) - \nu(y)),
\end{aligned}
$$

where $\phi(y) = c^{-1} \int_{\Lambda} f(w_\lambda(y)) \, dp(\lambda) \in \mathrm{Lip}_1(\mathbb{X})$ by the definition of c. Taking the supremum, we get

$$
d_{MK}(\mathbb{M}(\mu), \mathbb{M}(\nu)) \leq c \, d_{MK}(\mu, \nu)
$$

and the result follows.   □

We now prove that the invariant distribution is a continuous function of the "parameters" of the infinite IFSP.

Theorem 2.92. *Suppose that $p^{(n)}$ is a sequence of probability measures in $\mathcal{P}(\Lambda)$ that converges to p in the Monge-Kantorovich metric. Then $\mu_{p^{(n)}}$ also converges to μ_p in the Monge-Kantorovich metric.*

Proof. Let f be a bounded continuous function on $\mathbb{X}$. We calculate that

$$
\begin{aligned}
\int_{\mathbb{X}} f(x) \, d\left(\int_{\Lambda} \mu(w_\lambda^{-1}(x)) \, dp^{(n)}(\lambda) \right) &= \int_{\Lambda} \int_{\mathbb{X}} f(x) \, d\mu(w_\lambda^{-1}(x)) \, dp^{(n)}(\lambda) \\
&= \int_{\Lambda} \int_{\mathbb{X}} f(w_\lambda(x)) \, d\mu(x) \, dp^{(n)}(\lambda).
\end{aligned}
$$

Let $\phi(\lambda) = \int_{\mathbb{X}} f(w_\lambda(x)) \, d\mu(x)$. Clearly ϕ is bounded since f is bounded and μ is a probability measure. Let $\epsilon > 0$ be given. Now both f and w are uniformly continuous in $\mathbb{X}$ and Λ. Thus, there is a $\delta > 0$ such that if $d(\lambda, \lambda') < \delta$, then $|f(w_\lambda(x)) - f(w_{\lambda'}(x))| \leq \epsilon$ for all $x \in \mathbb{X}$. Therefore, for $\lambda, \lambda' \in \Lambda$ with $d(\lambda, \lambda') < \delta$, we have

$$|\phi(\lambda) - \phi(\lambda')| \leq \int_{\mathbb{X}} |f(w_\lambda(x)) - f(w_{\lambda'}(x))| \ d\mu(x)$$

$$\leq \int_{\mathbb{X}} \epsilon \ d\mu(x) = \epsilon$$

so $\phi \in C^*(\Lambda)$, and since $p^{(n)} \to p$ in the Monge-Kantorovich metric, we have that

$$\int_\Lambda \phi(\lambda) \ dp^{(n)}(\lambda) \quad \to \quad \int_\Lambda \phi(\lambda) \ dp(\lambda).$$

Since this is true for all $f \in C^*(\mathbb{X})$, we know that $\mathbb{M}_p$ is a continuous function of p (the distribution on Λ). To get continuity of μ_p (the fixed point of $\mathbb{M}_p$) as a function of p, we need to have a uniform bound for the contraction factor of the family $\mathbb{M}_{p^{(n)}}$. However, it suffices to get a bound for sufficiently large n. For fixed $x, y \in \mathbb{X}$, we have that $d(w_\lambda(x), w_\lambda(y))$ is a continuous function of λ, so we know that

$$\int_\Lambda d(w_\lambda(x), w_\lambda(y)) \ dp^{(n)}(\lambda) \quad \to \quad \int_\Lambda d(w_\lambda(x), w_\lambda(y)) \ dp(\lambda)$$

and thus the contraction factor of $\mathbb{M}_{p^{(n)}}$ converges to the contraction factor of $\mathbb{M}_p$. This gives us our uniform bound s on the contraction factors. Now, by the estimate

$$d_{MK}(\mu_p, \mu_q) \leq \frac{d_{MK}(\mathbb{M}_p(\mu_p), \mathbb{M}_q(\mu_p))}{1 - s},$$

we know that $\mu_{p^{(n)}} \Rightarrow \mu_p$ in the Monge-Kantorovich metric. $\qquad \square$

Finally we prove a result concerning the support of the invariant measure for the IFS $\{w_\lambda, p(\lambda)\}$. For $p \in \Lambda(p)$, let $\Omega_p \subseteq \Lambda$ be the support of p and A be the attractor of the IFS $\{w_\lambda : \lambda \in \Omega_p\}$. We see that since Λ is compact so is Ω_p, so this is well-defined.

Theorem 2.93. *Suppose that the contractivity of w_λ is uniformly bounded from above by $0 \leq c < 1$. Then the support of μ is equal to A.*

Proof. As the proof is similar to the proof of Theorem 2.63, we do not give the details. Alternatively, letting B be the support of μ, one can show that B is compact and invariant with respect to the IFS $\{w_\lambda : \lambda \in \Omega_p\}$ and thus $B = A$ by uniqueness. $\qquad \square$

Exercise 2.94. Give a complete proof of Theorem 2.93.

The previous two theorems do not imply that if $p^{(n)} \to p$ in $\mathcal{P}(\Lambda)$ then $\mathrm{supp}\mu_{p^{(n)}} \to \mathrm{supp}\mu_p$ in $\mathbb{H}(\mathbb{X})$. The following simple example illustrates what can go wrong.

Example 2.95. Let $\Lambda = \{0, 1\}$ and $w_i(x) = x/2 + i/2$. Let $p^{(n)}(\{0\}) = 1 - 1/n$ and $p^{(n)}(\{1\}) = 1/n$ so that $p^{(n)} \to p$ where p is the point mass at 0. Then the support of $\mu_{p^{(n)}}$ is the entire interval $[0, 1]$ but the support of μ_p is just $\{0\}$.

If in addition to $p^{(n)}$ converging to p in the Monge-Kantorovich metric we require the support of $p^{(n)}$ to converge to the support of p in $\mathbb{H}(\Lambda)$, then the support of $\mu_{p^{(n)}}$ will converge to the support of μ_p.

Exercise 2.96. Investigate which of the results for infinite IFSPs extend to the situation where $\mathbb{X}$ is not necessarily compact but is complete and separable. It will be necessary to place some assumptions on p, perhaps $p \in \mathcal{P}_1(\Lambda)$. Is this enough to guarantee that $\mathbb{M}\mu \in \mathcal{P}_1(\mathbb{X})$ if $\mu \in \mathcal{P}_1(\mathbb{X})$?

Chapter 3
IFS on Spaces of Functions

Geometric and measure-theoretic IFSs can easily be extended to IFS operators acting on functions. These operators are closely related to the IFS on measures from Sect. 2.5. Historically, one primary motivation for these operators was the desire to represent digital images by means of attractors of IFSs.

3.1 Motivation: Fractal imaging

The first practical method for using IFS fractals to represent and compress images was in the Ph.D. thesis of A. Jacquin (and was published in [81]). The basic idea is a very clever relaxation of the requirement for strict self-similarity. We describe a slightly simpler version that illustrates the idea.

For this we take $\mathbb{X} = [0, 1]^2$, the standard unit square in $\mathbb{R}^2$. We consider two different partitions of $\mathbb{X}$. The first one is into "parent" or "domain" blocks, and the second one is into "child" or "range" blocks. Of the two, the "parent" blocks are the larger, often twice as large in each direction. An illustration of these two different partitions is given in Fig. 3.1.

The compression algorithm proceeds block-by-block through the child block partition. For each such child block R_i, we search through all the parent blocks D_j, looking for the best match. The first problem is that D_j is larger than R_i, so we must downsample D_j in some way to match the size of R_i. This is usually accomplished by averaging four pixels in D_j to obtain a corresponding pixel in R_i and is simple to do. The second problem is that it is highly unlikely that any D_j will look very similar to the given R_i. To deal with this problem, we allow a sim-

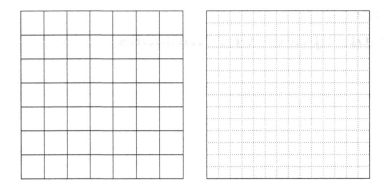

Fig. 3.1: "Parent" and "child" partitions.

ple transformation of the pixel values of D_j. What this often means is that we fix two values α_i, β_i (fixed for the given child block R_i) and transform the pixel values of D_j using the function $\phi_i(t) = \alpha_i t + \beta_i$. The best match between the given R_i and any D_j is then the best match between R_i and $\phi_i(D_j)$, with the values of α_i, β_i chosen in such a way as to optimize this match (i.e., minimize $\|R_i - \phi_i(D_j)\|$), usually using the Euclidean norm $\|x\| = (\sum_k |x_k|^2)^{1/2}$. The Euclidean norm has the obvious advantage that finding the optimal α_i, β_i is a linear least-squares problem, so it is very simple and fast to solve. Once the parent block D_j with the best match (smallest error $\|R_i - \phi_i(D_j)\|$) has been found, we record which D_j was the best match (by recording the index j of D_j) and record the values of the parameters α_i, β_i. This procedure is done for each range block. The data that comprises the compressed version of the original image is this list of triples (parent block index, α_i, β_i), with one triple for each child block R_i.

As an algorithm, compression goes as follows:

```
Loop over all child blocks R_i.
  Loop over all parent blocks D_j.
    Using least squares find α, β to minimize ||R_i − φ(D_j)||.
    Compute current error = ||R_i − φ(D_j)|| (using optimal α, β).
    If current error is the best for R_i, store index j, α, β.
  End of Loop over parent blocks D_j.
End of loop over child blocks R_i.
```

The data of triples (parent block index, α_i, β_i) defines a "fractal operator" that when iterated will yield a fixed-point image that is

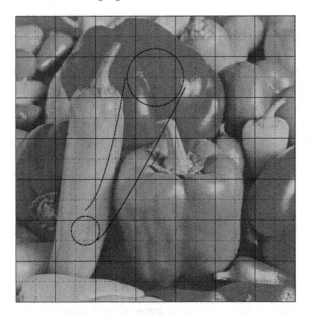

Fig. 3.2: "Parent" and "child" comparison in fractal block-coding.

the reconstructed approximation to the original image; usually the iteration is started off with a blank image. This fractal operator T is also defined in a block-by-block fashion, and T acts on an input image I and produces an output image J. To produce J from I, scan through the child block partition of J and for each child block R_i in J replace R_i with $\phi_i(D_{ind(i)})$, where $D_{ind(i)}$ refers to the parent block in I that was associated with R_i in the compression. Since we do this for each R_i in J, this fills up (or paints) each child block in J and thus constructs the output image J child block by child block.

For the final reconstructed image, we iterate T a number of times until we achieve convergence. As we expect T to be contractive in some sense, the iterates of T should converge to the fixed point of T. As an algorithm, image recovery goes as follows:

```
Loop over iterations.
    Loop over child blocks R_i in output image J.
        Set R_i = α_i D_ind(i) + β_i,
        where D_ind(i) is the appropriate parent block in I.
    End of loop on child blocks.
    Swap the roles of I and J.
End of loop on iterations.
```

The reconstruction is illustrated in Figs. 3.3 and 3.4. The first eight iterations and the final reconstructed image are given in both of these figures. In Fig. 3.3, the image is partitioned into an 8×8 grid of parent blocks, with each parent block being decomposed into four child blocks. In the first frame, you can see the 16×16 grid of child blocks. Figure 3.4 is similar, but here the image is partitioned into a 64×64 grid of parent blocks, with each parent block being decomposed into four child blocks. Again in the first frame, you can see the 128×128 grid of child blocks. Clearly the reconstructed image is much better in Fig. 3.4 than in Fig. 3.3, as we use more blocks and thus store more information.

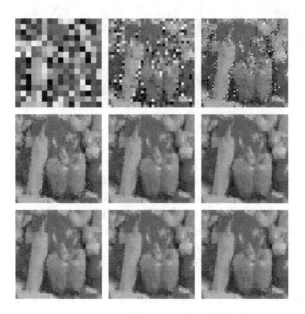

Fig. 3.3: Fractal block-coding reconstruction using a 16×16 grid of child blocks.

Viewed as subsets of $[0, 1]^2$, the mapping from $D_{ind(i)}$ to R_i is usually just a simple linear contraction; we call this map $w_i : D_{ind(i)} \to R_i$. For each i, we also have a function $\phi_i : \mathbb{R} \to \mathbb{R}$ (given above by $\phi_i(t) = \alpha_i t + \beta_i$). Thinking of the input image I and output image J as functions $I, J : [0, 1]^2 \to \mathbb{R}$, the action of the fractal operator T is

$$J(x) = T(I)(x) = \phi_i(I(w_i^{-1}(x))) \quad \text{for } x \in R_i.$$

Fig. 3.4: Fractal block-coding reconstruction using a 128×128 grid of child blocks.

Any type of data compression operates by finding correlations or structure in the given data and then using an appropriate model to represent this structure. A fractal block-coding algorithm exploits scaling relationships within a given image I in order to represent the image. It can be viewed as a type of vector coding (see [153]) but with the codebook generated from the image itself. Furthermore, contractivity ensures that only the parameters of the IFS transform need to be saved since the final reconstruction depends only on these parameters.

Since the introduction of the fractal block-coding algorithm, a huge number of variations and refinements have been implemented. These include using more complicated maps ϕ_i, adaptive partitions of various types, partitions developed through "region growing" heuristics, and many others. In terms of quality of the reconstructed image at a given compression rate, the best fractal coding techniques are very competitive. The main drawback of fractal image compression is usually the compression time, as the searching can take an enormous amount of time. The references [19, 60, 120, 151, 166] all contain more information about fractal imaging.

Fractal block-coding is the motivating example behind the technique of *collage coding*. Ideally we want to find a fractal transform whose fixed point is as close to the original image as possible. However, in the algorithm above, we do not compare the fixed point of the proposed transform to the original image. Instead we use the "collage error" (the difference between I and $T(I)$). This combination of a contractive operator and replacing the distance between the fixed point of T and I with the collage distance, the distance between I and $T(I)$, is the essence of the collage-coding method.

3.2 IFS on functions

The definition of the basic IFS operator on functions follows directly from the motivating example of IFS fractal block-coding. The general setup is one where we have *geometric* maps $w_i : \mathbb{X} \to \mathbb{X}$ and *grey-level maps* $\phi_i : \mathbb{R} \to \mathbb{R}$ for $i = 1, 2, \ldots, N$. For a function $f : \mathbb{X} \to \mathbb{R}$, we define

$$T(f)(x) = \sum_i \phi_i(f(w_i^{-1}(x))), \qquad (3.1)$$

where we use the convention that if $w_i^{-1}(x) = \emptyset$ we set the corresponding term equal to zero; that is, $\phi_i(f(w_i^{-1}(x))) = 0$ when $x \notin w_i(\mathbb{X})$. Sometimes the individual terms in the summation in (3.1) are referred to as the *fractal components* of the transform. This clearly defines a new function $T(f) : \mathbb{X} \to \mathbb{R}$. In order to prove anything about the operator T, it is necessary to have additional assumptions. We start in the framework with the least assumptions and then progressively add structure for the other models.

Notice that if $x \notin w_i(\mathbb{X})$ for any i, then by definition $T(f)(x) = 0$. Furthermore, if A is the attractor of the *geometric IFS* $\{w_i\}$ (that is, the IFS on sets), then the support of the fixed point of T, if there is one, is contained in A. For ease of reference, we record this as a proposition.

Proposition 3.1. *Suppose that $\bar{f}$ is the fixed point of T as defined in (3.1). Then the support of $\bar{f}$ is contained in the attractor A of the IFS $\{w_i\}$.*

3.2.1 Uniformly contractive IFSM

The first framework we consider requires only contractivity on the grey-level maps ϕ_i. For any set $\mathbb{X}$, the collection $\mathcal{F}^*(\mathbb{X})$ of all bounded functions $f : \mathbb{X} \to \mathbb{R}$ is complete in the supremum norm, $\|f\|_\infty :=$ $\sup_{x \in \mathbb{X}} |f(x)|$.

Theorem 3.2. *Suppose that each ϕ_i is contractive with contractivity c_i. Further suppose that for each $x \in \mathbb{X}$ we have*

$$\sum_{w_i^{-1}(x) \neq \emptyset} c_i < 1.$$

Then T defined as in (3.1) is contractive in the norm $\|f\|_\infty$ and therefore has a unique fixed point in $\mathcal{F}^*(\mathbb{X})$.

Proof. Let

$$c := \sup_{x \in \mathbb{X}} \sum_{w_i^{-1}(x) \neq \emptyset} c_i.$$

Since there are only finitely many w_i, the expression above takes on only finitely many different values. Since each one is strictly bounded above by 1 (by assumption), we know that $c < 1$. We show that T is contractive with contractivity bounded by c. To see this, suppose that $f, g : \mathbb{X} \to \mathbb{R}$. Then

$$
\begin{aligned}
|T(f)(x) - T(g)(x)| &= \left| \sum_i \phi_i(f(w_i^{-1}(x))) - \phi_i(g(w_i^{-1}(x))) \right| \\
&\leq \sum_{w_i^{-1}(x) \neq \emptyset} \left| \phi_i(f(w_i^{-1}(x))) - \phi_i(g(w_i^{-1}(x))) \right| \\
&\leq \sum_{w_i^{-1}(x) \neq \emptyset} c_i \left| f(w_i^{-1}(x)) - g(w_i^{-1}(x)) \right| \\
&\leq \sum_{w_i^{-1}(x) \neq \emptyset} c_i \|f - g\|_\infty \\
&\leq c\|f - g\|_\infty.
\end{aligned}
$$

$\square$

In the particular situation where $w_i(\mathbb{X}) \cap w_j(\mathbb{X}) = \emptyset$, the contractive factor then becomes $c := \max_i c_i$. Furthermore, in this case we can sometimes even have a "formula" for the invariant function, f. Recall from Chapter 2 the discussion on the code space and the address map. Suppose that $\mathbb{X}$ is a complete metric space and the w_i are also contractive. Let $\Sigma = \{1, 2, \ldots, N\}^{\mathbb{N}}$ be the code space and $a : \Sigma \to \mathbb{X}$ be the address map. Then since the IFS $\{w_i\}$ is nonoverlapping, we know that a is injective. Furthermore, each point of the attractor A of the geometric IFS $\{w_i\}$ (the support of f according to Proposition 3.1) corresponds to a unique address. If $a \in A$ corresponds to the address $\sigma = (\sigma_1, \sigma_2, \ldots)$, then

$$f(a) = \lim_{n \to \infty} \phi_{\sigma_1} \circ \phi_{\sigma_2} \circ \phi_{\sigma_3} \circ \cdots \circ \phi_{\sigma_n}(t) \tag{3.2}$$

for any $t \in \mathbb{R}$.

We can also easily adjust the setting to that of a *local* or *partitioned IFSM*. In this situation, we have a collection of maps $w_i : D_i \to \mathbb{X}$ where $D_i \subset \mathbb{X}$. We define the local IFSM operator in the same way as we previously defined the general IFSM operator; that is,

$$T(f)(x) = \sum_{w_i^{-1}(x) \neq \emptyset} \phi_i(f(w_i^{-1}(x))).$$

The contractivity and fixed-point conditions are exactly the same as for the nonlocal version from Theorem 3.2. Notice that if $\mathbb{X} \neq \bigcup_i w_i(D_i)$, then any $x \notin \bigcup_i w_i(D_i)$ has $T(f)(x) = 0$. For ease of reference, we record this as a theorem.

Theorem 3.3. *Let $\mathbb{X}$ be a set, $D_i \subset \mathbb{X}$, $w_i : D_i \to \mathbb{X}$, and $\phi_i : \mathbb{R} \to \mathbb{R}$ be contractive with contraction factor c_i. Then the local IFSM operator*

$$T(f)(x) = \sum_{w_i^{-1}(x) \neq \emptyset} \phi_i(f(w_i^{-1}(x)))$$

satisfies

$$\|T(f)(x) - T(g)(x)\|_\infty \leq c \|f - g\|_\infty,$$

where

$$c = \sup_{x \in \mathbb{X}} \sum_{x \in w_i(\mathbb{X})} c_i.$$

If $c < 1$, then T has a unique fixed point in $\mathcal{F}^*(\mathbb{X})$, the space of bounded real-valued functions on $\mathbb{X}$.

For the situation as given in the fractal block-coding algorithm, the sets D_i are the "domain" or "parent" blocks and $w_i(D_i)$ are the "range" or "child" blocks and form a partition of $\mathbb{X}$, and thus each $x \in \mathbb{X}$ is in exactly one $w_i(D_i)$. This also means that the contractivity for T is given by $c = \max_i c_i$. For $\phi_i(t) = \alpha_i t + \beta_i$, the fractal block-coding operator is contractive if $\max_i |\alpha_i| < 1$. Sometimes this condition is enforced in the compression phase; that is, the optimal α_i and β_i are found, then α_i is rescaled to enforce the condition $|\alpha_i| < 1$, and then a new β_i is found based on this new α_i. However, this restriction is not necessary for convergence and can result in a lower-quality reconstruction.

3.2.2 IFSM on $\mathcal{L}^p(\mathbb{X})$

The IFSM framework in the previous section has very few assumptions; in particular, it has no assumptions on $\mathbb{X}$. However, the conditions for convergence are quite strong and sometimes are too strong for a particular application (i.e., for fractal image compression). In this section, we add structure to $\mathbb{X}$ and gain in slightly weaker conditions on T for contractivity. These conditions allow for the possibility of some $c_i > 1$, as long as it is offset by having its corresponding geometric map w_i be "measure contractive" in a suitable sense.

For our first case, we take $\mathbb{X} \subset \mathbb{R}^n$ and μ to be the Lebesgue measure restricted to $\mathbb{X}$. For each map $w_i : \mathbb{X} \to \mathbb{X}$, we assume that

$$\text{there exist } s_i \geq 0 \text{ with } d\mu(w_i(x)) \leq s_i \, d\mu(x) \text{ for all } x \in \mathbb{X}. \quad (3.3)$$

An example of such a map is when w_i is affine, say $w_i(x) = A_i x + b_i$, in which case $s_i = |\det(A_i)|$.

Theorem 3.4. *Let $\mathbb{X} \subset \mathbb{R}^n$ and μ be the Lebesgue measure restricted to $\mathbb{X}$. Suppose that $w_i : \mathbb{X} \to \mathbb{X}$, $i = 1, 2, \ldots, N$, are such that there exist $s_i \geq 0$ with $d\mu(w_i(x)) \leq s_i \, d\mu(x)$ for all x. Finally, take $\phi_i : \mathbb{R} \to \mathbb{R}$ to have Lipschitz constant c_i, and let T be defined as in (3.1). Then, for $f, g \in \mathcal{L}^p(\mathbb{X})$, we have*

$$\|T(f) - T(g)\|_p \leq \left(\sum_i s_i^{1/p} c_i \right) \|f - g\|_p.$$

In particular, $T : \mathcal{L}^p(\mathbb{X}) \to \mathcal{L}^p(\mathbb{X})$.

Proof. We compute that

$$\|T(f) - T(g)\|_p \leq \sum_i \left(\int_{\mathbb{X}} |\phi_i(f(w_i^{-1}(x))) - \phi_i(g(w_i^{-1}(x)))|^p \, d\mu \right)^{1/p}$$

$$\leq \sum_i c_i \left(\int_{w_i(\mathbb{X})} |f(w_i^{-1}(x)) - g(w_i^{-1}(x))|^p \, d\mu(x) \right)^{1/p}$$

$$\leq \sum_i c_i \left(\int_{\mathbb{X}} |f(y) - g(y)|^p s_i \, d\mu(y) \right)^{1/p}$$

$$\leq \left(\sum_i c_i s_i^{1/p} \right) \|f - g\|_p.$$

Using this result with $g = 0$, we see that $\|T\|_p \leq \sum_i c_i s_i^{1/p}$ and thus $f \in \mathcal{L}^p(\mathbb{X})$ implies that $T(f) \in \mathcal{L}^p(\mathbb{X})$. $\square$

Condition (3.3) is key to getting the estimate in the theorem. Without this assumption, there is no way to bring any properties of the "geometric" maps w_i into the estimate of the $\mathcal{L}^p$ norm. This condition is clearly a condition on both the measure μ and the maps w_i.

A few simple observations are in order at this point:

• If the IFS $\{w_i\}$ is *measure disjoint* with respect to μ (i.e., if $\mu(w_i(\mathbb{X}) \cap w_j(\mathbb{X})) = 0$ for $i \neq j$), then the estimate in the theorem can be improved to

$$\|T(f) - T(g)\|_p \leq \left(\sum_i s_i c_i^p \right)^{1/p}.$$

To see this, just note that in this case $T(f)(x) = \phi_i(f(w_i^{-1}(x)))$ for $x \in w_i(\mathbb{X})$ and thus

$$\|T(f) - T(g)\|_p = \left(\sum_i \int_{w_i(\mathbb{X})} |\phi_i(f(w_i^{-1}(x))) - \phi_i(g(w_i^{-1}(x)))|^p \, d\mu \right)^{1/p},$$

and the rest follows as before.

• If A is the attractor of the IFS $\{w_i\}$ and $\mu(A) = 0$, then any fixed point $\bar{f}$ of T will be zero in $\mathcal{L}^p(\mathbb{X})$ by Proposition 3.1 since the support of $\bar{f}$ is contained in a set of measure zero. This means that we have to be careful in defining T if we are to get a nontrivial fixed point for it. For instance, if the w_i are all affine and $\bigcup_i w_i(\mathbb{X})$ does not cover $\mathbb{X}$, then $\mu(A) = 0$.

• Another situation that is similar to that of $\mathbb{X} \subset \mathbb{R}^n$ with μ Lebesgue measure is where $\mathbb{X}$ is a metric space and μ is Hausdorff d-dimensional measure on $\mathbb{X}$ and is σ-finite with $0 < \mu(\mathbb{X})$. In this context, if w_i is a similitude with contractivity s_i, then $d\mu(w_i(x)) \leq s_i^d \, d\mu(x)$.

• Finally, a nice broad class of examples is where the space $\mathbb{X}$ is the attractor of the geometric IFS $\{w_i\}$ and μ is the attractor of an IFS with probabilities $\{w_i, p_i\}$ with $\{w_i\}$ measure disjoint with respect to μ. In this case, the inequality $d\mu(w_i(x)) \leq p_i \, d\mu(x)$ is given by the self-similarity of μ, and thus condition (3.3) is satisfied with $s_i = p_i$. A similar thing can be possible even if the IFS $\{w_i\}$ is not measure disjoint with respect to μ. The constants s_i are slightly more complicated in this case and involve sums of the p_i.

3.2.2.1 The case of fractal block-coding and local IFSM

The situation for a general local IFSM on $\mathcal{L}^p(\mathbb{X})$ is rather complicated and depends on the precise setup of the IFS maps. We illustrate it with a special case. We take the situation illustrated in Fig. 3.1. We take $\mathbb{X} = [0, 1]^2$ and the two partitions in such a way that each parent block is formed from four child blocks. This also means that each w_i is affine with linear part $x \mapsto x/2$ and such that $s_i = 1/4$ for each i. Take the parent partition to be a grid of N^2 parent blocks (N in each direction), so there are $2N \times 2N$ child blocks. Then the Lipschitz constant of T is

$$\|Tf - Tg\|_p \leq \left(\sum_{i,j=1}^{2N} c_i^p/2 \right)^{1/p} \|f - g\|_p = (1/2)^{1/p} \left(\sum_{i,j=1}^{2N} c_i^p \right)^{1/p} \|f - g\|_p.$$

This can be simplified if some further assumptions are made. For instance, sometimes fractal block-coding is done without searching. Instead, each child block is matched with the unique parent block that contains it. In this case, the action of T on each parent block D_i is independent of the action of T on any other parent block $D_{i'}$ and thus T decomposes as a direct sum of operators T_i, one for each D_i:

$$T : \bigoplus \mathcal{L}^p(D_i) \to \bigoplus \mathcal{L}^p(D_i). \tag{3.4}$$

Each parent block D_i contains four child blocks, which we label $R_i^1, R_i^2, R_i^3, R_i^4$. Then each T_i has the form

$$T_i(f)(x) = \alpha_i^j f(w_j^{-1}(x)) + \beta_i^j, \quad \text{for } x \in R_i^j \subset D_i, \quad j = 1, 2, 3, 4,$$

and we have for $f, g \in \mathcal{L}^p(D_i)$

$$\|T_i(f - g)\|_p \leq \left(\frac{|\alpha_i^1|^p + |\alpha_i^2|^p + |\alpha_i^3|^p + |\alpha_i^4|^p}{4} \right)^{1/p} \|f - g\|_p$$

and so, for $f, g \in \mathcal{L}^p(\mathbb{X})$,

$$\|T(f - g)\|_p \leq \max_i \left(\frac{|\alpha_i^1|^p + |\alpha_i^2|^p + |\alpha_i^3|^p + |\alpha_i^4|^p}{4} \right)^{1/p} \|f - g\|_p.$$

One nice way of looking at this situation is to realize that the decomposition of $\mathcal{L}^p(\mathbb{X})$ given in (3.4) is given by a partition of unity. That is, for each D_i, there is a restriction operator $P_i : \mathcal{L}^p(\mathbb{X}) \to \mathcal{L}^p(D_i)$ given by $P_i(f) = f|_{D_i}$. These operators P_i are projections that form a partition of unity, $I = \sum_i P_i$, and $P_i P_j = 0$. The decomposition of T given in (3.4) is due to the fact that $T P_i = P_i T$. Many of the block-coding transforms can be viewed in this framework. The power of this viewpoint is evident when one uses other families of projections instead of simple restriction operators; see Sects. 3.3 and 3.4 for some related discussions.

3.2.3 Affine IFSM

The particular case of IFSM operators that are affine has nice and simple properties. The simplest version of this is where each $\phi_i(t) = \alpha_i t + \beta_i$, as in the case of simple fractal block-coding, and in which case T has the form

$$T(f)(x) = \sum_{x \in w_i(\mathbb{X})} \alpha_i f(w_i^{-1}(x)) + \sum_{x \in w_i(\mathbb{X})} \beta_i. \qquad (3.5)$$

The second term in the sum defines a function $\beta : \mathbb{X} \to \mathbb{R}$, and the first term defines a linear operator $A : \mathcal{F}^*(\mathbb{X}) \to \mathcal{F}^*(\mathbb{X})$ such that $T(f) = Af + \beta$. The contractivity of T is the same as the contractivity of A, so assuming that A is contractive we have that the fixed point $\bar{f}$ of T can be written as

$$\bar{f} = \sum_{n \geq 0} A^n \beta = (I - A)^{-1} \beta. \qquad (3.6)$$

Notice that (3.6) means that $\bar{f}$ depends linearly on β. A nice way to think about (3.6) is that β provides the details of $\bar{f}$ on the largest resolution scale, then $A\beta$ refines this on the next smaller level of resolution, then $A^2\beta$ fills in the finer details on the next level, and so on. In fractal block-coding, one can see this in the reconstruction as finer and finer details are added (see Figs. 3.3 and 3.4).

The fact that $\bar{f}$ is a linear function of β can be exploited. To show one possibility, take $\mathbb{X} = [0,1]$, μ to be the Lebesgue measure, $\{w_i\}$ an IFS that partitions $\mathbb{X}$ (i.e., $[0,1] = \bigcup_i w_i([0,1])$ and $\mu(w_i([0,1]) \cap w_j([0,1])) = 0$; we call such a partition an *IFS partition*), and let $X_i = w_i([0,1])$. We also take $w_i(x) = a_i x + b_i$ for simplicity.

To start with, take β to be piecewise constant on this partition with $\beta_1 \mu(X_1) + \beta_2 \mu(X_2) + \cdots \beta_N \mu(X_N) = 0$, so that the integral of β with respect to μ is zero. Because of the simple structure of T in this situation, we have that $\beta \circ w_i^{-1}$ is perpendicular to β for each i. To see this, just note that β is constant on X_i and $\beta \circ w_i^{-1}$ has integral equal to zero on X_i. Thus, $A\beta$ is perpendicular to β. Similarly, $A^2\beta$ is perpendicular to both β and $A\beta$. So, in fact, the sum in (3.6) is a sum of orthogonal components.

More is true, in fact. Let $\widehat{\beta}$ be another function that is piecewise constant on the X_i, has zero integral, and is perpendicular to β. Then $A^n\beta$ and $A^m\widehat{\beta}$ are perpendicular for all $n, m \in \mathbb{N}$. This means that if T and $\widehat{T}$ are the corresponding IFSM operators,

$$T(f) = Af + \beta \quad \text{and} \quad \widehat{T}(f) = Af + \widehat{\beta},$$

and $\bar{f}$ and $\bar{\widehat{f}}$ are their fixed points, then f and $\bar{\widehat{f}}$ are also perpendicular. With techniques like this, it is possible to design families of functions that are all self-affine and mutually orthogonal.

3.2.4 IFSM with infinitely many maps

We briefly show how the notion of an IFSM can be extended to one containing an infinite number of maps. The extension is simple and natural. The framework is analogous with the previous situation. We have an indexing set Λ, and for each $\lambda \in \Lambda$ we have a geometric map $w_\lambda : \mathbb{X} \to \mathbb{X}$ and a grey-level map $\phi_\lambda : \mathbb{R} \to \mathbb{R}$. Since we will have to integrate over Λ, we assume that Λ is a measure space and ν is a probability measure on Λ. For a function $f : \mathbb{X} \to \mathbb{R}$, we define

$$T(f)(x) = \int_\Lambda \phi_\lambda \left(f(w_\lambda^{-1}(x)) \right) \, d\nu(\lambda). \tag{3.7}$$

Clearly we will need some conditions on the function f in order for the integral to exist and thus for the transform T to be well-defined.

3.2.4.1 Uniformly contractive ϕ_λ

Our first case is when we assume that there is some $0 < c < 1$ such that for all λ we have ϕ_λ contractive with the contractivity factor bounded by c. We again take the complete space $\mathcal{F}^*(\mathbb{X})$ of all bounded real-valued functions on $\mathbb{X}$ with the supremum norm $\|f\|_\infty$. The proof of this result is exactly the same as that of Theorem 3.2.

Theorem 3.5. *Suppose that the contractivity factor of ϕ_λ is c_λ. Further suppose that for each $x \in \mathbb{X}$ we have*

$$\int_{x \in w_\lambda(\mathbb{X})} c_\lambda \, d\nu(\lambda) < 1.$$

Then T defined as in (3.7) is contractive in the norm $\|f\|_\infty$ and therefore has a unique fixed point in $\mathcal{F}^*(\mathbb{X})$.

3.2.4.2 IFSM in $\mathcal{L}^p(\mathbb{X})$

We again take $\mathbb{X} \subset \mathbb{R}^n$ and μ to be the Lebesgue measure. For each λ, we assume that

$$\exists s_\lambda \geq 0 \text{ such that } d\mu(w_\lambda(x)) \leq s_\lambda \, d\mu(x) \text{ for all } x \in \mathbb{X}. \qquad (3.8)$$

This is the same type of assumption we made in Sect. 3.2.2.

Theorem 3.6. *Let* $\mathbb{X} \subset \mathbb{R}^n$ *and* μ *be the Lebesgue measure restricted to* $\mathbb{X}$. *Suppose also that condition (3.8) is satisfied. Finally, suppose that the Lipschitz constant for* ϕ_λ *is* c_λ. *Then, for* T *as defined in (3.7) and* $f, g \in \mathcal{L}^p(\mathbb{X})$, *we have*

$$\|T(f) - T(g)\|_p \leq \left(\int_\Lambda s_\lambda^{1/p} c_\lambda \, d\nu(\lambda) \right) \|f - g\|_p.$$

Proof. The proof is the same as the proof of Theorem 3.4, with the slight modification that we use Minkowski's theorem with integrals instead of with sums. That is, we use

$$\left(\int_{\mathbb{X}} \left| \int_\Lambda \phi_\lambda(f(w_\lambda^{-1}(x))) - \phi_\lambda(g(w_\lambda^{-1}(x))) \, d\nu(\lambda) \right|^p d\mu(x) \right)^{1/p} \leq$$
$$\int_\Lambda \left(\int_{\mathbb{X}} |\phi_\lambda(f(w_\lambda^{-1}(x))) - \phi_\lambda(g(w_\lambda^{-1}(x)))|^p \, d\mu(x) \right)^{1/p} d\nu(\lambda).$$

$\square$

3.2.5 Progression from geometric IFS to IFS on functions

There is a step-by-step progression from IFS on sets to IFS on functions, which we now briefly outline; this was first described in [67].

1. **From IFS on $\mathbb{H}(\mathbb{X})$ to fractal transforms on black-and-white images in $\mathcal{F}_{BW}(\mathbb{X})$.** We first define the space of "black and white" image functions on $\mathbb{X}$,

$$\mathcal{F}_{BW}(\mathbb{X}) = \{u : \mathbb{X} \to \{0, 1\} \mid S_u = \text{supp}\,(u) \in \mathbb{H}(\mathbb{X})\}.$$

It is then natural to define the following metric in $\mathcal{F}_{BW}$:

$$d_{BW}(u, v) = d_{\mathbb{H}}(S_u, S_v).$$

An N-map IFS set-valued mapping $\mathbf{w}$ on $\mathbb{H}(\mathbb{X})$ is equivalent to the following fractal transform T on $\mathcal{F}_{BW}$:

$$(Tu)(x) = \sup_{1 \le k \le N} \{u(w_k^{-1}(x))\}, \quad x \in \mathbb{X}.$$

The range maps $\phi_k(t)$ are the identity maps on $[0, 1]$, and the combination operator $\mathcal{O}$ is the supremum operator.

Of course, most black-and-white images are not bitmap (binary) but have shades of grey running from black to white. The next step is to accommodate such greyscale images.

2. **From fractal transforms on $\mathcal{F}_{BW}(\mathbb{X})$ to iterated fuzzy set systems (IFZS) [36].** We modify the formalism above to accommodate greyscale images $u : \mathbb{X} \to [0, 1]$ with the addition of greyscale maps $\phi : [0, 1] \to [0, 1]$. Now consider the Hausdorff distance between level sets of two functions and (omitting some technical details) define the metric

$$d_\infty(u, v) = \sup_{c \in [0,1]} h([u]^c, [v]^c),$$

where $[u]^c$ denotes the c-level set of u:

$$[u]^c = \begin{cases} \{x \in \mathbb{X} \mid u(x) \ge c\}, \, c \in (0, 1] \\ \overline{\{x \in \mathbb{X} \mid u(x) > 0\}}, \, c = 0. \end{cases}$$

The complete metric space of functions is the space of (upper semicontinuous) *fuzzy set functions* on $\mathbb{X}$. The action of the fractal transform T defined by an N-map IFS $\{w_i\}_{i=1}^N$ associated with greyscale maps $\{\phi_i\}_{i=1}^N$ is given by

$$(Tu)(x) = \sup_{1 \le i \le N} \{\phi_i(\tilde{u}(w_i^{-1}(x)))\}, \quad x \in \mathbb{X},$$

where, for $B \subset \mathbb{X}$, (i) $\tilde{u}(B) = \sup_{z \in B}\{u(z)\}$ if $B \ne \emptyset$ and (ii) $\tilde{u}(\emptyset) = 0$.

3. **From IFZS to fractal transforms on $L^1(\mathbb{X})$.** The Hausdorff distance is very restrictive, however, from both visual as well as practical perspectives. In [64], two fundamental modifications of the IFZS were made:

   a. The Hausdorff distance between level sets $d_{\mathbb{H}}([u]^c, [v]^c)$ was replaced by the μ-measure (over $\mathbb{X}$) of the symmetric differences of the two level sets,

   $$G(u, v; c) = \mu([u]^c \triangle [v]^c).$$

   (For $A, B \subseteq \mathbb{X}$, $A \triangle B = (A \cup B) \setminus (A \cap B)$.)

   b. For a finite measure ν on the range space $\mathcal{R} = [0, 1]$, we then integrate over these symmetric differences,

   $$g(u, v; \nu) = \int_{\mathcal{R}} G(u, v; c)\, d\nu(c).$$

   In the special case where μ and ν are Lebesgue measures on $\mathbb{X}$ and $\mathcal{R}$, respectively, we have

   $$g(u, v, m) = \int_{\mathbb{X}} |u(x) - v(x)|\, dx = \|u - v\|_1,$$

   the L^1 distance between u and v. The restrictive Hausdorff metric d_∞ for the IFZS has been replaced by a weaker pseudometric. Furthermore, the supremum operator in the fractal transform T may be replaced by the summation operator.

4. **Fractal transforms on $L^p(\mathbb{X})$.** It is natural to extend the L^1 fractal transform T to L^p spaces. The result is (1.1), with the combination operator $\mathcal{O}$ being the summation operator. The resulting fractal transform is referred to as an *N-map IFS with greyscale maps* (IFSM), as discussed previously in this chapter.

3.3 IFS on wavelets

Wavelets have proven to be incredibly useful in their ability to describe and manipulate data. One of the primary reasons for this is that wavelets measure a combination of "frequency" information and "spatial" information. The spatial information comes because a wavelet is typically compactly supported, and thus a particular wavelet coefficient only reflects local information of the signal. The "frequency" information is reflected in the fact that a particular wavelet basis function has a given size. We cannot hope to do justice to this huge area in this short section, so we will give only the briefest general discussion of wavelets and then move on to describe the IFS transform in the discrete wavelet domain. There are many books that give a thorough accounting of the theory and applications of wavelets, of which we mention only [44, 123, 158].

3.3.1 Brief wavelet introduction

To begin, we restrict our attention to $\mathbb{X} = \mathbb{R}$. A *wavelet basis* for $\mathcal{L}^2(\mathbb{R})$ is an orthonormal basis $\{\psi_{i,j}\}$, with each $\psi_{i,j}$ generated as dilations and translations of a single *mother wavelet* Ψ as

$$\psi_{i,j}(x) = 2^{i/2}\Psi(2^i x - j).$$

The mother wavelet Ψ has zero integral, and thus all $\psi_{i,j}$ do as well. The condition that $\{\psi_{i,j}\}$ be an orthonormal basis is very strong and clearly puts restrictions on the mother wavelet Ψ. In fact, the usual way to construct Ψ is by first constructing the *scaling function* ϕ, and then Ψ is a finite linear combination of dilations of ϕ. Wavelets have very strong connections with IFSs since the scaling function satisfies a self-similarity condition. In fact, it is the fixed point of the *dilation equation*

$$\phi(x) = \sum_i h_i\phi(2x - i). \tag{3.9}$$

For compactly supported solutions to (3.9), it is only necessary that finitely many h_n be nonzero. In addition, ϕ is usually required to satisfy a self-orthogonality condition

$$\langle \phi(x), \phi(x - i)\rangle = 0, \quad i \neq 0 \in \mathbb{Z}.$$

The *scale space* V_n, for $n \in \mathbb{Z}$, is defined by

$$V_n = \overline{\mathrm{Span}\{\phi(2^n x - j) : j \in \mathbb{Z}\}}.$$

This space V_n has a nice interpretation in terms of image processing. One can view V_n as the level of detail you see at a given scale. That is, if you take an ideal image I as a function, then V_n projects this function down to a given "pixel" size. The finer the pixel size, the smaller the scale (and the larger the n). The wavelet functions give the "difference" information between these V_n, so that

$$V_n = \overline{\text{Span}\{\psi_{i,j} : j \in \mathbb{Z}, i \geq n\}} = \overline{\text{Span}\{\phi(2^n x - j) : j \in \mathbb{Z}\}}.$$

If the functions are all supported in a finite interval, then there is a coarsest resolution level that we can rescale to be the entire coarsest resolution level. We do this and assume that $\mathbb{X} = [0, 1]$. In this case, since $\int \Psi(x)\, dx = 0$, the coarsest resolution level is that of constant functions. Any function $f \in \mathcal{L}^2([0, 1])$ is decomposed as

$$f(x) = b_{0,0}\phi(x) + \sum_{i \geq 0, 0 \leq j \leq 2^i - 1} c_{i,j}\psi_{i,j}(x), \tag{3.10}$$

where $b_{0,0} = \langle f, 1 \rangle$ and $c_{i,j} = \langle f, \psi_{i,j} \rangle$ are the *wavelet coefficients* for f. The expansion coefficients are conveniently displayed in the form of an infinite *wavelet tree* (see Fig. 3.5). In this figure, each $B_{i,j}$ represents a tree of infinite depth. We also refer to such a tree with apex $c_{i,j}$ as the *block* $B_{i,j}$.

b_{00}							
c_{00}							
c_{10}				c_{11}			
c_{20}		c_{21}		c_{22}		c_{23}	
B_{30}	B_{31}	B_{32}	B_{33}	B_{34}	B_{35}	B_{36}	B_{37}

Fig. 3.5: Wavelet expansion tree

The wavelet tree displays the recursive structure of the wavelet coefficients. The position of a coefficient in the tree signals both the position and scale of the corresponding wavelet basis function. Moving down the tree corresponds to moving down to finer scales, and moving horizontally on a given level moves left to right in $[0, 1]$. We put b_{00} at the top of the tree, as it corresponds to the overall average value of the function f.

There are many different wavelet bases. In fact, the large variety of different wavelet bases is one of the strengths of wavelet analysis – you can choose a basis that has some convenient properties for your particular application.

In many ways, the simplest wavelet basis is the Haar basis. We restrict our description to a basis for $\mathcal{L}^2([0, 1])$. The scaling function is $\phi(x) = \chi_{[0,1]}(x)$, the indicator function of the interval. The mother wavelet Ψ is defined by

$$\Psi(x) = \begin{cases} 1 & \text{if } 0 \leq x < 1/2, \\ -1 & \text{if } 1/2 \leq x \leq 1. \end{cases} \tag{3.11}$$

Thus, the spaces V_n consist of those functions that are piecewise constant on intervals of the form $[i/2^n, (i+1)/2^n]$. The nested property of the scale spaces, $V_n \subset V_{n+1}$, is clear from this description. The projection of some $f \in \mathcal{L}^2([0,1])$ onto V_n consists of computing the average value of f over each of these intervals and putting these values together into a piecewise constant function.

Sometimes it is useful to restrict a wavelet basis of $\mathbb{R}$ to some bounded interval. There are several different ways of doing this, each with their own benefits and drawbacks. Perhaps the simplest is to use *periodic wavelets*. For a function $f \in \mathcal{L}^2(\mathbb{R})$, we define the periodized version of f to be the function

$$f^*(x) = \sum_{n \in \mathbb{Z}} f(x+n).$$

Since $f \in \mathcal{L}^2(\mathbb{R})$, the sum exists and $f^* \in \mathcal{L}^2([0,1])$. Because of the integer translations in the definition of $\psi_{i,j}$, it turns out that $\{\psi_{i,j}^* : i \geq 0, 0 \leq j < 2^i\}$ is an orthonormal basis for $\mathcal{L}^2([0,1])$.

3.3.2 IFS operators on wavelets (IFSW)

IFS operators map "information" from larger scales down to smaller scales. The recursive nature of the wavelet tree (Fig. 3.5) makes the idea of defining an IFS-type operation on a wavelet tree seem natural. Because of the tree structure, it is very easy to do this mapping. We give a couple of examples of such mappings to illustrate.

Example 3.7. Our first example is very simple. The transform M will preserve the average value of the input function f, and therefore keep b_{00} unchanged. We also keep c_{00} unchanged but modify every other coefficient. Graphically, we can represent it as

$$M \; : \; B_{00} \; \rightarrow \; \boxed{\begin{array}{c} c_{00} \\ \hline \alpha_0 B_{00} \mid \alpha_1 B_{00} \end{array}}, \quad |\alpha_i| < \frac{1}{\sqrt{2}}. \tag{3.12}$$

We need the restriction on α_i in order to ensure that the coefficients remain in $\ell^2(\mathbb{N})$. The representation above means that if we expand f as in (3.10) and $g = Mf$, then we can write

$$g(x) = b_{0,0}\phi(x) + c_{00}\psi_{00}(x) + \sum_{i>0, 0 \leq j \leq 2^i - 1} d_{i,j}\psi_{i,j}(x),$$

where for $i > 0$ we have

$$d_{i,j} = \begin{cases} \alpha_0 c_{i-1,j} & \text{if } 0 \leq j < 2^{i-1}, \\ \alpha_1 c_{i-1,j-2^{i-1}} & \text{if } 2^{i-1} \leq j < 2^i. \end{cases}$$

That is, we take a copy of the entire tree B_{00}, multiply each coefficient by α_0, and make it the "left" subtree $\widehat{B}_{10}$ for Mf. Similarly, we take the entire tree B_{00}, multiply each coefficient by α_1, and make it the right subtree $\widehat{B}_{11}$ for Mf. Repeated application of M will make the information flow from the top of the tree down toward the bottom. The top coefficient c_{00} stays fixed. After one application, the second row is $\alpha_0 c_{00}$ and $\alpha_1 c_{00}$, and then this second row will stay fixed for all subsequent iterations. After two iterations, the third row becomes

$$\alpha_0^2 c_{00}, \quad \alpha_0 \alpha_1 c_{00}, \quad \alpha_1 \alpha_0 c_{00}, \quad \alpha_1^2 c_{00}$$

and then will stay fixed for all subsequent iterations. In general, after the $n-1$th iteration, the nth row is fixed and will remain unchanged for all further iterations. Clearly, knowing only c_{00} and α_0, α_1 is sufficient for knowing the limiting tree and hence the fixed-point function of M.

$\square$

Example 3.8. Consider the fractal wavelet transform with four block maps

$$W_1 : B_{10} \to B_{20}, \ W_2 : B_{11} \to B_{21}, \ W_3 : B_{10} \to B_{22}, \ W_4 : B_{11} \to B_{23},$$
$$(3.13)$$

with associated multipliers α_i, $1 \leq i \leq 4$. Diagrammatically,

$$M \ : \ B_{00} \ \to \ \begin{array}{|c|c|} \hline \multicolumn{2}{|c|}{c_{00}} \\ \hline c_{10} & c_{11} \\ \hline \alpha_1 B_{10} \mid \alpha_2 B_{11} & \alpha_3 B_{10} \mid \alpha_4 B_{11} \\ \hline \end{array} \ . \qquad (3.14)$$

This is more like a local or partitioned IFSM than the previous example. The block B_{10} may be thought of as the projection of f onto part of the wavelet basis (that part whose coefficients are contained in this part of the wavelet tree); call this projection P_{10}. Similarly, we have a projection P_{11} associated with that part of the wavelet tree contained in block B_{11}. Then we see that (3.10) can be written as

$$f(x) = b_{00}\phi(x) + c_{00}\psi(x) + (P_{10}f)(x) + (P_{11}f)(x).$$

Here the decomposition of f given by P_{10} and P_{11} is not a simple restriction onto two disjoint parts of the space, as it is in the case of a local IFSM. It is an orthogonal splitting, projecting onto two orthogonal subspaces. The orthogonality of the subspaces (or of the projections) is the analogue to the disjointness of the parts in a local IFSM. The four block maps W_i from (3.13) correspond to linear isometries $\widehat{W}_i$. For example, $\widehat{W}_1$ has domain equal to $P_{10}(\mathcal{L}^2(\mathbb{X}))$, the range of P_{10} that also corresponds to the block B_{10}, and has range the subspace of $\mathcal{L}^2(\mathbb{X})$ that corresponds to block B_{20}. In terms of these projections and block maps, M as in (3.14) can be written as

$$M(f) = b_{00}\phi + c_{00}\psi + c_{10}\psi_{1,0} + c_{11}\psi_{1,1} + \alpha_1 \widehat{W}_1(P_{10}(f))$$
$$+ \alpha_2 \widehat{W}_2(P_{11}(f)) + \alpha_3 \widehat{W}_3(P_{10}(f)) + \alpha_4 \widehat{W}_4(P_{11}(f)).$$

In this example, the top two levels of the tree (in addition to the scaling coefficient b_{00}) are unchanged by M, and then M propagates the top of the tree down the levels of the tree, at each level multiplying by the appropriate α_i. It is again clear that M will have a limiting wavelet tree. As long as the coefficients belong to $\ell^2(\mathbb{N})$, this limiting tree will correspond to some function in $\mathcal{L}^2(\mathbb{X})$. The first few levels of the limiting tree are given in Fig. 3.6. □

b_{00}							
c_{00}							
c_{10}				c_{11}			
$\alpha_1 c_{10}$		$\alpha_2 c_{11}$		$\alpha_3 c_{10}$		$\alpha_4 c_{11}$	
$\alpha_1^2 c_{10}$	$\alpha_1\alpha_2 c_{11}$	$\alpha_2\alpha_3 c_{10}$	$\alpha_2\alpha_4 c_{11}$	$\alpha_3\alpha_1 c_{10}$	$\alpha_3\alpha_2 c_{11}$	$\alpha_4\alpha_3 c_{10}$	$\alpha_4^2 c_{11}$

Fig. 3.6: Limiting tree for Example 3.8.

From these examples, it is clear that an IFSW generates a wavelet tree with a very constrained type of geometric decay down the branches of the tree. The limiting tree of an IFSW operator M has a particular type of self-similarity. The associated function $\bar{f} \in \mathcal{L}^2(\mathbb{X})$ also has self-similarity, but in the wavelet domain. How does this self-similarity translate to the spatial domain?

3.3.3 Correspondence between IFSW and IFSM

Since the wavelet basis is made up of dilations and translations of a single function (the mother wavelet), it is not surprising that self-similarity in the wavelet domain might mean self-similarity in the spatial domain. This seems to have been first observed in [46] in the particular situation of Haar wavelets and was used to give a wavelet-based explanation/description of standard fractal block-coding. In fact, in the Haar wavelet basis, there is really no difference between an IFSM and an IFSW; they are direct translations of each other. In this section, we explain this relationship in the more general context. We do this by discussing each of the two example IFSW operators from the previous section. This section draws heavily from [134].

We mention that it is also possible to show this correspondence in the case of periodic wavelets. A discussion of this situation can also be found in [134].

3.3.3.1 IFSM for IFSW of Example 3.7

The goal is to find an IFSM operator T (as defined by (3.1)) acting in "physical space" (i.e., on functions supported on $\mathbb{X}$), that corresponds to an IFSW operator M acting on the wavelet coefficients. The dilation/translation relations within the wavelet basis provide the key to this goal. Let $w_1(x) = x/2$ and $w_2(x) = x/2 + 1/2$. Then

$$\sqrt{2}B_{00} \circ w_1^{-1} = B_{10} \quad \text{and} \quad \sqrt{2}B_{00} \circ w_2^{-1} = B_{11} \tag{3.15}$$

since

$$\sqrt{2}\psi_{i,j}\left(w_1^{-1}(x)\right) = \psi_{i+1,j}(x) \quad \text{and} \quad \sqrt{2}\psi_{i,j}\left(w_2^{-1}(x)\right) = \psi_{i+1,j+2^i}(x). \tag{3.16}$$

Thus, this simple IFSW operator will correspond to the following two-map IFSM with condensation function $\psi(x)$:

$$T(f)(x) = c_{00}\psi(x) + \sqrt{2}\alpha_0 f\left(w_1^{-1}(x)\right) + \sqrt{2}\alpha_1 f\left(w_2^{-1}(x)\right). \tag{3.17}$$

Note that the IFSM operator depends on the particular wavelet basis chosen, as indicated by the presence of the function ψ in the expression for T.

In the simple case of Haar wavelets, the mother wavelet $\psi(x)$ decomposes into nonoverlapping components:

$$\psi(x) = \chi_{[0,1/2)}(x) - \chi_{[1/2,1)}(x).$$

As such, the IFSM operator T in (3.17) corresponds to a simple two-map IFSM with IFS maps w_1 and w_2 and grey-level maps $\phi_1(t) = \sqrt{2}\alpha_1 t + 1$ and $\phi_2(t) = \sqrt{2}\alpha_2 t - 1$. If T is contractive, then its fixed-point attractor function $\bar{f}$ has [0,1] as support.

However, in the case of other compactly supported wavelets, no such spatial decomposition into separate grey-level maps is possible. As well, the support of the attractor function $\bar{f}$ is necessarily larger than $[0,1]$, as the support of the mother wavelet Ψ is usually larger than $[0,1]$ for other compactly supported wavelets. To illustrate, consider the particular IFSW in which $\alpha_1 = 0.4$ and $\alpha_2 = 0.6$. The IFSM operator T in (3.17) is contractive. The left and right diagrams of Figure 3.7 show the IFSM attractor functions for, respectively, the Haar wavelet and "Coifman-6" cases. In both cases, we have chosen $b_{00} = 0$ and $c_{00} = 1$.

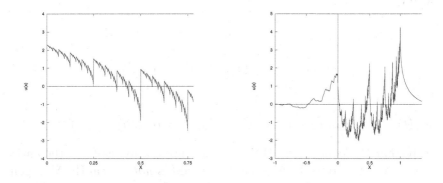

Fig. 3.7: Attractor functions $\bar{f}$ for Example 3.7: (left) Haar wavelet basis, (right) Coifman-6 wavelet basis.

3.3.3.2 IFSM for IFSW of Example 3.8

For this example, we see that the fixed point $\bar{u}$ has wavelet expansion

$$\bar{u} = c_{00}\psi_{00} + \bar{v}.$$

We need only focus on the function $\bar{v}$ that admits the wavelet expansion

c_{10}		c_{11}	
$\alpha_1 B_{10}$	$\alpha_2 B_{11}$	$\alpha_3 B_{10}$	$\alpha_4 B_{11}$

Because of the orthogonality of the functions $\psi_{i,j}$, we may write

$$\bar{v} = \bar{v}_1 + \bar{v}_2,$$

where the components $\bar{v}_i$ satisfy the relations

$$\bar{v}_1(x) = c_{10}\psi_{10}(x) + \alpha_1\sqrt{2}\bar{v}_1(2x) + \alpha_2\sqrt{2}\bar{v}_2(2x),$$
$$\bar{v}_2(x) = c_{11}\psi_{11}(x) + \alpha_3\sqrt{2}\bar{v}_1(2x-1) + \alpha_4\sqrt{2}\bar{v}_2(2x-1). \quad (3.18)$$

We may consider these equations as defining a kind of *vector IFSM with condensation*. The vector $\bar{v}$ is composed of the orthogonal components $\bar{v}_1$ and $\bar{v}_2$ that satisfy the fixed-point relations in (3.18). These equations may be written more compactly as

$$\bar{v}_i(x) = b_i(x) + \sum_{j=1}^{2} \Phi_{ij}(\bar{v}_j(w_{ij}^{-1}(x))), \quad i = 1, 2,$$

where

$$w_{11}(x) = w_{12}(x) = \frac{1}{2}x, \quad w_{21}(x) = w_{22}(x) = \frac{1}{2}x + \frac{1}{2},$$

$$b_1(x) = c_{10}\psi_{10}(x), \quad b_2(x) = c_{11}\psi_{11}(x),$$

and

$$\Phi_{11}(t) = \alpha_1\sqrt{2}t, \quad \Phi_{12}(t) = \alpha_2\sqrt{2}t, \quad \Phi_{21}(t) = \alpha_3\sqrt{2}t, \quad \Phi_{22}(t) = \alpha_4\sqrt{2}t.$$

Note that the contractive IFS maps w_{ij} are mappings from the entire base space $\mathbb{X}$ into itself. As such, this IFSM is *not* a local IFSM in general. What appeared to be a "local" transform in wavelet coefficient space is a normal IFSM in the base space. (Again, in the special case of the nonoverlapping Haar wavelets, the IFSM above may be written as a local IFSM.) The "locality" of the block transform has been passed on to the orthogonal components $\bar{v}_1$ and $\bar{v}_2$ of the function $\bar{v}$. These components may be considered as "nonoverlapping" elements of a vector.

This "vector IFSM" is really nothing more than a recurrent IFSM on B_{00}, where we split B_{00} as $B_{10} \oplus B_{11}$ and have the IFSM act "between" the components of this splitting.

3.4 IFS and integral transforms

An IFS on wavelets is just a particular case of a more general construction. Starting with an IFSM operator T on some function space $\mathcal{F}$ (such as $\mathcal{L}^p(\mathbb{X})$), we can attempt to transfer T to some other space of transforms $\mathcal{G}$. In the case of wavelets, the IFSM operator T induces an IFS operator on the wavelet coefficients.

The general situation is as depicted in Fig. 3.8, where $\mathcal{F}$ is the original function space, T is the IFSM operator on $\mathcal{F}$, $\mathcal{S} : \mathcal{F} \to \mathcal{G}$ is the transform, $\mathcal{G}$ is the "transform space" and M is the induced operator on $\mathcal{G}$. We will see that if the transform $\mathcal{S}$ is an integral transform with a self-similar kernel, then the induced operator M is also of IFS type.

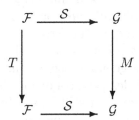

Fig. 3.8: Inducing an IFS operator on transforms

Examples of this appear in work on *generalized fractal transforms* acting on complete metric spaces, such as probability measures [75], moments of probability measures [65], fuzzy sets [36], $\mathcal{L}^p$ spaces [64], distributions [67], Fourier transforms [66], and discrete wavelet transforms [134, 166]. As we saw in Sect. 3.3, sometimes moving to a transform space has clear advantages; in that case in the quality of the reconstructed image. In some other situations, the transform space is the more natural place to do data processing. An example of this is in MRI imaging, where the data collected are already in the Fourier space.

Inverse problems of fractal approximation in $(\mathcal{F}, d_{\mathcal{F}})$ are then transformed to inverse problems in $(\mathcal{G}, d_{\mathcal{G}})$. Under the condition that T, and hence M, is contractive, solutions to the inverse problem in $(\mathcal{G}, d_{\mathcal{G}})$ using the collage theorem may then be formulated. Indeed, this was the procedure followed in [65, 66, 166].

We also mention that once function approximation/image compression is achieved in the space $\mathcal{G}$, it may not be necessary to return to the space $\mathcal{F}$. In fact, in some cases, such as moments of measures/images, a return may not be practically possible, nor would it be of interest. In such cases, one works exclusively in the transform domain. We emphasize, however, that when working in such "dual spaces" it is important to establish a number of properties, including:

1. completeness of the dual space $(\mathcal{G}, d_{\mathcal{G}})$ and
2. the "faithfulness" of $\mathcal{G}$ (i.e., is it an isomorphism of $\mathcal{F}$?) as well as the operator M (i.e., does M map $\mathcal{G}$ to itself?).

Assuming that T is contractive, the properties above are necessary to ensure the existence of a unique fixed point $\bar{g} \in \mathcal{G}$ (i.e., $\bar{g} = M\bar{g}$) by Banach's fixed-point theorem.

A natural question that arises in the study of induced operators is, "How does operating on $\mathcal{G}$ instead of on $\mathcal{F}$ relate to operating directly on $\mathcal{F}$?" For example, what is "self-similarity" in the transform domain if, in fact, this is a meaningful question?

Although it appears very natural to consider the induced operator M on the space $\mathcal{G}$, we are not constrained to use it. As we shall see below, it may be advantageous to use another operator, depending on the application. Nevertheless, the conditions of contractivity as well as completeness listed above must still be established.

In this section, $(\mathcal{G}, d_{\mathcal{G}})$ will be an appropriate space of *function transforms*, in particular *integral transforms*. There are two major motivations for this approach:

1. In many cases, the data that we seek to represent or compress are the result of an integral transform on some function space (e.g., MRI data, blurred images).
2. It may be more convenient to work in certain spaces of integral transforms. For example, as we show below, Lebesgue transforms of normalized nonnegative L^1 functions are nondecreasing and continuous functions. These latter functions may be easier to work with, especially in the sense of approximability.

In Sect. 3.4.1, we consider the integral transform, with kernel K, of a "fractally transformed" function (i.e., Tf) and relate it to the integral transform of f. This equation simplifies if K satisfies a general functional equation. It is then of interest to examine whether the kernel K itself can satisfy an IFS-type equation, for which it is necessary to examine the general space of kernels. A special class of solutions for this functional equation is considered. After this, we present some examples to illustrate the range of possibilities.

For ease of notation and clarity of discussion, the following discussion is restricted to the one-dimensional case. However, the extension to two (or more) dimensions is straightforward. This section draws substantially from [63].

3.4.1 Fractal transforms of integral transforms

3.4.1.1 Derivation of a functional equation for the kernel

We will denote our integral transform by $\mathcal{S} : \mathcal{F} \to \mathcal{G}$, and we assume it has kernel $K : \mathbb{X} \times \mathbb{R} \to \mathbb{R}$,

$$\widehat{f}(s) = (\mathcal{S}f)(s) = \int_{\mathbb{X}} K(t,s)f(t)\ dt. \qquad (3.19)$$

We shall also write this transform in inner product form as $\mathcal{S}f = \langle K, f \rangle$.

Let T be an affine IFSM operator as defined in (3.1). For an $f \in \mathcal{L}^p(\mathbb{X})$, let $g = Tf$. Then the transform $\widehat{g} = \mathcal{S}(g)$ is given by

$$\widehat{g}(s) = \int_{\mathbb{X}} K(t,s) \sum_{i=1}^{N} \left[\alpha_i f(w_i^{-1}(t)) + \beta_i \right] I_{\mathbb{X}_i}(t)\ dt$$

$$= \sum_{i=1}^{N} \alpha_i \int_{\mathbb{X}_i} K(t,s)f(w_i^{-1}(t))\ dt\ +\ \sum_{i=1}^{N} \beta_i \int_{\mathbb{X}_i} K(s,t)dt$$

$$= \sum_{i=1}^{N} \alpha_i c_i \int_{\mathbb{X}} K(c_i u + a_i, s)f(u)\ du\ +\ \widehat{\beta}(s), \qquad (3.20)$$

where

$$\widehat{\beta}(s) = \sum_{i=1}^{N} \beta_i \widehat{I_{\mathbb{X}_i}}(s). \qquad (3.21)$$

(Note that $\widehat{\beta}(s)$ depends only on the β_i – and, of course, the $\mathbb{X}_i$ – but not on f.)

Equation (3.20) may be written in the form

$$\langle K, Tf \rangle = \langle T^\dagger K, f \rangle\ +\ L(s), \qquad f \in \mathcal{F},$$

where the operator $T^\dagger$ may be interpreted as a kind of "adjoint" fractal operator on the kernel K,

$$(T^\dagger K)(t,s) = \sum_{i=1}^{N} \alpha_i c_i K(c_i t + a_i, s), \qquad (3.22)$$

and L as a kind of condensation function. However, the dilations in the spatial variable produced by $T^\dagger$ in (3.22) represent *expansions*. In contrast to IFSM fractal transforms on functions, the transform K is tiled with expanded copies of itself.

We now focus on (3.20) and attempt to rewrite the integrals involving K as *bona fide* integral transforms of f. First, we write

$$\int_{\mathbb{X}} K(c_i u + a_i, s) f(u) du = \int_{\mathbb{X}} \frac{K(c_i u + a_i, s)}{K(u, \zeta_i(c_i, a_i, s))} K(u, \zeta_i(c_i, a_i, s)) f(u)\, du,$$

where the ζ_i functions perform a renormalization or scaling of the transform variable s. It is desirable that the quotient in the integrand on the right be independent of the integration variable u (i.e., constant with respect to u). Most generally, this implies that

$$K(c_i u + a_i, s) = C_i(c_i, a_i, s) K(u, \zeta_i(c_i, a_i, s)), \qquad i = 1, 2, \ldots, N.$$

However, allowing each scaling relation to possess its own functions C_i and ζ_i may be too general since, for example, no "self-similarity" property is imposed on K. Therefore, we postulate the following functional relation to be satisfied by the kernel K and the functions C and ζ:

$$K(c_i u + a_i, s) \doteq C(c_i, a_i, s) K(u, \zeta(c_i, a_i, s)) \qquad (3.23)$$

for all $u \in \mathbb{X}$ and $i = 1, 2, \ldots, N$. Equation (3.23) may be considered in several ways, including:

1. as a functional relation between the kernel K, the constant C, and scaling function ζ,
2. as a functional equation in the unknown functions K and ζ, given C, and
3. as a functional equation in the unknown functions C and ζ, given K.

As in the case of differential equations, the solution of functional equations requires "initial conditions." In addition, however, an admissible space of functions in which solutions are sought must also be specified. This is a subject for future research. A few simple examples are presented in Sect. 3.4.4.

3.4.2 Induced fractal operators on fractal transforms

If the functional equation in (3.23) is satisfied by the kernel K, then the integrals in the first sum of (3.20) simplify to

$$\sum_{i=1}^{N} \alpha_i c_i \int_{\mathbb{X}} C(c_i, a_i, s) K\left(u, \zeta(c_i, a_i, s)\right) f(u) \, du$$

$$= \sum_{i=1}^{N} \alpha_i c_i C(c_i, a_i, s) \widehat{f}(\zeta(c_i, a_i, s)).$$

The net result is the relation

$$\widehat{g}(s) = (M\widehat{f})(s)$$

$$= \sum_{i}^{N} \alpha_i c_i C(c_i, a_i, s) \widehat{f}(\zeta(c_i, a_i, s)) \; + \; \widehat{\beta}(s), \qquad (3.24)$$

a kind of self-similarity equation defining the action of operator M in Fig. 3.8.

If we now assume that T is contractive in $\mathcal{L}^p(\mathbb{X})$ with fixed point $\bar{f}$, then $\widehat{\bar{f}} = \mathcal{S}(\bar{f})$ satisfies the fixed-point equation $\widehat{\bar{f}} = M\widehat{\bar{f}}$ or

$$\widehat{\bar{f}}(s) = \sum_{i=1}^{N} \alpha_i c_i C(c_i, a_i, s) \widehat{\bar{f}}(\zeta(c_i, a_i, s)) + \widehat{\beta}(s).$$

However, there remains the question of whether the operator M is *contractive* in the space of transforms $\mathcal{G}$ and with respect to what metric. Assuming that $\mathcal{G}$ is a Banach space with norm denoted $\|\cdot\|_{\mathcal{G}}$, we define the metric $d_{\mathcal{G}}(u, v) = \|u - v\|_{\mathcal{G}}$ for $u, v \in \mathcal{G}$. If M is contractive in this metric, then, by Banach's fixed-point theorem, $\widehat{\bar{f}}(s)$ may be generated by standard iteration: Start with any function $v_0 \in \mathcal{G}$, and define $v_{n+1} = Mv_n$. Then $v_n \to \widehat{\bar{f}}$ as $n \to \infty$ in the $d_{\mathcal{G}}$ metric.

In many practical examples, including Fourier and wavelet transforms, the coefficients $C(c_i, a_i, s)$ in (3.24) may be bounded with respect to s. In such cases, a straightforward calculation yields

$$d_{\mathcal{G}}(Mu, Mv) \leq D d_{\mathcal{G}}(u, v), \qquad u, v \in \mathcal{G},$$

where

$$D = \sum_{i=1}^{N} c_i |\alpha_i \bar{C}_i(c_i, a_i, s)| \|J_i\|^{-1}, \qquad (3.25)$$

where $\bar{C}_i = \max_s C(c_i, a_i, s)$ and $|J_i|$ denotes the (maximum of the) Jacobian of the transformation $s \to \zeta(c_i, a_i, s)$. Contractivity of M is guaranteed if $D < 1$.

3.4.3 The functional equation for the kernel

It is natural to inquire about the actual meaning of the functional equation in (3.23). Suppose that K is a solution. Furthermore, consider the particular case in which the sets $\mathbb{X}_i = w_i(\mathbb{X})$ (or the range blocks R_i), $1 \leq i \leq N$, form a partition of $\mathbb{X}$, herewith to be referred to as an *IFS partition* of $\mathbb{X}$, the case normally employed in practical fractal image and signal compression. In this nonoverlapping case, each point $x \in \mathbb{X}$ has only one fractal component (neglecting boundary points in the continuous case). As a result, we may "invert" (3.23) to give

$$K(t,s) = C(c_i, a_i, s)K(w_i^{-1}(t), \zeta(c_i, a_i, s)), \quad t \in \mathbb{X}_i, \quad i = 1, 2, \ldots, N.$$

The nonoverlapping nature of the $\mathbb{X}_i$ allows us to express this result as

$$
\begin{aligned}
K(t,s) &= (\mathcal{M}K)(t,s) \\
&= \sum_{i=1}^{N} C(c_i, a_i, s)K(w_i^{-1}(t), \zeta(c_i, a_i, s)).
\end{aligned}
\tag{3.26}
$$

Thus, as in the case of the IFSM fractal transform T (see (3.1)), K is now written as a linear combination of its own *fractal components* under the action of the IFS maps w_i. In other words, K is the fixed point of a fractal transform $\mathcal{M}$ that operates on kernels. Note that there is no restriction on the IFS partition of $\mathbb{X}$, implying that K satisfies a kind of *universal self-similarity*. Clearly, this is a special property.

The following proposition shows that the functional equation in (3.23) is equivalent to this type of universal self-similarity.

Proposition 3.9. *The function $K(t,s)$ satisfies the functional equation (3.23) for fixed functions $C(c,a,s)$ and $\zeta(c,a,s)$ if and only if for every IFS partition of $\mathbb{X}$ with IFS maps of the form $w_i(x) = c_i x + a_i$ there are functions $C_i(s)$ and $\zeta_i(s)$ such that K is the fixed point of the fractal transform operator*

$$(\mathcal{M}K)(t,s) = \sum_{i=1}^{N} C_i(s)K(w_i^{-1}(t), \zeta_i(s)). \tag{3.27}$$

Proof. By the comments immediately preceding the statement of the proposition, we know that if K satisfies the functional equation, then, for any IFSM partition of $\mathbb{X}$, K is the fixed point of the IFSM (3.27), where $C_i(s) = C(c_i, a_i, s)$ and $\zeta_i(s) = \zeta(c_i, a_i, s)$.

Conversely, suppose that for any IFSM partition of $\mathbb{X}$ there are functions $C_i(s)$ and $\zeta_i(s)$ such that K is the fixed point of the induced IFSM operator (3.27). In order to show that K is a solution to the functional equation, we must define the functions $C(c, a, s)$ and $\zeta(c, a, s)$.

To this end, let c and a be fixed such that $w_1(x) = cx + a$ defines a contractive map from $\mathbb{X}$ to itself. Choose $w_2, w_3, \ldots, w_n$ to be affine maps such that the IFS $\{w_1, w_2, \ldots, w_n\}$ is an IFS partition of $\mathbb{X}$. Then by hypothesis we know that there are functions $C_i(s)$ and $\zeta_i(s)$ such that K is the fixed point of the induced IFSM given by (3.27). Define

$$C(c, a, s) = C_1(s)$$

and

$$\zeta(c, a, s) = \zeta_1(s).$$

Then, for all s and t and for this specific choice of c and a, we have

$$K(ct + a, s) = C(c, a, s)K(t, \zeta(c, a, s))$$

and so K satisfies the functional equation (3.23) for this specific choice of c and a.

Clearly, since c and a were arbitrary, the procedure above can be performed for all c and a, thus constructing functions $C(c, a, s)$ and $\zeta(c, a, s)$ such that K satisfies the functional equation. $\qquad\square$

To repeat, the functional equation is equivalent to the property of universal self-similarity. The solution K is the fixed point of an IFSM-type operator on kernel functions. Note, however, that it is *not* guaranteed that the coefficients $C_i(s)$ are contractive. An additional complication arises from the fact that the operator $\mathcal{M}$ in (3.26) is linear in K. In order to avoid the trivial solution $K(t, s) = 0$, it may be necessary to restrict the solution space of the functional equation to appropriate "shells," as is done, for example, in the case of IFSP Markov operators and probability measures. These are open questions for further research. We now examine the functional equation for some very special cases.

Proposition 3.10. *Suppose that a kernel K satisfies the functional equation in (3.23) for $\zeta(c_i, a_i, s) = s$. Then K is independent of t (i.e., $K(t, s) = K(s)$).*

Proof. For simplicity of notation, let us drop the subscripts i. Choose fixed values of c and a. Then, for $K(t, s) \neq 0$, we have

$$C(c, a, s) = \frac{K(ct + a, s)}{K(t, s)}.$$

Since $t \in \mathbb{X}$ and $s \in \mathbb{R}$ are independent variables, it follows that both sides of the equation are independent of t (since the left-hand side is t-independent). Now, choose the value $t^* = a/(1 - c)$ such that $ct^* + a = t^*$. (The existence of such a $t^* \in \mathbb{X}$ is guaranteed by the assumption on the IFS maps that $w_i : \mathbb{X} \to \mathbb{X}$.) Inserting this value of t into the equation above yields

$$C(c, a, s) = \frac{K(t^*, s)}{K(t^*, s)} = 1.$$

This result is true for all values of c, a, s. Therefore, the functional equation reduces to

$$K(ct + a, s) = K(t, s),$$

the only solution of which is $K(t, s) = f(s)$, a function of s only. $\qquad \square$

The following result is obtained in a very similar fashion.

Proposition 3.11. *Suppose that the kernel K satisfying the functional equation in (3.23) is independent of s (i.e., $K(t, s) = K(t)$). Then K is a constant.*

These two simple results illustrate the importance of "mixing" between the spatial variable $t \in \mathbb{X}$ and the transform variable $s \in \mathbb{R}$. In the following, the particular consequences of separability of the kernel K are examined.

Proposition 3.12. *Suppose that the kernel K satisfying the functional equation in (3.23) is separable (i.e., $K(t, s) = K_1(t)K_2(s)$). Then K_1 is constant on $\mathbb{X}$, and K_2 satisfies the relation*

$$K_2(s) = C(c_i, a_i, s)K_2(\zeta(c_i, a_i, s)). \qquad (3.28)$$

Proof. Once again, for simplicity of notation, we omit the subscripts i and choose fixed values of c and a. Then, assuming separability (as well as $K(t, s) \neq 0$), a rearrangement of (3.23) yields

$$\frac{K_1(ct + a)}{K_1(t)} = C(c, a, s)\frac{K_2(\zeta(c, a, s))}{K_2(s)} = A,$$

where A is a real constant, since $t \in \mathbb{X}$ and $s \in \mathbb{R}$ are independent. For the particular value $t = t^* = a/(1 - c)$, we find that $A = 1$, which must hold true for all values of c, a, s. Therefore $K_1(cu + a) = K_1(u)$, implying that K_1 is constant on $\mathbb{X}$. The functional relation (3.28) for K_2 then follows immediately. $\qquad \square$

Proposition 3.13. *Let* $T : \mathcal{L}^p(\mathbb{X}) \to \mathcal{L}^p(\mathbb{X})$ *be the fractal transform operator associated with an N-map affine IFSM. For an $f \in \mathcal{L}^p(\mathbb{X})$, let $g = Tf$. Let $\hat{f}$ and $\hat{g}$ denote the integral transforms of f and g, respectively, assuming that the kernel K satisfies the functional equation in (3.23) and is separable (i.e., $K(t,s) = K_1(t)K_2(s)$). Then*

$$\hat{g}(s) = \left[\sum_{i=1}^{N} \alpha_i c_i\right] \hat{f}(s) + \widehat{\beta}(s), \tag{3.29}$$

where $\widehat{\beta}(s)$ is defined in (3.21).

Proof. From the previous proposition, it follows that $K_1(x) = B$, a constant, on $\mathbb{X}$. Therefore,

$$\hat{f}(s) = BK_2(s) \int_{\mathbb{X}} f(t)dt.$$

Substitution into (3.24) yields

$$\hat{g}(s) = \sum_{i=1}^{N} \alpha_i c_i BK_2(s) \int_{\mathbb{X}} f(t)dt + \widehat{\beta}(s),$$

which, when rearranged, gives the desired result. $\square$

The reader may compare (3.29) with (3.24). When the kernel K is separable, the resulting operator M relating $\hat{g}$ to $\hat{f}$ is rather simple in form, involving no dilations in the transform variable s. (This is a consequence of the fact that K is constant with respect to variations in the spatial variable $t \in \mathbb{X}$.) If we further assume that the IFSM operator T is contractive with fixed point $\bar{f}$, then, from (3.29), with $f = g = \bar{f}$, it follows that

$$\widehat{\bar{f}}(s) = \widehat{\beta}(s) \left[1 - \sum_{i=1}^{N} \alpha_i c_i\right]^{-1}.$$

3.4.4 Examples

We now present a few examples of integral transforms and IFS on these integral transform spaces to illustrate the theory presented above. In particular, we examine the kernel of the integral transform for any self-similarity properties.

3.4.4.1 Fourier transform

The kernel is $K(t, \omega) = e^{i\omega t}$, so that

$$K(c_i u + a_i, \omega) = e^{i(c_i u + a_i)\omega} = e^{ia_i\omega} e^{iuc_i\omega} = e^{ia_i\omega} K(u, c_i\omega).$$

Thus, $C(c_i, a_i, \omega) = e^{ia_i\omega}$ and $\zeta(c_i, a_i, \omega) = c_i\omega$. If $g = Tf$, then (3.24) becomes

$$\widehat{g}(\omega) = \sum_i \alpha_i c_i e^{ia_i\omega} \widehat{f}(c_i\omega) + \widehat{\beta}(\omega).$$

Notice that if T has contractive spatial maps (the w_i), then the induced operator will have expansive spatial maps. Intuitively, this happens because large frequencies correspond to small scales and small frequencies correspond to large scales. Computationally this happens because the kernel is of the form $K(t, s) = M(st)$. For a kernel of the form $K(t, s) = M(t/s)$, one would have a direct relationship between frequency and scale.

It is well-known that $f, g \in \mathcal{L}^2(\mathbb{X})$ implies that $\widehat{f}, \widehat{g} \in \mathcal{L}^2(\mathbb{R})$. Thus it is convenient to use the usual $\mathcal{L}^2$ metric in $\mathcal{G}$. Following the calculation of (3.25), we find

$$\|M\widehat{u} - M\widehat{v}\|_2 \leq \sum_{i=1}^{N} c_i^{1/2} |\alpha_i| \|\widehat{u} - \widehat{v}\|_2, \qquad \widehat{u}, \widehat{v} \in \mathcal{L}^2(\mathbb{R}).$$

From Theorem 3.4, with $p = 2$, contractivity of the IFSM operator T implies contractivity of M.

In the case of measures (i.e., $f, g \in \mathcal{P}(\mathbb{X})$), the set of Borel probability measures on $\mathbb{X}$, some care must be taken in constructing a suitable metric on the space of transforms $\mathcal{G}$ of measures [67]. It can then be shown that contractivity of T implies contractivity of M. We refer the reader to [67] for details.

Finally, from a historical viewpoint, we recall Zygmund's [175] analysis of the Fourier transform of a (uniform) Cantor-Lebesgue measure on the classical Cantor set. Not surprisingly, his analysis, which exploited the self-similarity of the problem, was quite analogous to the fractal transform method.

3.4.4.2 Wavelet transform

In this case, the kernel is given by $K(t, s, b) = \psi\left(\frac{t-b}{s}\right)$, where $\psi(x)$ denotes a mother wavelet function. There are two transform variables, s and b, corresponding to scaling and translation, respectively.

$$K(c_i u + a_i, s, b) = \psi \left(\frac{c_i u + a_i - b}{s} \right)$$
$$= \psi \left(\frac{u - c_i^{-1}(b - a_i)}{sc_i^{-1}} \right)$$

so that the functional equation satisfied by K is

$$K(c_i u + a_i, s, b) = K(u, sc_i^{-1}, c_i^{-1}(b - a_i)).$$

Here, $C(c_i, a_i, s, b) = 1$ and the scaling function for the parameter s is $\zeta(c_i, a_i, s, b) = sc_i^{-1}$. Thus, for $g = Tf$, we obtain

$$\widehat{g}(s, b) = \sum_i \alpha_i c_i \widehat{f} \left(sc_i^{-1}, c_i^{-1}(b - a_i) \right).$$

Numerous authors have studied the possibilities of mixing IFSs with wavelets (see, for example, [46, 66, 134, 163, 166] and references therein) with good results. The multiresolution structure of the wavelet transform makes it an ideal candidate for fractal analysis. See Sect. 3.3 for more details.

3.4.4.3 Lebesgue transform

Here the kernel is

$$K(t, s) = \begin{cases} 1 & \text{if } 0 \leq t \leq s, \\ 0 & \text{if } s \leq t \leq 1. \end{cases}$$

This kernel satisfies the functional equation with

$$K(c_i u + a_i, t) = K \left(u, \frac{t - a_i}{c_i} \right),$$

where we recall that $c_i > 0$. Another (perhaps more useful) way to write the Lebesgue transform of f is as

$$\widehat{f}(s) = \int_0^s f(t) \, dt.$$

If we restrict f to be positive (as is the case for image functions), then it may be viewed as a density function on $\mathbb{X} = [0, 1]$. Then $\widehat{f}(s)$ will be the cumulative distribution function (CDF) for f. For $f \in L^1$, $\widehat{f}(s)$ is nondecreasing and continuous, with $(\mathcal{S}f)(0) = 0$. If we assume f is normalized (i.e., $\|f\|_1 = 1$), then $\widehat{f}(1) = 1$.

If we relax the condition that the Lebesgue transform $\hat{f}(s)$ be continuous in s, then the space $\mathcal{G}$ is given by

$$\mathcal{G} = \{F : [0, 1] \to [0, 1] : F(0) = 0, F(1) = 1, F \text{ nondecreasing}\},$$

which is the set of CDFs for probability measures on [0,1]. A suitable choice of fractal-type transforms M on this space is as follows [140, 148]. We again assume that the sets $w_i(\mathbb{X})$ overlap only at their endpoints. Then, for an $f \in \mathcal{G}$,

$$(Mf)(x) = \alpha_i f(w_i^{-1}(x)) + \beta_i, \quad x \in \mathbb{X}_i,$$

where

$$\beta_1 = 0, \qquad \alpha_i + \beta_i \leq \beta_{i+1} \leq 1, \qquad \text{and} \qquad \alpha_N + \beta_N = 1.$$

(The reader may verify that the conditions above guarantee that M preserves the nondecreasing property. Technically, the equation above is not valid at any points of intersection of the $w_i(\mathbb{X})$. At those points, one would choose either the maximum or minimum of the two "fractal components" in order to preserve right or left continuity, respectively.) The papers [140, 148] contain an extended discussion of this example with some very nice applications to image representation.

This definition of M on $\mathcal{G}$ is slightly more general than the operator induced by the IFSM map $T : \mathcal{F} \to \mathcal{F}$, as it permits the introduction of point masses at the boundaries of the intervals $\mathbb{X}_i = w_i(\mathbb{X})$. It essentially represents a "grand unification" of some IFS-type schemes since $\mathcal{G}$ may now include the Lebesgue transforms of measures, functions, and distributions.

3.4.4.4 Moments of measures

This final example is perhaps more of a "nonexample" in two respects. First, the domain of the transform is the space of Borel probability measures on $[0, 1]$. Second, the transform space is a sequence space – the space of moments of measures on $[0, 1]$ (as discussed in [65]).

Let $\mathcal{P}(\mathbb{X})$ be the collection of probability measures on $\mathbb{X} = [0, 1]$ and $\mu \in \mathcal{P}(\mathbb{X})$. We define the moment sequence by

$$g_k(\mu) = \int_{\mathbb{X}} x^k \, d\mu(x) \quad \text{for } k = 0, 1, 2, \dots,$$

so the "kernel" for this transform is the function $K(x, n) = x^n$. Clearly, this kernel does not satisfy (3.23).

Note that $\mu_0 = 1$ since μ is a probability measure. Furthermore, $g_{k+1} \leq g_k$ since $x^{k+1} \leq x^k$ for $x \in [0,1]$. Thus, $(g_k) \in l^\infty$. We define the space

$$\bar{l}_0^2 = \{\mathbf{c} = (c_0, c_1, \ldots) \in l^\infty | c_0 = 1\}$$

with the weighted inner product

$$\langle \mathbf{c}, \mathbf{d} \rangle = 1 + \sum_{k=1}^{\infty} \frac{c_k d_k}{k^2}.$$

Then $\bar{l}_0^2$ is a complete metric space and the operator on $\bar{l}_0^2$ induced from the Markov operator on $\mathcal{P}(\mathbb{X})$ is a contraction (see [65]). However, this operator is not of IFS type since the kernel does not satisfy the functional equation (3.23).

Chapter 4
Iterated Function Systems, Multifunctions, and Measure-Valued Functions

In this chapter, we present some generalizations of iterated function systems to more general settings. The purpose of the first part of this chapter is to extend the earlier idea of iterated function systems by considering multifunctions which leads quite naturally to the definition of *iterated multifunction systems*(IMSs). In the second part, we consider a notion of iterated function systems on spaces of multifunctions, while in the third section we introduce a generalized fractal transform on the space of measure-valued functions. As one possible justification of these extensions, we provide potential applications of these approaches to fractal image coding.

4.1 Iterated multifunction systems and iterated multifunction systems with probabilities

Before introducing the notion of *iterated multifunction systems*, let us now cast the idea of families of N-map IFSs into a multifunction framework. We first consider the following example, from which the notion of iterated multifunction systems arises quite naturally from families of IFSs.

Example 4.1. Let $(\mathbb{X}, d)$ be a complete metric space and $\{w_1^i, \ldots w_N^i\}$ and $\{p_1^i, \ldots p_N^i\}$, $i = 1 \ldots M$, be M families of IFSs with probabilities. We define the *multifunctions* $F_j : \mathbb{X} \rightrightarrows \mathbb{X}$, $j = 1 \ldots N$, as

$$F_j(x) = \bigcup_{i \leq M} w_j^i(x).$$

Each multifunction F_j is a contraction since, for any $x, y \in \mathbb{X}$,

$$d_{\mathbb{H}}(F_j(x), F_j(y)) = d_{\mathbb{H}} \left(\bigcup_{i \leq M} w_j^i(x), \bigcup_{i \leq M} w_j^i(y) \right)$$

$$\leq \sup_{i \leq M} c_j^i d(x, y) := c_j d(x, y),$$

(4.1)

where $0 \leq c_j < 1$.

We shall consider the collection of multifunctions $\mathbf{F} = \{F_1, \cdots, F_N\}$ to define an *iterated multifunction system* (IMS) (see [41, 99, 96] for more mathematical details and applications). The next step is to introduce probabilities to this new set of (combined) multifunctions. Let $q_1, q_2, \ldots, q_M$ be a set of probabilities that we will use to mix the individual IFSs. We associate with the family of N-map IFSs above a set of probabilities

$$z_j = \sum_{i=1}^{M} q_i p_j^i, \quad 1 \leq j \leq N.$$

It is easy to see that $\sum_{j=1}^{N} z_j = 1$. To each multifunction F_j we shall associate the probability z_j defined above. The result is an *iterated multifunction with probabilities* (IMSP), $(\mathbf{F}, \mathbf{z})$. As with the classical case, we are interested in both the *geometric* and *probabilistic* versions of this type of IFS.

Let $(\mathbb{X}, d)$ be a compact metric space and $F_i : \mathbb{X} \to \mathbb{H}(\mathbb{X})$ be a set of multifunctions for $i = 1, 2, \ldots, N$. To each multifunction F_j there corresponds another multifunction $F_j^* : \mathbb{H}(\mathbb{X}) \to \mathbb{H}(\mathbb{X})$ defined as

$$F_j^*(A) = \bigcup_{a \in A} F_i(a) \quad \forall A \in \mathbb{H}(\mathbb{X}).$$

As before (see Sect. 2.2.3), F_j^* is contractive on the space $(\mathbb{H}(\mathbb{X}), d_{\mathbb{H}})$. As a result, we may associate with the IMS $\mathbf{F}$ the system of multifunctions $\mathbf{F}^* = (F_1^*, \ldots F_N^*)$, which is a set of contractions on $\mathbb{H}(\mathbb{X})$. Associated with this multifunction system is a mapping

$$F^{**} : \mathbb{H}(\mathbb{H}(\mathbb{X})) \to \mathbb{H}(\mathbb{H}(\mathbb{X}))$$

defined as follows. For $\mathcal{A}, \mathcal{B} \in \mathbb{H}(\mathbb{H}(\mathbb{X}))$,

$$F^{**}(\mathcal{A}) = \bigcup_{j=1}^{N} F_j^*(\mathcal{A}) = \bigcup_{j=1}^{N} \bigcup_{A \in \mathcal{A}} F_j^*(A).$$

We equip the space $\mathbb{H}(\mathbb{H}(\mathbb{X}))$ with the metric

$$d_{\mathbb{H}\mathbb{H}}(\mathcal{A}, \mathcal{B}) = \max \left\{ \sup_{A \in \mathcal{A}} \inf_{B \in \mathcal{B}} d_{\mathbb{H}}(A, B), \sup_{B \in \mathcal{B}} \inf_{A \in \mathcal{A}} d_{\mathbb{H}}(A, B) \right\},$$

where $\mathcal{A}, \mathcal{B} \in \mathbb{H}(\mathbb{H}(\mathbb{X}))$. It is easy to prove that $(\mathbb{H}(\mathbb{H}(X)), d_{\mathbb{H}\mathbb{H}})$ is a complete metric space. The following result shows that F^{**} is contractive on $\mathbb{H}(\mathbb{H}(\mathbb{X}))$.

Theorem 4.2. *If* $\mathcal{A}, \mathcal{B} \in \mathbb{H}(\mathbb{H}(\mathbb{X}))$, *then*

$$d_{\mathbb{H}\mathbb{H}}(F^{**}(\mathcal{A}), F^{**}(\mathcal{B})) \leq c \, d_{\mathbb{H}\mathbb{H}}(\mathcal{A}, \mathcal{B}),$$

where $c = \max_{1 \leq i \leq N} c_i$.

Proof. We have

$$d_{\mathbb{H}\mathbb{H}}(F^{**}(A), F^{**}(B)) = d_{\mathbb{H}\mathbb{H}} \left(\bigcup_{j=1}^{N} \bigcup_{A \in \mathcal{A}} F_i^*(A), \bigcup_{j=1}^{N} \bigcup_{B \in \mathcal{B}} F_i^*(B) \right)$$

$$\leq \max_{1 \leq j \leq N} d_{\mathbb{H}\mathbb{H}} \left(\bigcup_{A \in \mathcal{A}} F_i^*(A), \bigcup_{B \in \mathcal{B}} F_i(B) \right)$$

$$\leq \max_{1 \leq j \leq N} c_j d_{\mathbb{H}\mathbb{H}}(A, B),$$

proving the result.   □

From Banach's theorem, there exists a unique fixed point $\mathcal{A}^{**}$ of F^{**} that satisfies the equation

$$\mathcal{A}^{**} = F^{**}(\mathcal{A}^{**}) = \bigcup_{j=1}^{N} F_j^*(\mathcal{A}) = \bigcup_{j=1}^{N} \bigcup_{A \in \mathcal{A}} F_j^*(A).$$

Example 4.3. Let us denote by ru_i, rd_i, m_i, and k_i the reflect up, reflect down, maple leaf, and von Koch iterated function systems. We construct two IMSs with $M_i = \{ru_i, rd_i\}$, and $M_i = \{m_i, k_i\}$. Figure 4.1 shows the projections of IMS attractors.

Consider now the space $\mathbb{H}(\mathbb{X})$ and the σ-algebra of all Borel subsets of $\mathbb{H}(\mathbb{X})$, say $\mathcal{B}(\mathbb{H}(\mathbb{X}))$. Consider now the space $\mathcal{P}(\mathbb{H}(\mathbb{X}))$ of all probability measures on $(\mathbb{H}(\mathbb{X}), \mathcal{B}(\mathbb{H}(\mathbb{X})))$ equipped with the usual Monge-Kantorovich metric: For $\mu, \nu \in \mathcal{P}(\mathbb{H}(\mathbb{X}))$,

$$d_{MKMK}(\mu, \nu) := \sup_{f \in \mathrm{Lip}_1(\mathbb{H}(\mathbb{X}), \mathbb{R})} \left[\int_{\mathbb{H}(\mathbb{X})} f d\mu - \int_{\mathbb{H}(X)} f d\nu \right], \qquad (4.2)$$

Fig. 4.1: The (left) reflect up-reflect down and (right) maple leaf-von Koch IMS attractor projections.

where

$$\text{Lip}_1(\mathbb{H}(X), \mathbb{R}) = \tag{4.3}$$
$$\{f : \mathbb{H}(\mathbb{X}) \to \mathbf{R} \mid |f(x_1) - f(x_2)| \le d_{\mathbb{H}}(x_1, x_2), \ \forall x_1, x_2 \in \mathbb{H}(\mathbb{X})\}.$$

It is easy to prove that the space $(\mathcal{P}(\mathbb{H}(\mathbb{X})), d_{MKMK})$ is a complete metric space if a first-order condition is assumed. For $1 \le i \le N$, let $0 < p_i < 1$ be a partition of unity associated with the IFS maps F_i such that $\sum_{i=1}^{N} p_i = 1$. Associated with this IMSP $(\mathbf{F}, \mathbf{p})$ is the so-called *Markov operator*, $\text{MM} : \mathcal{P}(\mathbb{H}(\mathbb{X})) \to \mathcal{P}(\mathbb{H}(\mathbb{X}))$, the action of which is

$$(\text{MM}\mu)(\mathcal{A}) := \sum_{i=1}^{N} p_i \mu((F_i^*)^{-1}(\mathcal{A})), \quad \forall \mathcal{A} \in \mathcal{B}(\mathbb{H}(\mathbb{X})).$$

It is easy to prove the following proposition.

Proposition 4.4. MM *is a contraction mapping on* $\mathcal{P}(\mathbb{H}(\mathbb{X}))$ *under the metric* d_{MKMK}:

$$d_{MKMK}(\text{MM}\mu, \text{MM}\nu) \le c d_{MKMK}(\mu, \nu), \quad \mu, \nu \in \mathcal{P}(\mathbb{H}(\mathbb{X})).$$

Corollary 4.5. *There exists a unique measure* $\bar{\mu} \in \mathcal{P}(\mathbb{H}(\mathbb{X}))$, *the so-called* invariant measure *of the IMSP* $(\mathbf{F}, \mathbf{p})$, *such that* $\bar{\mu} = M^* \bar{\mu}$. *Moreover, for any* $\mu \in d_{MKMK}(\mathbb{H}(\mathbb{X}))$, $d_{MKMK}(\text{MM}^n \mu, \bar{\mu}) \to 0$ *as* $n \to \infty$.

Example 4.6. We assign the probabilities $q_1 = 0.7$ to the maple leaf maps m_i, and $q_2 = 0.3$ to the fern maps f_i. Next, we construct the 4-map IMSP with $M_i = \{m_i, f_i\}$ for $i = 1, 2, 3, 4$ with probabilities

$$p_1 = (0.7)(0.25) + (0.3)(0.01) = 0.178,$$
$$p_2 = 0.196, p_3 = 0.196, \text{ and } p_4 = 0.43.$$

The IMSP projections of the attractor have been coloured based upon the number of visits made to each pixel (see Fig. 4.2).

Fig. 4.2: The fern-maple leaf IMS attractor.

4.1.1 Code space

We now extend the classical results on code space to the case of iterated multifunction systems. Let Ω be the set of all semi-infinite sequences of symbols $\{\sigma_k\}$, where $\sigma_k \in \{1, \ldots, N\}$ endowed with the usual metric $d_\Omega(\sigma, \omega) = 0$ if $\sigma = \omega$ and $d_\Omega(\sigma, \omega) = 2^{-k}$, where k is the least index for which $\sigma_k \neq \omega_k$. Given an N-map iterated multifunction system $\mathbf{F} = \{F_1, \ldots F_N\}$ on $\mathbb{X}$, $\sigma \in \Omega$, and $x \in \mathbb{X}$, we can consider the address function

$$\phi_F(\sigma) = \lim_{k \to +\infty} \varphi_k^x(\sigma),$$

where $\varphi_k^x(\sigma) := F_{\sigma_1} \cdot F_{\sigma_2} \ldots \cdot F_{\sigma_k}(x)$. The address function ϕ_F maps an element σ of Ω to a set of points on the attractor A of the IMS on $\mathbb{X}$. It is easy to prove that $\varphi_k^x(\sigma) \in \mathbb{H}(\mathbb{X})$ for all $x \in \mathbb{X}$, $k \in \mathbb{N}$, and $\sigma \in \Omega$. We prove that $\phi_F(\sigma)$ is well-defined.

Theorem 4.7. *The sequence of compact sets $\{\varphi_k^x(\sigma)\}_{k \in \mathbb{N}}$ is Cauchy and so the limit in (4.1.1) exists. Furthermore, the limit does not depend on the choice of the point $x \in \mathbb{X}$.*

Proof. Let $c \in [0, 1)$ denote the maximum of the contraction factors of the multifunctions F_i, as usual, and, for $x \in \mathbb{X}$, let $L(x) = \max_{1 \leq j \leq N} d_{\mathbb{H}}\left(x, F_{\sigma_j}(x)\right)$. We have

$$d_{\mathbb{H}}\left(\varphi_i^x(\sigma), \varphi_{i+1}^x(\sigma)\right)$$
$$= d_{\mathbb{H}}\left(F_{\sigma_1} \circ F_{\sigma_2} \circ \cdots \circ F_{\sigma_i}(x), F_{\sigma_1} \circ F_{\sigma_2} \circ \cdots \circ F_{\sigma_{i+1}}(x)\right)$$

$$\leq c\, d_{\mathbb{H}} \left(F_{\sigma_2} \circ \cdots \circ F_{\sigma_i}(x), F_{\sigma_2} \circ \cdots \circ F_{\sigma_{i+1}}(x) \right)$$
$$\leq c^i d_{\mathbb{H}} \left(x, F_{\sigma_{i+1}}(x) \right)$$
$$\leq c^i L(x).$$

Thus, for $n > m$, we get

$$d_{\mathbb{H}} \left(\varphi_m^x(\sigma), \varphi_n^x(\sigma) \right) \leq \sum_{i=m}^{n-1} d_{\mathbb{H}} \left(\varphi_i^x(\sigma), \varphi_{i+1}^x(\sigma) \right)$$
$$\leq \sum_{i=m}^{n-1} c^i L(x) = \frac{c^m - c^n}{1 - c} L(x).$$

We conclude that $\{\varphi_k^x(\sigma)\}_{k \in \mathbb{N}}$ is Cauchy and define

$$\phi_F(\sigma) = \lim_{k \to +\infty} \varphi_k^x(\sigma),$$

as in (4.1.1), noting that the limit does not depend upon x. $\qquad\square$

4.2 Iterated function systems on multifunctions

Now we turn to IFS operators that act on spaces of multifunctions. This is the multifunction analogue of the IFSM from Sect. 3.2. The following discussion borrows heavily from [107, 105, 110, 111].

4.2.1 Spaces of multifunctions

We recall that a set-valued mapping or multifunction $F : \mathbb{X} \rightrightarrows \mathbb{Y}$ is a function from $\mathbb{X}$ to the power set $2^{\mathbb{Y}}$. In the following we will suppose that $\mathbb{Y}$ is compact and $F(x)$ is compact for each $x \in \mathbb{X}$. Define the following spaces of multifunctions:

$$\mathcal{F}(X, Y) = \{F : \mathbb{X} \to \mathbb{H}(\mathbb{Y})\}.$$

We place on $\mathcal{F}(X, Y)$ the metric

$$d_\infty(F, G) = \sup_{x \in X} d_{\mathbb{H}}(F(x), G(x)),$$

where $d_{\mathbb{H}}$ denotes, as usual, the Hausdorff distance between sets.

Proposition 4.8. *The space* $(\mathcal{F}(X, Y), d_\infty)$ *is a complete metric space.*

Proof. It is trivial to prove that $d_\infty(F, G) = 0$ if and only if $F = G$ and that $d_\infty(F, G) = d_\infty(G, F)$. Furthermore, for all $F, G, L \in \mathcal{F}(X, Y)$, we have

$$
\begin{aligned}
d_\infty(F, G) &= \sup_{x \in X} d_\mathbb{H}(F(x), G(x)) \\
&\leq \sup_{x \in X} d_\mathbb{H}(F(x), L(x)) + d_\mathbb{H}(L(x), G(x)) \\
&\leq \sup_{x \in X} d_\mathbb{H}(F(x), L(x)) + \sup_{x \in X} d_\mathbb{H}(L(x), G(x)) \\
&= d_\infty(F, L) + d_\infty(L, G).
\end{aligned}
$$

To prove that it is complete, let F_n be a Cauchy sequence of elements of $\mathcal{F}(X, Y)$; so for $\forall \epsilon > 0$ there exists $n_0(\epsilon) > 0$ such that for all $n, m \geq n_0(\epsilon)$ we have $d_\infty(F_n, F_m) \leq \epsilon$. So, for all $x \in X$ and for all $n, m \geq n_0(\epsilon)$, we have $\mathbb{H}(F_n(x), F_m(x)) \leq \epsilon$, and the sequence $F_n(x)$ is Cauchy in $\mathbb{H}(\mathbb{Y})$. Since it is complete, there exists $A(x)$ such that $d_\mathbb{H}(F_n(x), A(x)) \to 0$ when $n \to +\infty$. So, for all $x \in X$ and for all $n, m \geq n_0(\epsilon)$, we have $\mathbb{H}(F_n(x), F_m(x)) \leq \epsilon$, and sending $m \to +\infty$ we have $d_\mathbb{H}(F_n(x), A(x)) \leq \epsilon$; that is, $d_\infty(F_n, A) \leq \epsilon$. $\square$

Let $I \subset \mathbb{R}$ be a compact interval and $F : I \rightrightarrows \mathbb{R}^s$ be a given multifunction with compact values. A multifunction F is said to be *integrably bounded* if there exists a Lebesgue integrable function $m : I \to \mathbb{R}$ such that $\|F(t)\| \leq m(t)$ for all $t \in I$ where $\|F(t)\| := d_\mathbb{H}(F(t), \{0\})$. Define $\mathrm{Sel}(F)$ as the set of all Lebesgue integrable selectors of F. It can be proved that every measurable and integrably bounded multifunction F has $\mathrm{Sel}(F) \neq \emptyset$. A measurable multifunction F is said to be Aumann integrable on I if $\mathrm{Sel}(F) \neq \emptyset$. In this case, we define, as in Appendix C, the Aumann integral $\int_I F(t)dt = \{\int_I f(t)dt, f \in \S(F)\}$. In the following, we consider the space $\mathcal{A}(I, \mathbb{R}^s)$ of all Aumann integrable multifunctions $F : I \rightrightarrows \mathbb{H}(\mathbb{R}^s)$.

Lemma 4.9. *[86] The mapping* $d : \mathcal{A}(I, \mathbb{R}^s) \times \mathcal{A}(I, \mathbb{R}^s) \to \mathbb{R}_+$ *defined by*

$$
d_1(F, G) = \int_I d_\mathbb{H}(F(x), G(x)) \, d\mu(x)
$$

for every $F, G \in \mathcal{A}(I, \mathbb{R}^s)$ *is a metric on* $\mathcal{A}(I, \mathbb{R}^s)$.

Proposition 4.10. *[86] $(\mathcal{A}(I, \mathbb{R}^s), d_1)$ is a complete metric space. Furthermore, if F_k is a sequence of $\mathcal{A}(I, \mathbb{R}^s)$ converging in the metric d_1 to F then there is a subsequence, say F_{n_k} of F_k such that $d_\mathbb{H}(F_{n_k}(t), F(t)) \to 0$ for a.e. $t \in I$ as $n \to +\infty$.*

As in the classical L^p space theory one can extend the previous results to the space $\mathcal{A}^p(I, \mathbb{R}^s)$ of all Aumann L^p integrable multifunctions $F : I \rightrightarrows d_{\mathbb{H}}(\mathbb{R}^s)$.

Proposition 4.11. *[86] The space $(\mathcal{A}^p(I, \mathbb{R}^s), d_p)$ is a complete metric space.*

4.2.2 Some IFS operators on multifunctions (IFSMF)

Having these preliminaries out of the way, in the following sections we define two different IFS-type operators on $\mathcal{F}(X, Y)$. Each of these will be called an *iterated function system on multifunctions* or IFSMF. The difference between the two will be the way in which we combine the various fractal components. In the first case, we use the union, and in the second we use the Minkowski sum.

Let $w_i : \mathbb{X} \to \mathbb{X}$ be maps on $\mathbb{X}$ and $\phi_i : d_{\mathbb{H}}(\mathbb{Y}) \to d_{\mathbb{H}}(\mathbb{Y})$ with Lipschitz constants K_i. We define our first operator $T_1 : \mathcal{F}(X, Y) \to \mathcal{F}(X, Y)$ by

$$T_1(F)(x) = \bigcup_i \phi_i(F(w_i^{-1}(x))).$$

Proposition 4.12. *If $K = \max_i K_i < 1$, then T_1 is contractive in d_∞.*

Proof. We compute that

$$
\begin{aligned}
d_\infty(T_1(F), T_1(G)) &= \sup_x d_{\mathbb{H}}\left(\bigcup_i \phi_i(F(w_i^{-1}(x))), \bigcup_i \phi_i(G(w_i^{-1}(x))) \right) \\
&\leq \sup_x \max_i d_{\mathbb{H}}\left(\phi_i(F(w_i^{-1}(x))), \phi_i(G(w_i^{-1}(x))) \right) \\
&\leq \sup_x \max_i K_i d_{\mathbb{H}}\left(F(w_i^{-1}(x)), G(w_i^{-1}(x)) \right) \\
&\leq K \sup_z d_{\mathbb{H}}(F(z), G(z)) = K d_\infty(F, G).
\end{aligned}
$$

The result follows.   □

Proposition 4.13. *Assume that $d\mu(w_i(x)) \leq s_i d\mu(x)$, where $s_i \geq 0$. Then*

$$d_p(T_1(F), T_1(G)) \leq \left(\sum_i K_i^p s_i \right)^{1/p} d_p(F, G).$$

Proof. Computing, we get

$$
\begin{aligned}
& d_p(T_1(F), T_1(G)) \\
&= \left\{ \int_X d_{\mathbb{H}} \left[\bigcup_i \phi_i(F(w_i^{-1}(x))), \bigcup_i \phi_i(G(w_i^{-1}(x))) \right]^p d\mu(x) \right\}^{1/p} \\
&\leq \left\{ \int_X \max_i d_{\mathbb{H}} \left[\phi_i(F(w_i^{-1}(x))), \phi_i(G(w_i^{-1}(x))) \right]^p d\mu(x) \right\}^{1/p} \\
&\leq \left\{ \int_X \max_i K_i \, d_{\mathbb{H}} \left[F(w_i^{-1}(x)), G(w_i^{-1}(x)) \right]^p d\mu(x) \right\}^{1/p} \\
&= \left\{ \sum_i K_i^p \int_{M_i} d_{\mathbb{H}} \left[F(w_i^{-1}(x)), G(w_i^{-1}(x)) \right]^p d\mu(x) \right\}^{1/p} \\
&\leq \left\{ \sum_i K_i^p \int_{w_i(X)} d_{\mathbb{H}} \left[F(w_i^{-1}(x)), G(w_i^{-1}(x)) \right]^p d\mu(x) \right\}^{1/p} \\
&\leq \left\{ \sum_i K_i^p s_i \int_X d_{\mathbb{H}} \left[F(z), G(z) \right]^p d\mu(z) \right\}^{1/p} \\
&= \left[\sum_i K_i^p s_i \right]^{1/p} d_p(F, G).
\end{aligned}
$$

In the above, we have used the sets $M_i \subset w_i(X)$ defined by

$$
\begin{aligned}
M_i = \Big\{ x \in \mathbb{X} : & \, d_{\mathbb{H}}(F(w_i^{-1}(x)), G(w_i^{-1}(x))) \\
& \geq d_{\mathbb{H}}(F(w_j^{-1}(x)), G(w_j^{-1}(x))) \text{ for all } j \Big\}.
\end{aligned}
$$

That is, the set M_i consists of all those points for which the ith preimage gives the largest Hausdorff distance. □

Notice that if $\mathbb{X} \subset \mathbb{R}$ and μ is a Lebesgue measure, and $w_i(x)$ satisfy $|w_i'(x)| \leq s_i$, then the condition $d\mu(w_i(x)) \leq s_i d\mu(x)$ is satisfied. This is the situation that is used in image processing applications.

With a setup similar to that in the previous section, we define our second operator $T_2 : \mathcal{F}(X, Y) \to \mathcal{F}(X, Y)$ by

$$
T_2(F)(x) = \sum_i p_i(x) \phi_i(F(w_i^{-1}(x))),
$$

where the sum depends on x and is over those i such that $x \in w_i(\mathbb{X})$. We require that the functions p_i satisfy $\sum_i p_i(x) = 1$ (again with the dependence of the sum on x). The idea is to average the contributions of the various components in the areas where there is overlap.

Proposition 4.14. *We have*

$$d_\infty(T_2(F), T_2(G)) \leq \left[\sup_x \sum_i p_i(x)K_i\right] d_\infty(F, G).$$

Proof. We compute and see that

$$d_\infty(T_2(F), T_2(G))$$
$$= \sup_x d_{\mathbb{H}}\left(\sum_i p_i(x)\phi_i(F(w_i^{-1}(x))), \sum_i p_i(x)\phi_i(G(w_i^{-1}(x)))\right)$$
$$\leq \sup_x \sum_i p_i(x)K_i d_{\mathbb{H}}(F(w_i^{-1}(x)), G(w_i^{-1}(x)))$$
$$\leq \left[\sup_x \sum_i p_i(x)K_i\right] d_\infty(F, G).$$

$$\square$$

Lemma 4.15. *Let $a_i \in \mathbb{R}$, $i = 1\ldots n$. Then*

$$\left|\sum_i a_i\right|^p \leq C(n)^p \sum_i |a_i|^p,$$

with $C(n) = n^{(p-1)/p}$. Thus, if $p = 1$, we can choose $C(n) = 1$.

Proposition 4.16. *Let $p_i = \sup_x p_i(w_i(x))$ and $s_i \geq 0$ be such that $d\mu(w_i(x)) \leq s_i d\mu(x)$. Then we have*

$$d_p(T_2(F), T_2(G)) \leq C(n) \left(\sum_i K_i^p s_i^p p_i^p\right)^{1/p} d_p(F, G).$$

Proof. We compute and see that

$$d_p(T_2(F), T_2(G))^p$$
$$= \int_X \left(d_{\mathbb{H}}\left(\sum_i p_i(x)\phi_i(F(w_i^{-1}(x))), \sum_i p_i(x)\phi_i(G(w_i^{-1}(x)))\right)\right)^p d\mu(x)$$
$$\leq \int_X \left(\sum_i p_i(x) K_i d_{\mathbb{H}}\left(F(w_i^{-1}(x)), G(w_i^{-1}(x))\right)\right)^p d\mu(x)$$
$$\leq \int_{w_i(X)} C(n)^p \sum_i p_i(x)^p K_i^p \left(d_{\mathbb{H}}\left(F(w_i^{-1}(x)), G(w_i^{-1}(x))\right)\right)^p d\mu(x)$$

$$\leq C(n)^p \sum_i K_i^p s_i^p \int_X p_i(w_i(z))^p \, h\left(F(z), G(z)\right)^p \, d\mu(z)$$

$$\leq C(n)^p \left(\sum_i K_i^p s_i^p p_i^p\right) d_p(F, G)^p.$$

$\square$

Notice that it is easy (but messy) to tighten the estimate in the proposition. This is similar to our discussion in Section 3.2 for the standard IFSM.

4.2.3 An application to fractal image coding

We now present some practical realizations and applications of the IFSMF, with particular focus on the coding of signals and images. The idea of this section is that to each pixel of an image there is associated an interval that measures the "error" in the value for that pixel. In this situation, therefore, we restrict our set-valued functions so that they only take closed intervals as values. We also need to restrict the ϕ_i maps so that they map intervals to intervals.

Thus, we shall consider $\mathbb{X} = [0,1]^n$ for $n = 1$ or 2 and $\mathbb{Y} = [a, b]$. For each x, let $\beta(x) \in \mathbb{H}(\mathbb{R})$ be an interval in $\mathbb{Y}$. Then we define $T : \mathcal{F}(X, Y) \to \mathcal{F}(X, Y)$ by

$$T(F)(x) = \beta(x) + \sum_i p_i(x) \, \alpha_i F(w_i^{-1}(x)),$$

where $\alpha_i \in \mathbb{R}$.

Corollary 4.17. *We have the inequalities*

$$d_\infty(T(F), T(G)) \leq \left[\sup_x \sum_i \alpha_i p_i(x)\right] d_\infty(F, G)$$

and

$$d_p(T(F), T(G)) \leq C(n) \left(\sum_i \alpha_i^p s_i^p p_i^p\right)^{1/p} d_p(F, G),$$

where $p_i = \sup_x p_i(w_i(x))$ and $s_i \geq 0$ are such that $d\mu(w_i(x)) \leq s_i d\mu(x)$.

Proof. We only need to see that

$$d_{\mathbb{H}}\left(\beta(x) + \sum_i p_i(x)\alpha_i F(w_i^{-1}(x)), \beta(x) + \sum_i p_i(x)\alpha_i G(w_i^{-1}(x))\right)$$

$$= d_{\mathbb{H}}\left(\sum_i p_i(x)\alpha_i F(w_i^{-1}(x)), \sum_i p_i(x)\alpha_i G(w_i^{-1}(x)),\right)$$

after which the proof is the same as the proof of Proposition 4.14. □

The inverse problem can be formulated as follows. Given a multifunction (a statistical estimate of an image) $F \in \mathcal{F}(X,Y)$, find a contractive IFSMF operator $T : \mathcal{F}(X,Y) \to \mathcal{F}(X,Y)$ that admits a unique fixed point $\tilde{F} \in \mathcal{F}(X,Y)$ such that $d_\infty(F, \tilde{F})$ is small enough. As discussed before, it is in general a very difficult task to find such operators. A tremendous simplification is provided by the collage theorem, Theorem 2.6, which we now state with particular reference to the IFSMF.

Theorem 4.18. *(Collage theorem for IFSMF) Given $F \in \mathcal{F}(X,Y)$, assume there exists a contractive operator T such that $d_\infty(F, T(F)) < \epsilon$. If F^* is the fixed point of T and $c := \sup_x \sum_i \alpha_i p_i(x)$, then*

$$d_\infty(F, F^*) \le \frac{\epsilon}{1-c}.$$

The inverse problem then becomes one of finding some contractive IFSMF operator that maps the "target" multifunction F as close to itself as possible.

Corollary 4.19. *Under the assumptions of the collage theorem, Theorem 4.18, we have the inequality*

$$d_\infty(F, TF) \le \sum_i p_i \sup_{x \in X} \max\{\underline{A}_i(x), \bar{A}_i(x)\},$$

where $\underline{A}_i(x) = |\min F(x) - \min(\beta(x) + \alpha_i F(w_i^{-1})(x))|$, $\bar{A}_i(x) = |\max F(x) - \max(\beta(x) + \alpha_i F(w_i^{-1}(x)))|$ and $p_i = \sup_{x \in X} p_i(w_i(x))$.

Proof. In fact, using a previous result on the Hausdorff distance and recalling that F is a closed interval multifunction,

$$d_\infty(F, TF) = d_\infty(F(x), \beta(x) + \sum_i p_i(x)\alpha_i F(w_i^{-1}(x)))$$

$$\le d_\infty(F(x), \sum_i p_i(x)(\beta(x) + \alpha_i F(w_i^{-1}(x))))$$

$$\leq \sum_i p_i \dot{d}_\infty(F(x), \beta(x) + \alpha_i F(w_i^{-1}(x)))$$

$$\leq \sum_i p_i \sup_{x \in X} \max\{\underline{A}_i(x), \bar{A}_i(x)\}$$

where $\underline{A}_i(x) = |\min F(x) - \min(\beta(x) + \alpha_i F(w_i^{-1})(x))|$, $\bar{A}_i(x) = |\max F(x) - \max(\beta(x) + \alpha_i F(w_i^{-1}(x)))|$ and $p_i = \sup_{x \in X} p_i(w_i(x))$. $\square$

We now prove a similar result for the d_p metric.

Corollary 4.20. *Under the assumptions of the collage theorem, we have the following inequality*

$$d_p(F, TF)^p \leq \|\min F - \min TF\|_p^p + \|\max F - \max TF\|_p^p.$$

Proof. Computing, we have

$$d_p(F, TF)^p$$

$$= \int_X \left(h(F(x), \beta(x) + \sum_i p_i(x)\alpha_i F(w_i^{-1}(x))) \right)^p d\mu(x)$$

$$\leq \int_X \left| \min F(x) - \min(\beta(x) + \sum_i p_i(x)\alpha_i F(w_i^{-1}(x))) \right|^p d\mu(x)$$

$$+ \int_X \left| \max F(x) - \max(\beta(x) + \sum_i p_i(x)\alpha_i F(w_i^{-1}(x))) \right|^p d\mu(x)$$

$$= \|\min F - \min TF\|_p^p + \|\max F - \max TF\|_p^p.$$

$\square$

The pixel array defining the image is partitioned into a set of nonoverlapping *range subblocks* R_i. Associated with each R_i is a larger *domain subblock* D_i, chosen so that the image function $u(R_i)$ supported on each R_i is well-approximated by a greyscale-modified copy of the image function $u(D_i)$. In practice, affine greyscale maps,

$$u(R_i) \approx \phi_i(u(w_i(D_i)) = \alpha_i u(w_i(D_i)) + \beta_i, 1 \leq i \leq N,$$

are used, where $w_i(x)$ denotes the contraction that maps R_i to D_i (in discrete pixel space, the w_i maps will have to include a decimation that reduces the number of pixels in going from R_i to D_i). The greyscale map coefficients α_i and β_i are usually determined by least squares. The domain blocks D_i are usually chosen from a common *domain pool*

$\mathcal{D}$. The domain block yielding the best approximation to $u(R_i)$ (i.e., the lowest *collage error*)

$$\Delta_{ij} = \| \, u(R_i) - \phi_{ij}(u(w_{ij}(D_j))) \, \|, \quad 1 \le j \le M,$$

is chosen for the fractal coding (the L^2 norm is usually chosen). Figure 4.3 presents the fixed-point approximation $\bar{u}$ to the standard 512×512 *Lena* image (8 bits per pixel, or 256 greyscale values) using a partition of 8×8 nonoverlapping pixel blocks ($64^2 = 4096$ in total). The domain pool for each range block was the set of $32^2 = 1024$ 16×16 nonoverlapping pixel blocks. (This is not an optimal domain pool – nevertheless it works quite well.) The image $\bar{u}$ was obtained by starting with the seed image $u_0 = 255$ (plain white image) and iterating $u_{n+1} = T u_n$ to $n = 15$.

Fig. 4.3: The *Lena* attractor $\bar{u}$.

We now consider a simple IFSMF version of image coding using the partition described above. Since the range blocks R_i are nonoverlapping, all coefficients $p_i(x)$ in our IFSMF operator will have value 1. From the *Lena* image function $u(x)$ used above, we shall construct a multifunction $U(x)$ such that

$$U(x) = [u^-(x), u^+(x)].$$

The approximation of the multifunction range block $U(R_i)$ by $U(D_i)$ then takes the form of two coupled problems,

$$u^-(R_i) \approx \alpha_i u^- (w_i(D_i)) + \beta_i^-(R_i),$$
$$u^+(R_i) \approx \alpha_i u^+ (w_i(D_i)) + \beta_i^+(R_i), \quad 1 \le i \le N.$$

For simplicity, we assume that the $\beta^+(x)$ and $\beta^-(x)$ functions are piecewise constant over each block R_i. For a given domain-range block pair D_i/R_i, we then have a system of three equations in the unknowns α_i, β_i^-, and β_i^+. The domain block yielding the best total L^2 collage distance,

$$\Delta_{ij} = \quad \| u^-(R_i) - \alpha_i u^- (w_{ij}(D_j)) - \beta_i^-(R_i) \|$$
$$+ \| u^+(R_i) - \alpha_i u^+ (w_{ij}(D_j)) - \beta_i^+(R_i) \|, \quad 1 \le j \le M,$$

is selected for the fractal code. Corresponding to this fractal code will be the multifunction attractor $\bar{U}(x) = [\bar{u}^-(x), \bar{u}^+(x)]$.

To illustrate, we consider the multifunction constructed from the *Lena* image defined as

$$U_{ij} = [u_{ij} - \delta_{ij}, u_{ij} + \delta_{ij}],$$

where

$$\delta_{ij} = \begin{cases} 0, & 1 \le i,j \le 255, \\ 40, & 256 \le i,j \le 512, \\ 20, & \text{otherwise.} \end{cases}$$

In other words, the error or uncertainty in the pixel values is zero for the upper left quarter of the image, 20 for the upper right and lower left quarters, and 40 for the lower right quarter. In Fig. 4.4, we show the lower and upper functions, $\bar{u}^-(x)$ and $\bar{u}^+(x)$, respectively, produced by a fractal coding of this multifunction.

Fig. 4.4: $\bar{u}^-(x)$ and $\bar{u}^+(x)$.

4.3 Iterated function systems on measure-valued images

We first set up our space of measure-valued images. More details and interesting applications of this measure-valued function method to nonlocal image processing, in particular nonlocal means denoising and fractal image coding, can be found in [107, 105, 108, 112]. In what follows, $\mathbb{X} = [0,1]^n$ will denote the "base space," the support of the images. $\mathbb{R}_g \subset \mathbb{R}$ will denote a compact "greyscale space" of values that our images can assume at any $x \in \mathbb{X}$. (The following discussion is easily extended to $\mathbb{R}_g \subset \mathbb{R}^m$ to accommodate colour images, etc.) And $\mathcal{B}$ will denote the Borel σ algebra on $\mathbb{R}_g$ with λ the Lebesgue measure. Let $\mathcal{P}$ denote the set of all Borel probability measures on $\mathbb{R}_g$ and d_{MK} the Monge-Kantorovich metric on this set. For a given $M > 0$, let $\mathcal{M}_1 \subset \mathcal{M}$ be a complete subspace of $\mathcal{M}$ such that $d_{MK}(\mu, \nu) \leq M$ for all $\mu, \nu \in \mathcal{M}_1$. We now define

$$Y = \{\mu(x) : \mathbb{X} \to \mathcal{M}_1, \mu(x) \text{ is measurable}\}$$

and consider on this space the metric

$$d_Y(\mu, \nu) = \int_{\mathbb{X}} d_{MK}(\mu(x), \nu(x)) d\lambda.$$

We observe that d_Y is well-defined, and since μ and ν are measurable functions, d_{MK} is bounded and so the function $\xi(x) = d_{MK}(\mu(x), \nu(x))$ is integrable on X.

Theorem 4.21. *The space (Y, d_Y) is complete.*

Proof. It is trivial to prove that this is a metric when we consider that $\mu = \nu$ if $\mu(x) = \nu(x)$ a.e. $x \in X$. To prove the completeness, we follow the trail of the proof of Theorem 1.2 in [86]. Let μ_n be a Cauchy sequence in Y. So, for all $\epsilon > 0$, there exists n_0 such that for all $n, m \geq n_0$ we have $d_Y(\mu_n, \mu_m) < \epsilon$. Let $\epsilon = 3^{-k}$, so you can choose an increasing sequence n_k such that $d_Y(\mu_n, \mu_{n_k}) < 3^{-k}$ for all $n \geq n_k$. So, choosing $n = n_{k+1}$, we have $d_Y(\mu_{n_{k+1}}, \mu_{n_k}) < 3^{-k}$. Let

$$A_k = \{x \in [0,1] : d_{MK}(\mu_{n_{k+1}}(x), \mu_{n_k}(x)) > 2^{-k}\}.$$

Then

$$\lambda(A_k) 2^{-k} \leq \int_{A_k} d_{MK}(\mu_{n_{k+1}}(x), \mu_{n_k}(x)) d\lambda \leq 3^{-k},$$

so that $\lambda(A_k) \leq \left(\frac{2}{3}\right)^k$. Let $A = \bigcap_{m=1}^{\infty} \bigcup_{k \geq m} A_k$. We observe that

$$\lambda \left(\bigcup_{k \geq m} A_k \right) \leq \sum_{k \geq m} \lambda(A_k)$$

$$\leq \sum_{k \geq m} \left(\frac{2}{3} \right)^k = \frac{\left(\frac{2}{3} \right)^m}{1 - \left(\frac{2}{3} \right)} \ . \tag{4.4}$$

Therefore

$$\lambda(A) \leq 3 \left(\frac{2}{3} \right)^m$$

for all m, which implies that $\lambda(A) = 0$. Now, for all $x \notin X\backslash A$, there exists $m_0(x)$ such that for all $m \geq m_0$ we have $x \notin A_m$, and so $d_{MK}(\mu_{n_{m+1}}(x), \mu_{n_m}(x)) < 2^{-m}$. This implies that $\mu_{n_m}(x)$ is Cauchy for all $x \notin X\backslash A$, and so $\mu_{n_m}(x) \to \mu(x)$ using the completeness of M_1. This also implies that $\mu : X \to Y$ is measurable; that is, $\mu \in Y$. To prove $\mu_n \to \mu$ in Y, we have that

$$\begin{aligned}
d_Y(\mu_{n_k}, \mu) &= \int_X d_{MK}(\mu_{n_k}(x), \mu(x)) d\lambda \\
&= \int_X \lim_{i \to +\infty} d_{MK}(\mu_{n_k}(x), \mu_{n_i}(x)) d\lambda \\
&\leq \liminf_{i \to +\infty} \int_X d_{MK}(\mu_{n_k}(x), \mu_{n_i}(x)) d\lambda \\
&= \liminf_{i \to +\infty} d_Y(\mu_{n_k}, \mu_{n_i}) \leq 3^{-k}
\end{aligned} \tag{4.5}$$

for all k. So $\lim_{k \to +\infty} d_Y(\mu_{n_k}, \mu) = 0$. Now we have

$$d_Y(\mu_n, \mu) \leq d_Y(\mu_n, \mu_{n_k}) + d_Y(\mu_{n_k}, \mu) \to 0$$

when $k \to +\infty$. $\qquad\square$

4.3.1 A fractal transform operator on measure-valued images

In this section, we construct and analyze a fractal transform operator M on the space (Y, d_Y) of measure-valued functions. We now list the ingredients for a fractal transform operator in the space Y. The reader will note that they form a kind of blending of IFS-based methods on measures (IFSP) and functions (IFSM). For simplicity, we assume that $X = [0, 1]$. The extension to $[0, 1]^n$ is straightforward. We have:

1. a set of N one-to-one contraction affine maps $w_i : \mathbb{X} \to \mathbb{X}$, $w_i(x) = s_i x + a_i$, with the condition that $\cup_{i=1}^{N} w_i(X) = X$,
2. a set of N greyscale maps $\phi_i : [0, 1] \to [0, 1]$, assumed to be Lipschitz (i.e., for each i, there exists a $\alpha_i \geq 0$ such that $|\phi_i(t_1) - \phi_i(t_2)| \leq \alpha_i |t_1 - t_2|$, $\forall t_1, t_2 \in [0, 1]$),
3. and for each $x \in \mathbb{X}$, a set of probabilities $p_i(x)$, $i = 1, \cdots, N$ with the following properties:

   a. $p_i(x)$ are measurable,
   b. $p_i(x) = 0$ if $x \notin w_i(\mathbb{X})$, and
   c. $\sum_{i}^{N} p_i(x) = 1$ for all $x \in \mathbb{X}$.

The action of the fractal transform operator $M : Y \to Y$ defined by the above is as follows. For a $v \in Y$ and any subset $S \subset [0, 1]$,

$$\nu(x)(S) = (M\mu(x))(S) = \sum_{i=1}^{N} p_i(x)\mu(w_i^{-1}(x))(\phi_i^{-1}(S)). \qquad (4.6)$$

Theorem 4.22. Let $p_i = \sup_{x \in X} p_i(x)$. Then, for $\mu_1, \mu_2 \in Y$,

$$d_Y(M\mu_1, M\mu_2) \leq \left(\sum_{i=1}^{n} |s_i| \alpha_i p_i\right) d_Y(\mu_1, \mu_2).$$

Proof. Computing, we have

$$d_Y(M\mu_1, M\mu_2) = \int_{\mathbb{X}} d_{MK}(M\mu_1(x), M\mu_2(x)) d\lambda$$

$$= \int_{\mathbb{X}} d_{MK}\left(\sum_{i=1}^{N} p_i(x)\mu_1(w_i^{-1}(x)) \circ \phi_i^{-1}, \sum_{i=1}^{N} p_i(x)\mu_2(w_i^{-1}(x)) \circ \phi_i^{-1}\right) d\lambda$$

$$\leq \int_{\mathbb{X}} \sum_{i=1}^{n} p_i(x) d_{MK}(\mu_1(w_i^{-1}(x)) \circ \phi_i^{-1}, \mu_2(w_i^{-1}(x)) \circ \phi_i^{-1}) d\lambda$$

$$\leq \int_{\mathbb{X}} \sum_{i=1}^{n} \alpha_i p_i(x) d_{MK}(\mu_1(w_i^{-1}(x)), \mu_2(w_i^{-1}(x))) d\lambda$$

$$\leq \int_{\mathbb{X}} \left(\sum_{i=1}^{n} |s_i| \alpha_i p_i\right) d_{MK}(\mu_1(x), \mu_2(x)) d\lambda$$

$$= \left(\sum_{i=1}^{n} |s_i| \alpha_i p_i\right) d_Y(\mu_1, \mu_2).$$

$\square$

Corollary 4.23. *Let $p_i = \sup_{x \in X} p_i(x)$. Then M is a contraction on (Y, d_Y) if*

$$\sum_{i=1}^{n} |s_i| \alpha_i p_i < 1.$$

Consequently there exists a measure-valued mapping $\bar{\mu} \in Y$ such that $\bar{\mu} = M\bar{\mu}$.

Example 4.24. The fractal transform M is defined by the following two-IFS-map system on $\mathbb{X} = [0, 1]$:

$$w_1(x) = \frac{1}{2}x, \quad \phi_1(t) = \frac{1}{2}t,$$
$$w_2(x) = \frac{1}{2}x + \frac{1}{2}, \quad \phi_2(t) = \frac{1}{2}t + \frac{1}{2}.$$

The sets $w_1(\mathbb{X})$ and $w_2(\mathbb{X})$ overlap at the single point $x = \frac{1}{2}$, so we let

$$p_1(x) = 1, \quad p_2(x) = 0 \quad x \in \left[0, \frac{1}{2}\right),$$

$$p_1(x) = 0, \quad p_2(x) = 1 \quad x \in \left(\frac{1}{2}, 1\right],$$

$$p_1\left(\frac{1}{2}\right) = p_2\left(\frac{1}{2}\right) = \frac{1}{2}.$$

It is easy to confirm that M is contractive. Its fixed point $\bar{\mu}$ is given by

$$\bar{\mu}(x) = \delta(t - x), \quad x \in [0, 1],$$

where $\delta(s)$ denotes the "Dirac delta function" at $s \in [0, 1]$.

Example 4.25. A "perturbation" of the fractal transform M in Example 4.24 is produced by adding the following IFS and associated greyscale maps:

$$w_3(x) = \frac{1}{2}x, \quad \phi_3(t) = \frac{1}{2}t + 0.1.$$

The sets $w_1(\mathbb{X})$ and $w_3(\mathbb{X})$ overlap over the entire subinterval $[0, \frac{1}{2}]$, so we let

$$p_1(x) = p_3(x) = \frac{1}{2}, \quad p_2(x) = 0 \quad x \in \left[0, \frac{1}{2}\right),$$

$$p_1(x) = p_3(x) = 0, \quad p_2(x) = 1 \quad x \in \left(\frac{1}{2}, 1\right],$$

$$p_1\left(\frac{1}{2}\right) = p_2\left(\frac{1}{2}\right) = p_3\left(\frac{1}{2}\right) = \frac{1}{3}.$$

Once again, it is easy to confirm that M is contractive. Its fixed point $\bar{\mu}(x)$ is sketched in Fig. 4.5. The darkness of a point is proportional to the measure $\bar{\mu}(S_1, S_2)$ of the region in $[0, 1]^2$ represented by the point. Note that the overlapping of the w_1 and w_3 maps over $[0, \frac{1}{2}]$ is responsible for the self-similar "splitting" of the measures $\bar{\mu}(x)$ over this interval since ϕ_3 produces an upward shift in the greyscale direction. Since $w_2(x)$ maps the support $[0, 1]$ of the entire measure-valued function onto $[\frac{1}{2}, 1]$, the self-similarity of the measure over $[0, \frac{1}{2}]$ is carried over to $[\frac{1}{2}, 1]$.

Fig. 4.5: A sketch of the invariant measure $\bar{\mu}(x)$ for the three-IFS map fractal transform in Example 4.25, $x \in X = [0, 1]$, $y \in \mathbb{R}_g = [0, 1]$.

4.3.2 Moment relations induced by the fractal transform operator

In this section we show that the moments of measures in the space (Y, d_Y) also satisfy recursion relations when the greyscale maps ϕ_i are affine. We now consider the local or x-dependent moments of a measure $\mu(x) \in Y$, defined as

$$g_n(x) = \int_{\mathbb{R}_g} s^n d\mu_x(s), \quad m = 0, 1, 2, \cdots,$$

where we use the notation $\mu_x = \mu(x)$ in the Lebesgue integral for simplicity. By definition, $g_0(x) = 1$ for $x \in \mathbb{X}$. Obviously the functions g_m are measurable on $\mathbb{X}$ (since $\mu(x)$ are measurable) and bounded so that $g_m \in \mathcal{L}^1(\mathbb{X}, \mathcal{L})$. We now derive the relations between the moments of a measure $\mu \in Y$ and the moments of $\nu = M\mu$, where M is the fractal transform operator defined in (4.6).

Let h_n denote the moments of $\nu = M\mu$. Computing, we have

$$h_n(x) = \int_{\mathbb{R}_g} s^n d(M\mu)_x(s)$$

$$= \int_{\mathbb{R}_g} s^n d \left(\sum_{i=1}^{N} p_i(x)\mu_{w_i^{-1}(x)} \circ \phi_i^{-1} \right) (s)$$

$$= \int_{\mathbb{R}_g} \sum_{i=1}^{N} p_i(x) s^n d(\mu_{w_i^{-1}(x)} \circ \phi_i^{-1})(s)$$

$$= \int_{\mathbb{R}_g} \sum_{i=1}^{N} p_i(x)[\phi_i(s)]^n d(\mu_{w_i^{-1}(x)})(s).$$

For affine greyscale maps of the form $\phi(s) = \alpha_i s + \beta_i$, we have

$$h_n(x) = \int_{\mathbb{R}_g} \sum_{i=1}^{N} p_i(x)(\alpha_i + s\beta_i)^n d\left(\mu_{w_i^{-1}(x)}\right)(s)$$

$$= \sum_{i=1}^{N} \sum_{j=0}^{n} p_i(x) \int_{\mathbb{R}_g} c_{nj}(\alpha_i s)^j \beta_i^{n-j} d\left(\mu_{w_i^{-1}(x)}\right)(s)$$

$$= \sum_{i=1}^{N} \sum_{j=0}^{n} p_i(x) c_{nj}\alpha_i^j \beta_i^{n-j} \int_{\mathbb{R}_g} s^j d\left(\mu_{w_i^{-1}(x)}\right)(s)$$

$$= \sum_{j=0}^{n} \left[\sum_{i=1}^{N} p_i(x) c_{nj}\alpha_i^j \beta_i^{m-j} \right] g_j(w_i^{-1}(x)), \tag{4.7}$$

where

$$c_{nj} = \binom{n}{j}.$$

The reader may compare the result above to that of (2.26) for the IFSP case. The place-dependent moments $h_n(x)$ are related to the moments g_n evaluated at the preimages $w_i^{-1}(x)$. And it is the greyscale $\phi_i(s)$ maps that now "mix" the measures, as opposed to the spatial IFS maps $w_i(x)$ in (2.26).

In the special case where $\mu = \bar{\mu} = M\bar{\mu}$, the fixed point of M, $h_n(x) = g_n(x)$ and we have

$$g_n(x) = \sum_{j=0}^{n} \left[\sum_{i=1}^{N} p_i(x) c_{nj} \alpha_i^j \beta_i^{n-j} \right] g_j(w_i^{-1}(x)).$$

In other words, the moments $g_n(x)$ satisfy recursion relations that involve moments of all orders up to n evaluated at preimages $w_i^{-1}(x)$. Note that this does not yield a rearrangement analogous to (2.26) that will permit a simple recursive computation of the moments $g_n(x)$. Nevertheless, the *moment functions* can be computed recursively, as we now show.

First, note that for the particular case $n = 1$, the moment function $g_1(x)$ is a solution of the fixed-point equation

$$g_1(x) = \sum_{i=1}^{N} p_i(x)[\alpha_i g_1(w_i^{-1}(x)) + \beta_i]. \tag{4.8}$$

(Note that $g_1(x)$ is the expectation value of $\mu(x)$.) But this implies that g_1 is the unique fixed point in $\mathcal{L}^1(\mathbb{X})$ of the contractive place-dependent IFSM operator defined by

$$(Mu)(x) = \sum_{i=1}^{N} p_i(x)[\alpha_i u(w_i^{-1}(x)) + \beta_i].$$

This provides a method for computing g_1 – at least approximately – by means of Banach's theorem. Starting at any $u_0 \in \mathcal{L}^1$, the sequence $M^n u = M(M^{n-1})u_0$ converges to g_1 as $n \to +\infty$.

Higher-order moments can be computed in a similar recursive manner. To illustrate, consider the case $n = 2$. From (4.8), the moment $g_2(x)$ satisfies the fixed-point equation

$$g_2(x) = \sum_{i=1}^{N} p_i(x)[\alpha_i^2 g_2(w_i^{-1}(x)) + 2\alpha_i \beta_i g_1(x) + \beta_i^2]. \tag{4.9}$$

In other words, g_2 is the fixed point of a contractive IFSM operator (see Sect. (3.2)) with condensation function

$$b(x) = \sum_{i=1}^{N} p_i(x)[2\alpha_i\beta_i g_1(x) + \beta_i^2].$$

From a knowledge of g_1, the moment function g_2 may be computed via iteration. The process may now be iterated to produce g_3, etc.

Finally, note that g_1 and g_2 determine the pointwise variance $\sigma^2(x)$ of the measure $\bar{\mu}(x)$:

$$\sigma^2(x) = \int_{\mathbb{R}_g} (s - g_1(s))^2 d\bar{\mu}_x(s)$$
$$= g_2(x) - [g_1(x)]^2. \tag{4.10}$$

Chapter 5
IFS on Spaces of Measures

In this chapter, we expand on the construction of an IFS with probability (as seen in Sect. 2.5 in Chapter 2). We do this in many different directions, but all with the same overall scheme in mind.

As previously mentioned, there are several motivations for moving from an IFS on strictly geometric objects (such as compact sets) to measures. One of these comes from the area of geometric measure theory, in which geometric objects are generalized to measures; it is mathematically richer to deal with the sum of two measures than the (geometric) sum of two sets. Another motivation is more applied and comes from thinking of using an IFS to represent images; a set can only represent a binary image (black/white), whereas a measure can represent a greyscale image.

We first discuss the case of signed measures. The main technical difficulty is finding the right constraints on a set of signed measures in order to obtain a complete metric space. The usual (positive) measures behave in a simpler manner: if the total measure is bounded, the measure is bounded. This is not the case for signed measures. Another complication comes when we wish to consider measures with possibly unbounded support.

The next level of generality is the case of vector-valued measures. For measures with values in $\mathbb{R}^n$, this is a rather simple extension of the signed measure case. However, it allows us to deal with some very interesting applications. The one we focus on is in defining both a "tangent measure" and a "normal measure" to a self-affine fractal curve. This allows us to define line integrals on such curves and hence do explicit computations in vector calculus on these curves. This framework also allows the construction of a host of differential geometric objects. Unfortunately, many of the natural examples of "tangent measures" for fractal curves result in measures of unbounded variation, so it is necessary to modify our basic framework in order to accommodate these measures.

Finally, the most general case we deal with is that of measures whose values are compact subsets of some metric space. In this case, we have two different types of such measures. The first type consists of those measures that are additive with respect to the Minkowski addition of sets $A+B = \{a+b : a \in A, b \in B\}$. This type of set-valued measure (or multimeasure) has many properties in common with the usual types of measures. We give a brief review of the definitions and basic properties of these multimeasures. For the second type, we use the union operator as the addition. These union-additive multimeasures are a little bit unusual, but we include them since they give a natural framework in which to discuss very general kinds of "fractal partitions." The theory is also much simpler than that of the Minkowski-additive multimeasures.

For each new type of measure, we define a metric (usually a very natural one) and prove completeness for some space of measures under this metric. We then define IFS operators, give conditions for contractivity, and give some examples.

The technical details are much simpler in the case where the space $\mathbb{X}$ is compact, so in each instance we deal with this case first. Our primary motivation for this is that the compact case contains all the important ideas, and the extra details just tend to obscure them. Furthermore, for the purposes of IFS fractal constructions, the compact case is often sufficient since if $\{w_i\}$ is a finite set of contractions, the attractor of this IFS is always a compact set (see Sect. 2.2.3); we can often restrict our attention to this compact set. However, for completeness, we deal with the general case in a separate section.

5.1 Signed measures

It is simple to generalize the IFS Markov operator (2.14) from Sect. 2.5 to signed measures. In fact, as (2.14) is linear, it is the natural operator to use in any of the contexts we consider here. We will essentially use this same operator for all the types of measures, modified slightly to fit within the given framework. The ideas in this section appear implicitly in [58, 20] and independently also in [135].

We assume that all signed measures have finite positive and negative parts, and thus finite variation. This means that if we construct a signed measure, we need to show that it has finite variation (in order to fit within our standing assumption). We fix a metric space $\mathbb{X}$ and use the Borel σ-algebra on $\mathbb{X}$; that is, we will assume that all our measures are Borel measures. Often we will make some additional assumptions on $\mathbb{X}$, such as completeness or compactness.

Definition 5.1. Let $\mathbb{X}$ be a metric space. We denote the space of all finite signed Borel measures on $\mathbb{X}$ by $\mathcal{M}(\mathbb{X}, \mathbb{R})$.

5.1.1 Complete space of signed measures

The first issue is to find a metric and a space of signed measures that is complete with respect to the chosen metric. The metric we use is the same as for probability measures, the Monge-Kantorovich metric; the definition is exactly the same as (2.15). Recall that $\text{Lip}(\mathbb{X})$ is the collection of real-valued Lipschitz functions on $\mathbb{X}$ and $\text{Lip}_1(\mathbb{X})$ is the subset of $\text{Lip}(\mathbb{X})$ whose Lipschitz constant is bounded by one (see Appendix A).

Definition 5.2. For two signed measures μ, ν, we define the quantity

$$d_{MK}(\mu, \nu) = \sup \left\{ \int_{\mathbb{X}} f \, d(\mu - \nu) : f \in \text{Lip}_1(\mathbb{X}) \right\}. \qquad (5.1)$$

Note that $d_{MK}(\mu, \nu) \in [0, \infty]$, but in general it is possible for $d_{MK}(\mu, \nu) = \infty$. In fact, if $\mu(\mathbb{X}) \neq \nu(\mathbb{X})$, then $d_{MK}(\mu, \nu) = \infty$ since if $f(x) = c > 0$ we have

$$\int_{\mathbb{X}} c \, d(\mu - \nu) = c(\mu(\mathbb{X}) - \nu(\mathbb{X}))$$

which can be made as large as we like. Thus, one necessary condition for d_{MK} to be a metric is that all the measures under consideration have the same total mass. This naturally happens in the case of probability measures.

There are two other possible problems. The first arises when $\mathbb{X}$ is unbounded because then $d_{MK}(\mu, \nu)$ can be unbounded even for two probability measures. As an example, on the real line if we take the point mass at 0 for μ and let ν have density proportional to $1/(1 + t^2)$ on $[0, \infty)$. Then

$$\int_{\mathbb{R}} x \, d\nu(x) = \infty \quad \Rightarrow \quad d_{MK}(\mu, \nu) = \infty.$$

The second problem is a little bit more subtle. Basically, it is possible for a sequence of finite signed measures to "converge" in the d_{MK} distance to a signed measure of unbounded variation. As an example, let $x_n = 2^{-n}$ and consider the sequence of finite signed measures

$$\mu_n = \sum_{k=1}^{n} k(\delta_{x_k} - \delta_{-x_k}),$$

for which $\mu_n(\mathbb{R}) = 0$ and $\|\mu_n\| = 2n$. It is simple to show that (μ_n) is d_{MK}-Cauchy, but the "limit" is not a finite signed measure.

For the rest of this section, we assume that $\mathbb{X}$ *is compact.* This implies that

$$\int_{\mathbb{X}} d(x, a) \, d|\mu|(x) < \infty$$

for any $a \in \mathbb{X}$.

Definition 5.3. Suppose $q \in \mathbb{R}$ and $k > 0$. Let

$$\mathcal{M}_{q,k}(\mathbb{X}, \mathbb{R}) = \{ \mu \in \mathcal{M}(\mathbb{X}, \mathbb{R}) : \mu(\mathbb{X}) = q, \ \|\mu\| \leq k \}.$$

The following theorem is our first completeness result. Restricting $\mathbb{X}$ to be compact is not much of a restriction for our purposes, as any finite IFS with strictly contractive maps will yield a compact attractor, and thus the attractor of any IFS on measures will also be compactly supported (see Theorem 2.63 and Proposition 5.16). Thus, if necessary, we can always restrict any discussion to the attractor of the IFS and hence obtain a compact space.

Theorem 5.4. *Let* $\mathbb{X}$ *be a compact metric space and* $k, q \in \mathbb{R}$ *with* $k > 0$. *Then* $\mathcal{M}_{q,k}(\mathbb{X}, \mathbb{R})$ *is complete under the metric* d_{MK}.

Proof. Let $a \in \mathbb{X}$. Then, for any $f \in \mathrm{Lip}_1(\mathbb{X})$ and $\mu, \nu \in \mathcal{M}_{q,k}(\mathbb{X}, \mathbb{R})$, we have

$$\int f \, d(\mu - \nu) = \int (f - f(a)) \, d(\mu - \nu) + \int f(a) \, d(\mu - \nu)$$
$$= \int (f - f(a)) \, d(\mu - \nu)$$
$$\leq \int |f - f(a)| \, d(|\mu| + |\nu|)$$
$$\leq 2k \ \mathrm{diam}(\mathbb{X}),$$

and so $d_{MK}(\mu, \nu) \leq 2k \ \mathrm{diam}(\mathbb{X})$ and is thus finite. The other properties of a metric follow in the same way as for probability measures, and we leave out the details. The only thing really left to show is completeness.

Suppose that $(\mu_n)_{n\geq 1}$ is a d_{MK}-Cauchy sequence in $\mathcal{M}_{q,k}(\mathbb{X}, \mathbb{R})$. Since subsets of $\mathcal{M}(\mathbb{X}, \mathbb{R})$ which are bounded in the variation norm have weak* compact closures (see Proposition B.27), there is a subsequence $(\mu_{n_\ell})_{\ell \geq 1}$ and a $\mu \in \mathcal{M}(\mathbb{X}, \mathbb{R})$ with $\|\mu\| \leq k$ such that

$$\int f \, d\mu_{n_\ell} \to \int f \, d\mu \text{ for all } f \in C(\mathbb{X}).$$

Taking $f = 1$, it follows that $\mu(\mathbb{X}) = q$ and thus $\mu \in \mathcal{M}_{q,k}(\mathbb{X}, \mathbb{R})$.

Now suppose $\epsilon > 0$. Since (μ_n) is d_{MK}-Cauchy, for all $n, n_\ell \geq N(\epsilon)$ we have for each $f \in \mathrm{Lip}_1(\mathbb{X})$ that

$$\left| \int_{\mathbb{X}} f \, d\mu_{n_\ell} - \int_{\mathbb{X}} f \, d\mu_n \right| \leq \epsilon.$$

Taking the limit as $\ell \to \infty$, we get

$$\left| \int_{\mathbb{X}} f \, d\mu - \int_{\mathbb{X}} f \, d\mu_n \right| \leq \epsilon,$$

independent of $f \in \mathrm{Lip}_1(\mathbb{X})$. Thus $d_M(\mu, \mu_n) \to 0$. $\qquad\square$

5.1.2 IFS operator on signed measures

As previously mentioned, the definition of the IFS operator on signed measures is completely parallel to that of an IFS on probability measures.

Definition 5.5. Let the N maps $w_i : \mathbb{X} \to \mathbb{X}$ be Lipschitz with constant s_i and $p_i \in \mathbb{R}$ for $i = 1, 2, \ldots, N$ satisfy $\sum_i p_i = 1$. We define the operator $\mathbb{M}$ on $\mathcal{M}(\mathbb{X}, \mathbb{R})$ by

$$\mathbb{M}\mu(B) = \sum_{i=1}^{N} p_i \mu(w_i^{-1}(B))$$

for each $B \in \mathcal{B}$.

Notice that since $\sum_i p_i = 1$, $\mu(\mathbb{X}) = q$ implies that $\mathbb{M}\mu(\mathbb{X}) = q$ as well.

Proposition 5.6. *The operator* $\mathbb{M}$ *satisfies*

$$d_{MK}(\mathbb{M}\mu, \mathbb{M}\nu) \leq \left(\sum_i s_i |p_i| \right) d_{MK}(\mu, \nu).$$

Proof. For a given Lipschitz f, we have

$$\int_{\mathbb{X}} f(x)\, d(\mathbb{M}\mu(x) - \mathbb{M}\nu(x))$$

$$= \sum_i p_i \int_{\mathbb{X}} f(x)\, d(\mu \circ w_i^{-1}(x) - \nu \circ w_i^{-1}(x))$$

$$= \int_{\mathbb{X}} \left(\sum_i p_i f(w_i(x)) \right) d(\mu - \nu)(x).$$

Now, we see that the function $\tilde{f} = \sum_i p_i f \circ w_i$ has Lipschitz factor at most $\sum_i s_i |p_i|$. Thus, when we take the supremum over all Lipschitz functions, we get that

$$d_{MK}(\mathbb{M}\mu, \mathbb{M}\nu) \leq \left(\sum_i s_i |p_i| \right) d_{MK}(\mu, \nu),$$

as was desired.   □

In particular, if $\mathbb{M}$ maps $\mathcal{M}_{q,k}(\mathbb{X}, \mathbb{R})$ to itself and $\sum_i s_i |p_i| < 1$, then $\mathbb{M}$ is contractive and thus has a unique attractive fixed point in $\mathcal{M}_{q,k}(\mathbb{X}, \mathbb{R})$. Since $\mu(\mathbb{X}) = q$ implies $\mathbb{M}\mu(\mathbb{X}) = 1$, the only things to find are conditions under which $\|\mu\| \leq k$ implies that $\|\mathbb{M}\mu\| \leq k$. Unfortunately, these conditions are rather stringent.

The conditions on $\mathbb{X}$ and the IFS in the following proposition are a type of nontriviality condition. Typically these conditions are easily met.

Proposition 5.7. *Suppose that* $\mathbb{X}$ *and the IFS* $\{w_i\}$ *satisfy the property that there are* $x_1, x_2 \in \mathbb{X}$ *such that* $w_i(x_k) \neq w_j(x_l)$ *for* $i \neq j$ *or* $k \neq l$. *Then, if* $\mathbb{M}\mu \in \mathcal{M}_{q,k}(\mathbb{X}, \mathbb{R})$ *for all* $\mu \in \mathcal{M}_{q,k}(\mathbb{X}, \mathbb{R})$, *we must have* $\sum_i |p_i| \leq 1$ *and thus* $p_i \geq 0$ *for each* i.

Proof. Take $\mu = (q - k)/2\delta_{x_1} + (q + k)/2\delta_{x_2}$. Then $\|\mu\| = k$ and $\mu(\mathbb{X}) = q$. Furthermore,

$$\mathbb{M}\mu = (q - k)/2 \sum_i p_i \delta_{w_i(x_1)} + (q + k)/2 \sum_i p_i \delta_{w_i(x_2)},$$

and so $\|\mathbb{M}\mu\| = k \sum_i |p_i|$. So, if $\mathbb{M}\mu \in \mathcal{M}_{q,k}(\mathbb{X}, \mathbb{R})$, we must have $\sum_i |p_i| \leq 1$. However, $\sum_i p_i = 1$ and $\sum_i |p_i| \leq 1$ imply that $p_i \geq 0$ for each i.   □

The condition $p_i \geq 0$ implies that any fixed point of an $\mathbb{M} :$ $\mathcal{M}_{q,k}(\mathbb{X}, \mathbb{R}) \to \mathcal{M}_{q,k}(\mathbb{X}, \mathbb{R})$ is simply a multiple of a fixed point of an IFS with probabilities (that is, an IFS on probability measures). So this framework is too restrictive to give any other examples of fractal measures.

Another way to describe the situation is that if $\sum_i |p_i| > 1$, then there will be some measures in $\mathcal{M}_{q,k}(\mathbb{X}, \mathbb{R})$ for which the variation is increased and thus the "fixed point" of $\mathbb{M}$ will be a signed "measure" of infinite variation, so a theory of integration against such a "measure," if it exists, is more complicated than for a finite signed measure. In fact, in general you cannot expect to integrate even a continuous function against such a "measure."

One way around this problem is to modify the IFS operator by including an additive term. This type of IFS is often called an *IFS with condensation*.

Definition 5.8. Let the N maps $w_i : \mathbb{X} \to \mathbb{X}$ be Lipschitz with constant s_i, and $p_i \in \mathbb{R}$ for $i = 0, 1, \ldots, N$ satisfy $\sum_i p_i = 1$. Finally, choose some $\nu \in \mathcal{M}(\mathbb{X}, \mathbb{R})$ with $\nu(\mathbb{X}) = q$ for some $q \in \mathbb{R}$. We define the operator $\mathbb{M}$ on $\mathcal{M}(\mathbb{X}, \mathbb{R})$ by

$$\mathbb{M}\mu(B) = p_0\nu(B) + \sum_{i=1}^{N} p_i\mu(w_i^{-1}(B)) \tag{5.2}$$

for each $B \in \mathcal{B}$.

If $\mu(\mathbb{X}) = q$, then

$$\mathbb{M}\mu(\mathbb{X}) = p_0\nu(\mathbb{X}) + \sum_{i=1}^{N} p_i\mu(\mathbb{X}) = q(p_0 + p_1 + \cdots + p_N) = q,$$

and thus $\mathbb{M}$ preserves total mass. Furthermore,

$$\|\mathbb{M}\mu\| \leq |p_0|\|\nu\| + \sum_{i=1}^{N} |p_i|\|\mu\|,$$

so one way to ensure $\|\mathbb{M}\mu\| \leq k$ is to enforce the condition

$$|p_0|\|\nu\| \leq \left(1 - \sum_{i=1}^{N} |p_i|\right) k, \tag{5.3}$$

which implies that we must have $\sum_i |p_i| < 1$.

Proposition 5.9. *If $\sum_{i=1}^{N}|p_i| < 1$, then for a fixed ν and q there exists a large enough k that $\mathbb{M}$ from Definition 5.8 maps $\mathcal{M}_{q,k}(\mathbb{X},\mathbb{R})$ to itself.*

Proof. From (5.3), we have that $1 - \sum_{i=1}^{N}|p_i| < 1$, and thus if we make k large enough the right-hand side will exceed the left-hand side. $\square$

As the map $\mu \mapsto \sum_i p_i \mu \circ w_i^{-1}$ is linear in μ, we can write the IFS operator in (5.2) as

$$\mathbb{M}\mu = \mathbb{A}\mu + p_0\nu,$$

and thus the fixed point of $\mathbb{M}$, if it exists, is the measure

$$\mu = p_0\nu + p_0\mathbb{A}\nu + p_0\mathbb{A}^2\nu + \cdots + p_0\mathbb{A}^n\nu + \cdots = p_0(I - \mathbb{A})^{-1}\nu.$$

In particular, this means that the fixed point of the IFS with condensation $\mathbb{M}$ is a linear function of the translational part ν. This simple observation is sometimes useful.

5.1.3 "Generalized measures" as dual objects in $\mathbf{Lip}(\mathbb{X}, \mathbb{R})^*$

We now turn to a framework that deals with this problem of $\mathbb{M}\mu$ having larger variation than μ. The key lies in the observation that if $\sum_i s_i|p_i| < 1$, then $\mathbb{M}$ is contractive in some sense, even if $\sum_i |p_i| > 1$. The paper [58] uses this framework to construct fractal vector measures, and we borrow heavily from this paper. These ideas are also implicit in [135].

Following the usual construction, we put the norm

$$\|f\|_{\text{Lip}} = |f(a)| + \text{Lip}(f) \tag{5.4}$$

on $\text{Lip}(\mathbb{X})$. Here $a \in \mathbb{X}$, and

$$\text{Lip}(f) = \sup_{x \neq y} \frac{|f(x) - f(y)|}{d(x,y)}$$

is the Lipschitz constant of f. The inequalities

$$|f_n(x) - f_m(x)| \leq |(f_n(x) - f_m(x)) - (f_n(a) - f_m(a))|$$
$$+ |f_n(a) - f_m(a)|$$
$$\leq d(x,a)\text{Lip}(f_n - f_m) + |f_n(a) - f_m(a)|$$

can be used to easily show that $\mathrm{Lip}(\mathbb{X})$ is complete under the norm $\|\cdot\|_{\mathrm{Lip}}$ (use the inequality to show that a $\|\cdot\|_{\mathrm{Lip}}$-Cauchy sequence is uniformly convergent on bounded sets).

Now clearly every $\mu \in \mathcal{M}(\mathbb{X},\mathbb{R})$ induces a bounded linear functional on $\mathrm{Lip}(\mathbb{X})$ by

$$\hat{\mu}(f) = \int_{\mathbb{X}} f \, d\mu.$$

Furthermore, since $\|f\|_{\mathrm{Lip}} = \|f - f(a) + f(a)\|_{\mathrm{Lip}}$, for any $\mu \in \mathcal{M}(\mathbb{X},\mathbb{R})$ with $\mu(\mathbb{X}) = 0$ we have

$$\sup\left\{\int_{\mathbb{X}} f \, d\mu : f \in \mathrm{Lip}(\mathbb{X}), \|f\|_{\mathrm{Lip}} \leq 1\right\} = \sup\left\{\int_{\mathbb{X}} f \, d\mu : f \in \mathrm{Lip}_1(\mathbb{X})\right\}. \tag{5.5}$$

Now the left-hand side of (5.5) is the definition of the induced norm on $\mathrm{Lip}(\mathbb{X})^*$. Denoting this norm by $\|\cdot\|_{\mathrm{Lip}^*}$, we have that if $\mu(\mathbb{X}) = \nu(\mathbb{X})$, then

$$d_{MK}(\mu,\nu) = \sup\left\{\int_{\mathbb{X}} f \, d(\mu - \nu) : \|f\|_{\mathrm{Lip}} \leq 1\right\} = \|\mu - \nu\|_{\mathrm{Lip}^*}, \tag{5.6}$$

and so d_{MK}-Cauchy is equivalent to $\|\cdot\|_{\mathrm{Lip}^*}$-Cauchy for such measures.

Definition 5.10. Let $q \in \mathbb{R}$ be fixed, and define

$$\mathcal{S}_q(\mathbb{X}) = \{\varphi \in \mathrm{Lip}(\mathbb{X})^* : \varphi(1) = q\}.$$

Being a dual space, $\mathrm{Lip}(\mathbb{X})^*$ is complete under its norm, and since the set $\mathcal{S}_q(\mathbb{X})$ is a closed subset of $\mathrm{Lip}(\mathbb{X})^*$, we know that $\mathcal{S}_q(\mathbb{X})$ is also complete in this norm.

Now, as $\mathbb{M}$ in Definition 5.5 is formulated, it does not directly act on elements of $\mathrm{Lip}(\mathbb{X})^*$. The simplest way to describe an action of $\mathbb{M}$ on $\mathrm{Lip}(\mathbb{X})^*$ is through its dual action on elements of $\mathrm{Lip}(\mathbb{X})$. If $\mu \in \mathcal{M}(\mathbb{X},\mathbb{R})$ and $f \in \mathrm{Lip}(\mathbb{X})$, then we have

$$\int_{\mathbb{X}} f \, d\mathbb{M}\mu = \int_{\mathbb{X}} f \sum_i p_i d\mu \circ w_i^{-1} = \int_{\mathbb{X}} \sum_i p_i f \circ w_i \, d\mu = \int_{\mathbb{X}} \mathbb{T} f \, d\mu,$$

which defines the adjoint $\mathbb{T}$ to $\mathbb{M}$. We use this same idea to define the action of $\mathbb{M}$ on $\mathrm{Lip}(\mathbb{X})^*$.

Definition 5.11. Let the N maps $w_i : \mathbb{X} \to \mathbb{X}$ be Lipschitz with constant s_i and $p_i \in \mathbb{R}$ satisfy $\sum_i p_i = 1$. Furthermore, let $q \in \mathbb{R}$ be fixed. Define the IFS operator $\mathbb{M}$ on $\mathcal{S}_q(\mathbb{X})$ by setting for each $\varphi \in \mathcal{S}_q(\mathbb{X})$ and $f \in \mathrm{Lip}(\mathbb{X})$

$$\mathbb{M}\varphi(f) = \sum_i p_i \varphi(f \circ w_i). \tag{5.7}$$

Proposition 5.12. *The operator* $\mathbb{M}$ *defined as in (5.7) satisfies*

$$\|\mathbb{M}\varphi - \mathbb{M}\psi\|_{Lip^*} \leq \left(\sum_i s_i |p_i|\right) \|\varphi - \psi\|_{Lip^*}.$$

Proof. The proof is basically the same as the proof of contractivity for a usual IFS with probabilities (see Theorem 2.60). However, first we have to observe that if $\varphi, \psi \in \mathcal{S}_q(\mathbb{X})$, then $\varphi(c) - \psi(c) = 0$ and thus

$$\|\varphi - \psi\|_{\text{Lip}^*} = \sup\{\varphi(f) - \psi(f) : \|f\|_{\text{Lip}} \leq 1\}$$
$$= \sup\{\varphi(f) - \psi(f) : f \in \text{Lip}_1(\mathbb{X})\}.$$

Next we observe that if $f \in \text{Lip}_1(\mathbb{X})$, then the Lipschitz constant for $\sum_i p_i f \circ w_i$ is at most $\sum_i s_i |p_i|$. The rest of the details are a simple modification of those in Theorem 2.60. $\square$

Corollary 5.13. *If the operator* $\mathbb{M}$ *defined in (5.7) satisfies* $\sum_i s_i |p_i| < 1$, *then for each* $q \in \mathbb{R}$ *there is a unique "generalized signed measure"* $\varphi \in \mathcal{S}_q(\mathbb{X})$ *for which*

$$\|\mathbb{M}^n \mu - \varphi\|_{Lip^*} \to 0$$

for any $\mu \in \mathcal{M}(\mathbb{X})$ *with* $\mu(\mathbb{X}) = q$.

Basically, convergence occurs because if $\sum_i s_i |p_i| = C < 1$, then $\text{Lip}(\sum_i p_i f \circ w_i) \leq C \, \text{Lip}(f)$, and so, if we define $\mathbb{T}$ on $\text{Lip}(\mathbb{X})$ by $\mathbb{T}f = \sum_i p_i f \circ w_i$, then $\mathbb{T}^n f$ converges to a constant function. This view is "dual" to thinking about the action of $\mathbb{M}$ on $\text{Lip}(\mathbb{X})^*$.

Notice that $\varphi \in \text{Lip}(\mathbb{X})^*$ is not really a signed measure. By definition, we can "integrate" (i.e., evaluate) φ against any Lipschitz f, but there is no reason to expect that we can "integrate" φ against a general continuous function, much less a measurable one. One intuitive reason for this is that the "measure" represented by φ might oscillate so much that "integration" with respect to a smooth (Lipschitz) function will cancel out these oscillations but an arbitrary continuous function could reinforce them, with the consequence that no consistent result is possible.

Example 5.14. As the first example of an IFS on signed measures, take $\mathbb{X} = [0,1]$, $w_i(x) = x/3 + i/3$ for $i = 0,1$ and $p_0 \in (1,2)$ and $p_1 = 1 - p_0 \in (-1,0)$. Then $p_0 + p_1 = 1$ and $\sum_i s_i |p_i| = 1/3(2p_0 - 1) < 1$. Taking $q = 1$, we see that $\mathbb{M}$ defined as in (5.5) is contractive in $\text{Lip}(\mathbb{X})^*$ and thus has a unique fixed point in $\mathcal{S}_1(\mathbb{X})$. To get an idea of the attractor, we start with μ the uniform probability measure on $[0,1]$ and consider the iterate $\mathbb{M}^n \mu$. The nth iterate of the geometric

IFS has generated 2^n intervals of size 3^{-n}, which are all labeled in a natural way by the binary sequences of length n. If $\sigma = b_1 b_2 \ldots b_n$ is such a binary sequence and I_σ is the associated interval, then we see that

$$\mathbb{M}^n \mu(I_\sigma) = p_{b_1} p_{b_2} \cdots p_{b_n} = p_0^l p_1^{n-l},$$

where l is the number of zeros in the binary sequence σ. Clearly, as n becomes large, these numbers oscillate wildly, and the corresponding iterate $\mathbb{M}^n \mu$ is quite complicated. Nevertheless, for any $f \in \mathrm{Lip}([0,1])$, the integrals $\int_0^1 f \, d\mathbb{M}^n \mu$ converge! However, because of the oscillations, there is no reason to believe that integrals with respect to a generic continuous function would converge. $\square$

Example 5.15. As another example of an IFS on signed measures and its attractor, let $\mathbb{X} = [0,1]$, $w_i(x) = x/3 + i/3$ for $i = 0,1$ and $p_0 = -1$ and $p_1 = 1$. Then $\sum_i s_i |p_i| = 1/3 + 1/3 = 2/3 < 1$, so this $\mathbb{M}$ is contractive in $\mathrm{Lip}(\mathbb{X})^*$. Now, if we let $q = 0$, we see that $\mathbb{M}$ maps $\mathcal{S}_0(\mathbb{X})$ to itself, and therefore must have a fixed point in $\mathcal{S}_0(\mathbb{X})$. Since $\mathbb{M}$ is linear and $0 \in \mathcal{S}_0(\mathbb{X})$, the only possible fixed point is the zero measure. However, notice that if we start with the measure μ on $[0,1]$ defined by

$$\mu([a,b]) = \begin{cases} a - b & \text{if } 0 < a < b < 1/2, \\ b - a & \text{if } 1/2 < a < b < 1 \end{cases}$$

then $\|\mu\| = 1$ (total variation) and $\|\mathbb{M}^n \mu\| = 2^n$. $\square$

In the usual case of an invariant probability measure μ for an IFS with probabilities, the support of μ is (usually) the geometric attractor of the IFS (at least it is in the case where all the probabilities p_i are strictly positive). In the current case of "generalized measures," we have to redefine what we mean by support because if $\mu \in \mathrm{Lip}(\mathbb{X})^*$ it does not necessarily mean anything to ask for the value of $\mu(B)$ for many sets B. Instead we say that for any $f \in \mathrm{Lip}(\mathbb{X})$ whose support is separated from the attractor A of the geometric IFS, we have $\mu(f) = 0$, where μ is the fixed point.

Proposition 5.16. *Suppose that each w_i in Definition 5.5 is strictly contractive and that $\sum_i s_i |p_i| < 1$. Let A be the geometric attractor of the IFS $\{w_i\}$, U be an open neighbourhood of A, and $\mu \in \mathrm{Lip}(\mathbb{X})^*$ be the fixed point of $\mathbb{M}$. Then, for every $f \in \mathrm{Lip}(\mathbb{X})$ with $f|_U = 0$, we have $\mu(f) = 0$.*

Proof. Let $x \in \mathbb{X}$. Then there is some $n \in \mathbb{N}$ such that for all 2^n of the n-fold compositions we have $w_{i_1} \circ w_{i_2} \circ \cdots \circ w_{i_n}(x) \in U$. However, this means that $\mathbb{T}^n f(x) = 0$. Since this is true for all x, we know that $\mu(f) = 0$. $\square$

5.1.4 Noncompact case

We briefly indicate how the Monge-Kantorovich metric can be extended to the situation where $\mathbb{X}$ is not compact. We do, however, assume that $\mathbb{X}$ is a locally compact, complete, and separable metric space.

As usual, $C_0(\mathbb{X})$ denotes the collection of all continuous $f : \mathbb{X} \to \mathbb{R}$ such that for every $\epsilon > 0$ there is some compact $K \subseteq \mathbb{X}$ for which $f(\mathbb{X} \setminus K) \subset (-\epsilon, \epsilon)$. We also use $C^*(\mathbb{X})$ to denote the collection of all bounded continuous $f : \mathbb{X} \to \mathbb{R}$. By the Riesz representation theorem (Theorem B.26), $(\mathcal{M}(\mathbb{X}, \mathbb{R}), \|\cdot\|)$ is the dual space of $C_0(\mathbb{X})$ endowed with the supremum norm $\|f\|_\infty = \sup\{|f(x)| : x \in \mathbb{X}\}$.

Definition 5.17. Suppose $q \in \mathbb{R}$ and $k > 0$. Let $\mathcal{M}_{q,k}(\mathbb{X}, \mathbb{R})$ be the set of all $\mu \in \mathcal{M}(\mathbb{X}, \mathbb{R})$ such that

1. $\mu(\mathbb{X}) = q$;
2. $\|\mu\| = |\mu|(\mathbb{X}) \leq k$;
3. $\int_{\mathbb{X}} d(x, a)\, d|\mu|(x) < \infty$ for some, and hence any, $a \in \mathbb{X}$.

Note that we used the same notation, $\mathcal{M}_{q,k}(\mathbb{X}, \mathbb{R})$, in Definitions 5.3 and 5.17, but this should not cause confusion, as the third condition from Definition 5.17 is automatically satisfied when $\mathbb{X}$ is compact.

We use the exact same definition (Definition 5.2) for d_{MK} as in the case where $\mathbb{X}$ is compact. That is,

$$d_{MK}(\mu, \nu) = \sup\left\{ \int_{\mathbb{X}} f\, d(\mu - \nu) : f \in \text{Lip}_1(\mathbb{X}) \right\}.$$

Theorem 5.18. $(\mathcal{M}_{q,k}(\mathbb{X}, \mathbb{R}), d_{MK})$ *is a complete metric space.*

Proof. First we show that $d_{MK}(\mu, \nu)$ is finite on $\mathcal{M}_{q,k}(\mathbb{X}, \mathbb{R})$. Take $f \in \text{Lip}_1(\mathbb{X})$, and let a be as in condition 3 of Definition 5.17. Then

$$\int f\, d(\mu - \nu) = \int (f - f(a))\, d(\mu - \nu) \quad \text{(since } \mu(\mathbb{X}) = \nu(\mathbb{X}))$$
$$\leq \int |f - f(a)|\, d(|\mu| + |\nu|)$$
$$\leq \int d(\cdot, a)\, d(|\mu| + |\nu|) < \infty.$$

Since the bound is independent of $f \in \text{Lip}_1(\mathbb{X})$, it follows that $d_{MK}(\mu, \nu)$ is finite.

The other properties of a metric follow in a straightforward manner. The only thing left to prove is completeness.

Suppose that (μ_n) is a d_{MK}-Cauchy sequence of signed measures in $\mathcal{M}_{q,k}(\mathbb{X}, \mathbb{R})$. Then, by definition, for each $f \in \mathrm{Lip}_1(\mathbb{X})$, we have that

$$\phi(f) = \lim_n \int_{\mathbb{X}} f \, d\mu_n$$

exists. It is easy to see that ϕ is a linear functional on $\mathrm{Lip}_1(\mathbb{X})$ and thus can be extended to be a linear functional on $\mathrm{Lip}(\mathbb{X})$, the space of all Lipschitz functions on $\mathbb{X}$.

Our strategy is to show that ϕ can be (uniquely) extended to a bounded linear functional on $C_0(\mathbb{X})$ and thus corresponds to a signed measure μ on $\mathbb{X}$. Then we will show that $\mu \in \mathcal{M}_{q,k}(\mathbb{X}, \mathbb{R})$ and is thus the desired limit.

The second condition in Definition 5.17 implies that for all bounded Lipschitz f and all n we have

$$\left| \int_{\mathbb{X}} f \, d\mu_n \right| \leq k \|f\|_\infty,$$

and thus taking limits we see that $|\phi(f)| \leq k \|f\|_\infty$ as well and so ϕ is a bounded linear functional on the dense subspace $\mathrm{Lip}(\mathbb{X}) \cap C_0(\mathbb{X}) \subset C_0(\mathbb{X})$. Hence it has an extension (which we also denoted by ϕ) to all of $C_0(\mathbb{X})$. Since $\mathrm{Lip}(\mathbb{X}) \cap C_0(\mathbb{X})$ is dense in $C_0(\mathbb{X})$, the extension is in fact unique. This extension has the same bound of k. Let μ be the signed measure on $\mathbb{X}$ that corresponds to this extension ϕ.

Since the constant function $f = 1$ is in $\mathrm{Lip}_1(\mathbb{X})$, we have

$$\mu(\mathbb{X}) = \int_{\mathbb{X}} 1 \, d\mu = \phi(1) = \lim_n \int_{\mathbb{X}} 1 \, d\mu_n = \mu_n(\mathbb{X}) = q,$$

and thus the first condition for membership in $\mathcal{M}_{q,k}(\mathbb{X}, \mathbb{R})$ will be satisfied.

For the third condition, we first show that this condition is equivalent to the assumption that for some $a \in \mathbb{X}$ (and hence for any a) we have

$$\sup \left\{ \int_{\mathbb{X}} f d\mu : f \in \mathrm{Lip}_1(\mathbb{X}), f(a) = 0 \right\} < \infty. \tag{5.8}$$

In the first part of the proof above, we showed that the third condition implies (5.8). So, suppose that (5.8) holds. In particular, since $f(x) = d(x, a) \in \mathrm{Lip}_1(\mathbb{X})$, we have

$$\int_{\mathbb{X}} d(x, a) d\mu < \infty,$$

which implies that

$$\int_{\mathbb{X}} d(x,a) d\mu^+ < \infty \quad \text{and} \quad \int_{\mathbb{X}} d(x,a) d\mu^- < \infty,$$

which in turn implies

$$\int_{\mathbb{X}} d(x,a) d|\mu| < \infty.$$

Now, to show that the third condition holds, suppose to the contrary that it does not hold. Then there is some $a \in \mathbb{X}$ and sequence of functions $f_m \in \text{Lip}_1(\mathbb{X})$ with $f_m(a) = 0$ and such that $\phi(f_m) \to +\infty$. Since μ_n is d_{MK}-Cauchy, there is some N such that for $n, m \geq N$ we have $d_{MK}(\mu_n, \mu_m) < 1$. Let L be such that

$$-L < \int_{\mathbb{X}} f \, d\mu_N < L$$

for all $f \in \text{Lip}_1(\mathbb{X})$ with $f(a) = 0$. This implies that for all $n \geq N$ we have

$$-L - 1 < \int_{\mathbb{X}} f \, d\mu_n < L + 1.$$

But then, since $\phi(f_m) \to \infty$, there must be some $m \geq N$ such that $\phi(f_m) > L + 10$. Finally, by the definition of ϕ, there is some $n > N$ such that

$$\left| \int_{\mathbb{X}} f_m \, d\mu_n - \phi(f_m) \right| < 1.$$

However, this implies that

$$\int_{\mathbb{X}} f_m \, d\mu_n > L + 8,$$

which is a contradiction. This means that (5.8) holds and thus condition 3 also holds.

The argument above shows that, for any $f \in C_0(\mathbb{X})$, we have

$$\int_{\mathbb{X}} f \, d\mu_n \to \int_{\mathbb{X}} f \, d\mu,$$

and so μ is the $C_0(\mathbb{X})$-weak limit of the sequence μ_n. This implies that $\|\mu\| \leq \limsup \|\mu_n\| \leq k$ and thus condition 2 of Definition 5.17 holds for μ. Therefore, $\mu \in M_{k,q}(\mathbb{X}, \mathbb{R})$.

Since μ_n is d_{MK}-Cauchy, for large n, m we have for each $f \in \text{Lip}_1(\mathbb{X})$

$$\left| \int_{\mathbb{X}} f \, d\mu_m - \int_{\mathbb{X}} f \, d\mu_n \right| \leq \sup_{g \in Lip_1} \left| \int_{\mathbb{X}} g \, d(\mu_n - \mu_m) \right| < \epsilon.$$

Taking a limit as $m \to \infty$, we get

$$\left| \int_{\mathbb{X}} f \, d\mu - \int_{\mathbb{X}} f \, d\mu_n \right| \leq \epsilon$$

independent of f. Thus $d_{MK}(\mu, \mu_n) \to 0$. $\qquad\qquad\qquad\square$

5.2 Vector-valued measures

We now turn to considering spaces of vector-valued measures on $\mathbb{X}$. Again, for simplicity we will first deal with the case of compact $\mathbb{X}$ and also the case where the measure takes values in $\mathbb{R}^m$. The latter assumption reduces many of the technical complications introduced by having an infinite-dimensional range space for the measure. The general structure of the constructions and results remains the same in the simpler context and is more clearly seen than in the more general situation. For some general discussions on vector-valued measures, see [51, 48]. The results in this section are based on [135].

After some preliminary definitions and basic results on vector measures, the outline of the section is much the same as the previous section. That is, we set a space of vector measures along with a metric on this space of measures and prove completeness. We then construct an IFS operator on this space and show contractivity conditions. Again we have the situation that many naturally occurring examples lead to unbounded vector measures, so we move to a framework involving generalized vector measures. Finally, we end with some generalities involving noncompact $\mathbb{X}$ and measures with values in an infinite-dimensional space. Along the way, we illustrate the theory with several examples.

The material in this section appeared in [58, 20]. It was also independently discovered in [135], from which the presentation borrows very heavily, particularly the discussion of the coloured chaos game and the application to vector calculus on fractal domains.

Definition 5.19. Let $(\Omega, \mathcal{A})$ be a measurable space. By a *vector measure* μ on Ω with values in $\mathbb{R}^m$, we mean a set function $\mu : \mathcal{A} \to \mathbb{R}^m$ for which

1. $\mu(\emptyset) = 0$ and
2. for each disjoint sequence $E_i \in \mathcal{A}$, $\mu(\cup_i E_i) = \sum_i \mu(E_i)$.

The *variation* of such a vector measure μ is the nonnegative set function $|\mu|$ defined by

$$|\mu|(E) = \sup\left\{\sum_i \|\mu(E_i)\| : \text{ finite measurable partitions } \{E_i\} \text{ of } E\right\},$$

where the norm in the sum is the norm in $\mathbb{R}^m$.

Notice that the infinite sum $\sum_i \mu(E_i)$ converges unconditionally. Notice also that we assume that $\mu(E) \in \mathbb{R}^m$ for all E. This will imply that any such μ will be of bounded variation.

If μ is a vector measure with values in $\mathbb{R}^m$ and for all $E \in \mathcal{A}$ each component of $\mu(E)$ is nonnegative (that is, $\mu(E)_i \geq 0$ for all $E \in \mathcal{A}$ and $i = 1, 2, \ldots, m$), then we will say that μ is a *nonnegative vector measure*. In this situation, we have $|\mu|(E) = |\mu(E)|$ for all E, just the same as for normal (nonnegative) measures, and so computing the variation for such measures is simple.

Some elementary properties of vector measures are given in the next proposition. Note that we are assuming that the range space is $\mathbb{R}^m$. If the range space is some infinite-dimensional Banach space, some of these statements need to be modified.

Proposition 5.20.

1. *A vector measure has bounded range and hence has bounded variation.*
2. *The variation $|\mu|$ of a vector measure is a (positive) finite measure.*
3. *If μ_n is a sequence of vector measures such that $\mu(E) = \lim_n \mu_n(E)$ exists for all $E \in \mathcal{A}$, then μ is also a vector measure.*
4. *Let μ be a vector measure and λ a measure. Then $\lambda(E) = 0$ implies $\mu(E) = 0$ for all $E \in \mathcal{A}$ if and only if*

$$\lim_{\lambda(E)\to 0} \mu(E) = 0.$$

5. *For each $y \in \mathbb{R}^m$, the set function $\mu^y(E) = \langle y, \mu(E)\rangle$ is a finite signed measure and $|\mu^y|(E) \leq |\mu|(E)$, so $\mu^y \ll |\mu|$.*

If property 4 from this proposition holds, then μ is said to be λ *continuous*.

Let $\{\mathbf{e}_i\}$ be the standard basis for $\mathbb{R}^m$. Notice from property 5 that $\mu^{\mathbf{e}_i} \ll |\mu|$ for each i, so by the Radon-Nikodym theorem (Theorem B.23) there are g_i such that

$$\mu^{\mathbf{e}_i}(E) = \int_E g_i \, d|\mu|,$$

which then means that

$$\mu(E) = \left(\int_E g_2 \, d|\mu|, \int_E g_2 \, d|\mu|, \ldots, \int_E g_m \, d|\mu| \right). \qquad (5.9)$$

If we allow ourselves to integrate the vector function $\mathbf{g} = (g_1, \ldots, g_m)$ with respect to the measure $|\mu|$, (5.9) becomes

$$\mu(E) = \int_E \mathbf{g} \, d|\mu|. \qquad (5.10)$$

The simplest and most natural way to define the integral of a measurable $f : \Omega \to \mathbb{R}$ with respect to a vector measure μ is component-by-component. That is, we define the ith component of the integral

$$\int_\Omega f \, d\mu$$

(whose value should be a vector in $\mathbb{R}^m$) to be the number

$$\int_\Omega f \, d\mu^{\mathbf{e}_i},$$

provided these all exist for $i = 1, \ldots, m$. Since our vector measures have finite-dimensional values, there are no real complications in defining the integral in this way. Notice that, as in the case of a signed measure, it is possible to allow the components to take the value $+\infty$ or $-\infty$.

Using the Radon-Nikodym theorem and (5.10), we see that

$$\int_\Omega f \, d\mu = \left(\int_\Omega f g_1 \, d|\mu|, \int_\Omega f g_2 \, d|\mu|, \ldots, \int_\Omega f g_m \, d|\mu| \right) = \int_\Omega f \mathbf{g} \, d|\mu|.$$

A measurable function $f : \Omega \to \mathbb{R}$ will be *integrable* with respect to the vector measure μ if f is integrable with respect to μ^y for all $y \in \mathbb{R}^m$ with $\|y\| = 1$. An equivalent condition involves only y from an orthonormal basis for $\mathbb{R}^m$, such as the standard basis $\{\mathbf{e}_i\}$. In particular, f is integrable with respect to μ if and only if $|f|$ is integrable with respect to $|\mu|$ since for each $j = 1, \ldots, m$ we have

$$|\mu^{\mathbf{e}_j}|(E) \le |\mu|(E) \le \sum_{i=1}^m |\mu^{\mathbf{e}_i}|(E).$$

For a measurable function $\mathbf{f} : \Omega \to \mathbb{R}^m$ and a vector measure μ with values in $\mathbb{R}^m$, we can also define the integral using components. Using $f_i(x) = \langle \mathbf{f}(x), \mathbf{e}_i \rangle$, we have

$$\int_\Omega \mathbf{f} \cdot d\mu := \sum_i \int_\Omega f_i \, d\mu^{\mathbf{e}_i},$$

again provided that all the individual integrals in the sum exist and the sum is meaningful (i.e., it does not involve terms like $\infty - \infty$). The way to think about this integral is that we use the standard inner product $\langle \cdot, \cdot \rangle$ on $\mathbb{R}^m$ to pair the vector-valued function $\mathbf{f}(x)$ with the vector measure $d\mu(x)$ and obtain a scalar. Alternatively, again using (5.10), we can write

$$\int_\Omega \mathbf{f} \, d\mu = \int_\Omega \langle \mathbf{f}, \mathbf{g} \rangle \, d|\mu|,$$

where the inner product is much more explicit.

Finally, for a measurable vector function $\mathbf{f} : \Omega \to \mathbb{R}^m$, we say that $\mathbf{f}$ is *integrable* with respect to μ if $\|\mathbf{f}\|$ is integrable with respect to $|\mu|$.

Definition 5.21. Let $\mathbb{X}$ be a metric space. We denote the space of all Borel measures of finite variation on $\mathbb{X}$ with values in $\mathbb{R}^m$ by $\mathcal{M}(\mathbb{X}, \mathbb{R}^m)$.

We see that, under the obvious isomorphism using the standard basis, we have

$$\mathcal{M}(\mathbb{X}, \mathbb{R}^m) = \mathcal{M}(\mathbb{X}, \mathbb{R}) \oplus \mathcal{M}(\mathbb{X}, \mathbb{R}) \oplus \cdots \oplus \mathcal{M}(\mathbb{X}, \mathbb{R}).$$

Furthermore, $\mathcal{M}(\mathbb{X}, \mathbb{R}^m)$ is a normed linear space under the norm $\|\mu\| = |\mu|(\mathbb{X})$. In fact, it can be shown that it is a Banach space under this norm.

5.2.1 Complete space of vector measures

Again we use a slight modification of the Monge-Kantorovich metric, and we have the same requirements of fixed total mass and uniform bound on variation as in the signed measure case (this is clearly necessary, as a signed measure is a special case of a vector measure).

We assume that $\mathbb{X}$ is a compact metric space.

With an obvious adaptation of notation, $\mathrm{Lip}(\mathbb{X}, \mathbb{R}^m)$ denotes the Lipschitz functions $f : \mathbb{X} \to \mathbb{R}^m$ and $\mathrm{Lip}_1(\mathbb{X}, \mathbb{R}^m)$ those functions with Lipschitz constant bounded by 1. In this context, for $\mu, \nu \in \mathcal{M}(\mathbb{X}, \mathbb{R}^m)$, we define

$$d_{MK}(\mu, \nu) = \sup \left\{ \int_{\mathbb{X}} f \cdot d(\mu - \nu) : f \in \mathrm{Lip}_1(\mathbb{X}, \mathbb{R}^m) \right\}. \qquad (5.11)$$

Again we need some conditions on the collection of vector measures in the space in order to ensure that d_{MK} is finite and that the resulting space is complete.

Definition 5.22. Suppose $q \in \mathbb{R}^m$ and $k > 0$. Let

$$\mathcal{M}_{q,k}(\mathbb{X}, \mathbb{R}^m) = \{\mu \in \mathcal{M}(\mathbb{X}, \mathbb{R}^m) : \mu(\mathbb{X}) = q, \ \|\mu\| \leq k\}.$$

Theorem 5.23. *Let* $\mathbb{X}$ *be a compact metric space,* $q \in \mathbb{R}^m$, *and* $k > 0$. *Then* $\mathcal{M}_{q,k}(\mathbb{X}, \mathbb{R}^m)$ *is complete under the metric* d_{MK}.

Proof. First note that if $\mu(\mathbb{X}) = q$, then $\mu^{\mathbf{e}_i}(\mathbb{X}) = \langle q, \mathbf{e}_i \rangle := q_i$ and conversely. Furthermore, because of the way that the integral is defined and the fact that we can choose the components of $g \in \mathrm{Lip}(\mathbb{X}, \mathbb{R}^m)$ independently, we have

$$\sup\left\{\int_{\mathbb{X}} g \cdot d\mu : g \in \mathrm{Lip}_1(\mathbb{X}, \mathbb{R}^m)\right\} = \sum_{i=1}^m \sup\left\{\int_{\mathbb{X}} f \, d\mu^{\mathbf{e}_i} : f \in \mathrm{Lip}_1(\mathbb{X})\right\},$$

which means that $\mu_n \to \mu$ in the d_{MK} metric on vector measures if and only if for each $i = 1, 2, \ldots, m$ we have $\mu_n^{\mathbf{e}_i} \to \mu^{\mathbf{e}_i}$ in the d_{MK} metric on signed measures.

Suppose that (μ_n) is a d_{MK}-Cauchy sequence in $\mathcal{M}_{q,k}(\mathbb{X}, \mathbb{R}^m)$. Then for each i we know that $(\mu_n^{\mathbf{e}_i})$ is a d_{MK}-Cauchy sequence in $\mathcal{M}_{q_i,k}(\mathbb{X}, \mathbb{R})$ (notice that we can use the same k as $|\mu^{\mathbf{e}_i}|(\mathbb{X}) \leq |\mu|(\mathbb{X})$ for all i). By Theorem 5.4, we then have that $\mu_n^{\mathbf{e}_i} \to \mu_i \in \mathcal{M}_{q_i,k}(\mathbb{X}, \mathbb{R})$. We must prove that $\mu := (\mu_1, \mu_2, \ldots, \mu_m) \in \mathcal{M}_{q,k}(\mathbb{X}, \mathbb{R}^m)$. Clearly we have $\mu(\mathbb{X}) = q$, as $\langle \mu(\mathbb{X}), \mathbf{e}_i \rangle = q_i$, and thus we only have to show that $\|\mu\| \leq k$.

To do this we recall that for a variation norm bounded sequence $\nu_n \in \mathcal{M}(\mathbb{X}, \mathbb{R})$, $\nu_n \to \nu$ in the Monge-Kantorovich metric implies that $\nu_n \to \nu$ weakly. However, $\mathcal{M}(\mathbb{X}, \mathbb{R})$ is a normed space under the variation norm, and for weak* convergence in a normed space the norm of the limit cannot be larger than the limit of the norms. In the same way, since $\mathcal{M}(\mathbb{X}, \mathbb{R}^m)$ is the Banach dual to $C(\mathbb{X}, \mathbb{R}^m)$ by the Riesz representation theorem (Theorem B.26), the norm of the limit cannot be larger than the limit of the norms. Since $\lim_n \|\mu_n\| = k$, this means that $\|\mu\| \leq k$, as desired.   □

Notice that the measures having values in $\mathbb{R}^m$ (a finite-dimensional space) made the proof rather simple, as we could then essentially reduce it to the case of a signed measure.

5.2.2 IFS on vector measures

Before we give the (simple) form of an IFS operator on vector measures, we begin with a motivating example.

First, consider a nice smooth curve, such as that shown in Fig. 5.1. We see that the unit tangent vector $\overrightarrow{T}(s)$ exists at each point of the curve, where s is the arclength parameter for the curve. Furthermore, for any two points x, y along the curve, we have

$$\int_x^y \overrightarrow{T}(s)\, ds = \overrightarrow{x\,y},$$

where $\overrightarrow{x\,y}$ is the displacement vector from x to y.

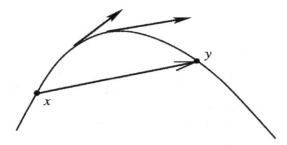

Fig. 5.1: A portion of a smooth curve.

The von Koch curve $\mathcal{K}$ in Fig. 5.2 is the attractor of a 4-map IFS. It is a continuous curve with infinite length and without a tangent at any point along the curve. However, we would like to have some way of defining a tangent vector to $\mathcal{K}$, with the particular goal of computing line integrals. There are two problems with this. First, not only does $\mathcal{K}$ have infinite length, but if $x \in \mathcal{K}$ is any point and $\epsilon > 0$, then the curve segment $\mathcal{K} \cap B_\epsilon(x)$ also has infinite length! That is, $\mathcal{K}$ is not only unrectifiable but is everywhere locally unrectifiable. This property is not surprising since $\mathcal{K}$ exhibits self-similarity. However, this means that the arclength parameter does not have any real meaning for $\mathcal{K}$. The second problem is that $\mathcal{K}$ does not have a tangent anywhere, so $\overrightarrow{T}$ also makes no sense.

The solution is to combine $\overrightarrow{T}$ and ds together in a measure. That is, to a segment $S \subseteq \mathcal{K}$, we assign the "measure" $\mu(S) = \overrightarrow{x\,y}$, where x and y are the endpoints of S. The fact that this can be done in a

Fig. 5.2: The von Koch curve.

consistent manner is at first surprising. However, the recursive nature of the construction of $\mathcal{K}$ carries all the work.

Figure 5.3 illustrates this by showing the first three stages in a recursive construction of $\mathcal{K}$ and the "tangent vector measure" μ to $\mathcal{K}$. The geometric IFS that generates $\mathcal{K}$ uses four shrunken copies of $\mathcal{K}$, joined together at the endpoints, to construct $\mathcal{K}$. Similarly, the IFS on vector measures $\mathbb{M}$ will use four scaled and rotated copies of μ to construct μ. The geometry of $\mathcal{K}$ ensures that $\mathbb{M}$ preserves total mass (that is, the overall displacement will be preserved) from one stage of the construction to the next. This preservation of mass is a consistency condition that will result in a well-defined limiting object. This will also give a natural total fixed mass, as is required by the Monge-Kantorovich metric.

Before we give the general definition, we illustrate the idea with the IFS operator for the tangent measure to $\mathcal{K}$. First, we need four geometric IFS maps to define the curve $\mathcal{K}$. Let Rot_θ be the rotation in $\mathbb{R}^2$ around the origin by an angle of θ. Then the four IFS maps are given by

$$
\begin{aligned}
w_1(x,y) &= (1/3)\mathrm{Rot}_0(x,y), \\
w_2(x,y) &= (1/3)\mathrm{Rot}_{\pi/6}(x,y) + (1/3,0), \\
w_3(x,y) &= (1/3)\mathrm{Rot}_{-\pi/6}(x,y) + (1/2,\sqrt{3}/6), \\
w_4(x,y) &= (1/3)\mathrm{Rot}_0(x,y) + (2/3,0).
\end{aligned}
$$

The von Koch curve $\mathcal{K}$ is the attractor of the IFS $\{w_1, w_2, w_3, w_4\}$, which can be easily seen since $\mathcal{K}$ is invariant with respect to this IFS. Next we have four linear maps $\alpha_i : \mathbb{R}^2 \to \mathbb{R}^2$, given by $\alpha_1 = (1/3)\mathrm{Rot}_0$, $\alpha_2 = (1/3)\mathrm{Rot}_{\pi/6}$, $\alpha_3 = (1/3)\mathrm{Rot}_{-\pi/6}$, and finally $\alpha_4 = (1/3)\mathrm{Rot}_0$. Notice that we have taken the linear part of w_i to be the map α_i.

Look at the first two stages of Fig. 5.3. The first "part" of the second stage is clearly just a copy of the first stage that is smaller by a factor

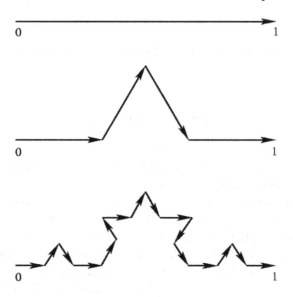

Fig. 5.3: First three stages for the tangent measure for the von Koch curve.

of $1/3$. Thinking of each stage in this diagram as a vector measure, we can describe the first stage as the vector measure μ_1 supported on the interval $[0,1] \times \{0\} \subset \mathbb{R}^2$ and defined as

$$\mu_1(S) = \overrightarrow{V} \text{ Lebesgue measure of } (S \cap [0,1] \times \{0\}),$$

where the vector $\overrightarrow{V} = \overrightarrow{\mathbf{i}}$, the unit vector in the direction of the horizontal axis. That is, μ_1 is the uniform distribution supported on $[0,1] \times \{0\}$ and with total mass equal to $\overrightarrow{V}$. Using this description, the first "part" of the second-stage vector measure μ_2 is equal to the "uniform" distribution on $[0,1/3] \times \{0\}$ and with total mass $(1/3)\overrightarrow{V}$. However, this is exactly the same as the vector measure

$$\alpha_1 \, \mu_1 \circ w_1^{-1}.$$

Similarly, the second "part" is equal to the vector measure

$$\alpha_2 \, \mu_1 \circ w_2^{-1},$$

and so on. That is, we have that the second-stage vector measure μ_2 satisfies

$$\mu_2 = \alpha_1 \, \mu_1 \circ w_1^{-1} + \alpha_2 \, \mu_1 \circ w_2^{-1} + \alpha_3 \, \mu_1 \circ w_3^{-1} + \alpha_4 \, \mu_1 \circ w_4^{-1}.$$

The maps w_i move the support of this "tangent field" to the appropriate places; the linear maps α_i do both the rotation and the scaling of the vector part of the tangent field.

We mention here that it turns out that $\alpha_1 + \alpha_2 + \alpha_3 + \alpha_4 = I$, the identity transformation on $\mathbb{R}^2$.

The general definition follows in a very straightforward manner. Notice that it is linear and is very similar both to the IFS on probability measures and that on signed measures.

Definition 5.24. Let $q \in \mathbb{R}^m$ be fixed. Let the N maps $w_i : \mathbb{X} \to \mathbb{X}$ be Lipschitz with constant s_i and $\alpha_i : \mathbb{R}^m \to \mathbb{R}^m$ be linear maps. Further suppose that the condition

$$\sum_i \alpha_i q = q \tag{5.12}$$

is satisfied. Define the IFS operator $\mathbb{M}$ on $\mathcal{M}(\mathbb{X}, \mathbb{R}^m)$ by

$$\mathbb{M}\mu(B) = \sum_i \alpha_i \mu(w_i^{-1}(B)). \tag{5.13}$$

The condition (5.12) ensures that if $\mu(\mathbb{X}) = q$, then $\mathbb{M}\mu(\mathbb{X}) = q$ as well and thus is a mass preservation condition. As we have seen, some type of mass preservation condition is necessary for our framework.

Proposition 5.25. *The operator $\mathbb{M}$ from (5.13) satisfies*

$$d_{MK}(\mathbb{M}\mu, \mathbb{M}\nu) \leq \left(\sum_i s_i \|\alpha_i^*\| \right) d_{MK}(\mu, \nu)$$

for two $\mu, \nu \in \mathcal{M}(\mathbb{X}, \mathbb{R}^m)$ with $\mu(\mathbb{X}) = \nu(\mathbb{X}) = q$.

Proof. For a given Lipschitz function f, we have

$$\int_{\mathbb{X}} f(x) \, d(\mathbb{M}\mu(x) - \mathbb{M}\nu(x))$$

$$= \sum_i \int_{\mathbb{X}} f(x) \cdot d\left(\alpha_i(\mu \circ w_i^{-1}(x) - \nu \circ w_i^{-1}(x)) \right)$$

$$= \sum_i \int_{\mathbb{X}} f(w_i(y)) \cdot \alpha_i \, d(\mu(y) - \nu(y))$$

$$= \int_{\mathbb{X}} \left(\sum_i \alpha_i^* f(w_i(y)) \right) \cdot d(\mu(y) - \nu(y))$$

$$= \left(\sum_i s_i \|\alpha_i^*\| \right) \int_{\mathbb{X}} \phi(y) \cdot d(\mu(y) - \nu(y)),$$

where the function $\phi(y) = \frac{1}{\sum_i s_i \|\alpha_i^*\|} \sum_i \alpha_i^* f \circ w_i \in \mathrm{Lip}_1(\mathbb{X}, \mathbb{R}^m)$ since s_i is the Lipschitz factor of w_i (for more, see below). Taking the supremum, we obtain

$$d_{MK}(\mathbb{M}\mu, \mathbb{M}\nu) \leq \left(\sum_i s_i \|\alpha_i^*\| \right) d_{MK}(\mu, \nu).$$

$\square$

For completeness, we now show that $\phi(x) \in \mathrm{Lip}_1(\mathbb{X}, \mathbb{R}^m)$. Let $x, y \in \mathbb{X}$ and $s = \sum_i s_i \|\alpha_i^*\|$. Then

$$\|\phi(x) - \phi(y)\| = \|s^{-1} \sum_i \alpha_i^* \left(f(w_i(x)) - f(w_i(y)) \right) \|$$

$$\leq s^{-1} \sum_i \|\alpha_i^*\| \|f(w_i(x)) - f(w_i(y))\|$$

$$\leq s^{-1} \sum_i \|\alpha_i^*\| d(w_i(x), w_i(y))$$

$$\leq s^{-1} \sum_i s_i \|\alpha_i^*\| d(x, y)$$

$$\leq d(x, y).$$

Corollary 5.26. *Suppose that $\mathbb{M}$ is given as in (5.13) and satisfies*

$$\sum_i s_i \|\alpha_i^*\| < 1.$$

Suppose further that there is a $k > 0$ such that $\mathbb{M} : \mathcal{M}_{q,k}(\mathbb{X}, \mathbb{R}^m) \to \mathcal{M}_{q,k}(\mathbb{X}, \mathbb{R}^m)$. Then there exists a unique measure $\mu \in \mathcal{M}_{q,k}(\mathbb{X}, \mathbb{R}^m)$ such that $\mathbb{M}\mu = \mu$. Furthermore, the support of μ is contained in the attractor of the IFS $\{w_i\}$.

Proof. The existence part is clear from the preceding proposition. The only thing to prove is the statement about the support of μ. Let A be the attractor of the IFS $\{w_i\}$ and $x \notin A$ and let $2\epsilon = d(x, A)$. Since $\mathbb{X}$ is compact, there is an n such that $\widehat{W}^n(\mathbb{X}) \subset A_\epsilon$, which means that $(\widehat{W}^n)^{-1}(B_\epsilon(x)) = \emptyset$, which means that $\mathbb{M}^n \mu(B_\epsilon(x)) = 0$ so x is not in the support of μ. Thus the support of μ is contained in A. $\square$

There is a simple necessary condition for $\mathbb{M}$ not to increase the variation norm of μ. This condition also works for signed measures, but in that case it is too strong a condition since $\sum_i |p_i| \leq 1$ and $\sum_i p_i = 1$ imply that $p_i \geq 0$ for each i. However, for linear functions α_i, the conditions

$$\sum_i \alpha_i q = q \quad \text{and} \quad \sum_i \|\alpha_i\| \leq 1$$

do not imply the same sort of triviality of the operator $\mathbb{M}$.

Proposition 5.27. *Suppose that $\mathbb{M}$ is given as in (5.13) and satisfies $\sum_i \|\alpha_i\| \leq 1$. Then $\|\mu\| \leq k$ implies that $\|\mathbb{M}\mu\| \leq k$.*

Proof. We simply see that

$$|\mathbb{M}\mu(\mathbb{X})| = \left| \sum_i \alpha_i \mu(\mathbb{X}) \right| \leq \sum_i \|\alpha_i\| |\mu(\mathbb{X})|.$$

$\square$

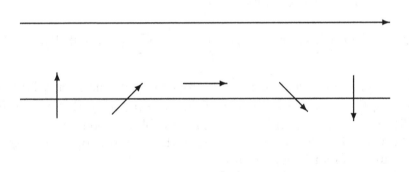

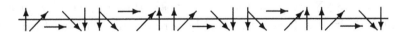

Fig. 5.4: Example of an $\mathbb{R}^2$-valued fractal vector measure on $[0, 1]$.

Example 5.28. As another example of an IFS on vector measures, consider the IFS maps $w_i : \mathbb{R} \to \mathbb{R}$ given by

$$w_i(x) = \frac{(-1)^n}{5} x + \frac{2}{5} \lceil \frac{i}{2} \rceil \quad \text{for } i = 0, 1, 2, 3, 4$$

(where $\lceil y \rceil$ is the smallest integer greater than or equal to y) and the scaled rotations

$$\alpha_i = (1/(\sqrt{2}+1))\mathrm{Rot}_{(i-2)\pi/4}.$$

The geometric attractor of the IFS is $[0,1]$, and the operator on vector measures acts on measures with values in $\mathbb{R}^2$. The first few iterations of the IFS operator are illustrated in Fig. 5.4. In this case, we have $\sum_i \alpha_i = I$. Since $s_i = 1/5$, we have $\sum_i s_i \|\alpha_i\| = p_1 < 1.$ ☐

This example illustrates that we do not have to match the dimension of $\mathbb{X}$ to the dimension of the output of the vector measure.

In the case of signed measures, the same operator $\mathbb{M}$ acts on $\mathcal{M}_{q,k}(\mathbb{X},\mathbb{R})$ and $\mathcal{M}_{q',k}(\mathbb{X},\mathbb{R})$ for different q, q' since $\mathbb{M}$ preserves the measure no matter what the total measure because $q = cq'$ for some $c \in \mathbb{R}$. This is no longer the case for vector measures, as it is no longer true that given any two $q, q' \in \mathbb{R}^m$ there is some $c \in \mathbb{R}$ with $q = cq'$ (obviously). However, it is simple to modify an IFS operator $\mathbb{M}$ that is designed to operate on μ with total mass q into $\widehat{\mathbb{M}}$, which operates on measures ν with total mass $\widehat{q}$. Just let R be any rotation in $\mathbb{R}^m$ for which $R(q) = (\|q\|/\|\widehat{q}\|)\,\widehat{q}$ and define

$$\widehat{\mathbb{M}}\mu = \frac{\|q'\|}{\|q\|} R\left(\sum_i \alpha_i R^{-1}\mu \circ w_i^{-1}\right) = \frac{\|q'\|}{\|q\|}\sum_i (R \circ \alpha_i \circ R^{-1})\mu \circ w_i^{-1}.$$

That is, assuming that $\mathbb{M}$ has the property that whenever $\mu(\mathbb{X}) = q$ then $\mathbb{M}\mu(\mathbb{X}) = q$, we have that $\widehat{\mathbb{M}}$ has the property that whenever $\nu(\mathbb{X}) = q'$, then $\widehat{\mathbb{M}}\nu(\mathbb{X}) = q'$ as well. This observation will be useful in Sect. 5.2.4 when we want to modify the construction of a tangent measure to obtain a normal measure.

The case $m = 2$ is special. If all the α_i are scaled rotations (so $\alpha_i = p_i R_i$) and $\sum_i \alpha_i q = q$ for some $q \in \mathbb{R}^2$, then in fact $\sum_i \alpha_i w = w$ for all $w \in \mathbb{R}^2$. To show this, let R be a rotation such that $R(q) = w$ (with no loss in generality, $\|q\| = \|w\|$). Then

$$\left(\sum_i \alpha_i\right) w = \sum_i \alpha_i Rv = R\left(\sum_i \alpha_i v\right) = Rv = w$$

since rotations commute in $\mathbb{R}^2$. Thus we have the condition on the p_i and the rotations R_i

$$\sum_i p_i R_i = I,$$

the identity matrix. In this case, the functional forms of $\mathbb{M}$ and $\widehat{\mathbb{M}}$ are identical – the only differences between $\mathbb{M}$ and $\widehat{\mathbb{M}}$ are their domains and ranges. This property will be used in Sect. 5.2.4. The operator that constructs the tangent vector measure of a fractal curve $\mathcal{C}$ will be modified appropriately in order to construct the normal vector measure of $\mathcal{C}$.

It is also possible to define an IFS with condensation on spaces of vector measures just as we did for signed measures and probability measures.

Definition 5.29. Fix $q \in \mathbb{R}^m$, and let the N maps $w_i : \mathbb{X} \to \mathbb{X}$ be Lipschitz with constant s_i and α_i be linear functions on $\mathbb{R}^m$ for $i = 1, \ldots, N$. Finally, choose some $\nu \in \mathcal{M}(\mathbb{X}, \mathbb{R})$ and suppose that we have the condition

$$\nu(\mathbb{X}) + \sum_i \alpha_i q = q. \tag{5.14}$$

Define the operator $\mathbb{M}$ on $\mathcal{M}(\mathbb{X}, \mathbb{R}^m)$ by

$$\mathbb{M}\mu = \nu + \sum_{i=1}^{N} \alpha_i \mu \circ w_i^{-1}. \tag{5.15}$$

Again it is easy to see that if $\mu(\mathbb{X}) = q$, then $\mathbb{M}\mu(\mathbb{X}) = q$ and

$$\|\mathbb{M}\mu\| \leq \|\nu\| + \sum_{i=1}^{N} \|\alpha_i\| \|\mu\|,$$

so to ensure $\|\mathbb{M}\mu\| \leq k$ for large k it is sufficient to have $\sum_i \|\alpha_i\| < 1$. Furthermore, we can write $\mathbb{M}$ as

$$\mathbb{M}\mu = \mathbb{A}\mu + \nu,$$

so that the fixed point of $\mathbb{M}$, if it exists, is the measure

$$\mu = \nu + \mathbb{A}\nu + \mathbb{A}^2\nu + \cdots + \mathbb{A}^n\nu + \cdots = (I - \mathbb{A})^{-1}\nu.$$

5.2.3 Coloured fractals

The first application of our IFS operator on vector measures will be to the case where the range of the measure is a set of probability measures.

As motivation, let us begin with a variation of the standard random iteration algorithm or "chaos game" (see Chapter 4) for rendering the attractor of an IFS with probabilities. (This variation was told to the authors by J. Anderson.) Given a set of N IFS maps w_i with probabilities p_i, $\sum_i p_i = 1$, recall that the standard chaos game algorithm is as follows (see Sect. 2.4):

1. Pick an initial point x_0 on the IFS attractor. One way to do this is to let x_0 be the fixed point of w_1. Set $n = 0$.
2. With probabilities defined by the p_i's, choose one of the w_i's, say w_j.
3. Let $x_{n+1} = w_j(x_n)$. Plot x_{n+1}.
4. If a sufficient number of points x_n have been generated, stop. Otherwise, let $n \to n + 1$ and go to Step 2.

The variation consists of adding "colour" to the algorithm. Assign a colour C_i to each map w_i and a probability $0 \leq pc_i \leq 1$ of changing to this colour. Then, the new algorithm is as follows:

1. Pick x_0 to be the fixed point of w_1, and set the current colour to $c_1 = C_1$. Set $n = 0$.
2. Choose a map w_j according to the probabilities p_i.
3. Change the current colour to colour C_j with probability pc_j. That is, with probability pc_j, set $c_{n+1} = C_j$, and otherwise $c_{n+1} = c_n$.
4. Set $x_{n+1} = w_j(x_n)$, and plot x_{n+1} using colour c_{n+1}.
5. If a sufficient number of points have been generated, stop. Otherwise, let $n \to n + 1$ and go to Step 2.

This algorithm yields a plot of the attractor "coloured" in a self-similar way by the colours associated with the IFS maps w_i. Figure 5.5 is an illustration of such a coloured version of the Sierpinski gasket (a 3-map IFS) that was generated with $pc_i = 1/2$ for all i. Since a black-and-white copy of the attractor obviously will not show the true colouring, we provide a very brief description. Let w_i, $i = 1, 2, 3$ denote the IFS maps for this attractor with fixed points $(0, 0)$, $(1, 0)$ and $(0, 1)$, respectively. Associated with these maps are the colours *red*, *green*, and *blue*, respectively. As one travels toward a vertex, the colour of the attractor points approaches the colour associated with the IFS map with that vertex as the fixed point. For example, as one approaches $(1, 0)$, the colour of the attractor points becomes more green.

We now present an IFS operator that can generate a "coloured" attractor as its fixed point. Let Λ be a finite set with $n = |\Lambda|$ and M_i be Markov transition matrices on Λ. Since we wish to multiply the

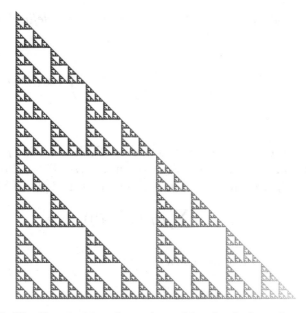

Fig. 5.5: The Sierpinski gasket coloured by the "colour chaos game."

M_i by probability vectors on the right, we require them to be *column stochastic*, that is, the column sum of each matrix M_i is 1.

For our set of N IFS maps w_i with probabilities p_i, $\sum_i p_1 = 1$, we define the "colour Markov" operator $\mathbb{M}_C : \mathcal{M}(\mathbb{X}, \mathbb{R}^m) \to \mathcal{M}(\mathbb{X}, \mathbb{R}^m)$ by

$$\mathbb{M}_C(\mu)(B) = \sum_i p_i M_i \mu \left(w_i^{-1}(B) \right)$$

for all Borel sets $B \subseteq \mathbb{X}$. In other words, we use Markov transition matrices for the linear operators that define $\mathbb{M}$. Now let $\mathbf{v}$ be an invariant probability distribution for the stochastic matrix $\sum_i p_i M_i$. (Being a convex combination of column stochastic matrices, this matrix is column stochastic.) We now not only restrict the total mass of the vector measures to be equal to $\mathbf{v}$ but also require them to be "positive", that is, each component must be nonnegative. So, we restrict our attention to the space

$$\mathcal{S}_{\mathbf{v}}(\mathbb{X}) = \{\mu \in \mathcal{M}(\mathbb{X}, \mathbb{R}^n) \ : \ \mu(B) \geq 0, \ \forall B \subseteq \mathbb{X}, \mu(\mathbb{X}) = \mathbf{v}\}.$$

Notice that this set is automatically bounded in variation because of the positivity condition. Clearly $\mathbb{M}_C$ maps $\mathcal{S}_{\mathbf{v}}(\mathbb{X})$ to itself. The condition that each M_i be stochastic also implies that $\|\mathbb{M}_C(\mu)\| \leq C$ if $\|\mu\| \leq C$. Also note that if A is a Markovian matrix, then $\|A\| \geq 1$.

Proposition 5.30. *Let* $\mathbb{M}_C$ *be the IFS operator defined above on* $\mathcal{S}_\mathbf{v}(\mathbb{X})$. *Then*

$$d_{MK}(\mathbb{M}_C(\mu), \mathbb{M}_C(\eta)) \leq \left(\sum_i s_i p_i \|M_i^*\| \right) d_{MK}(\mu, \eta)$$

for all $\mu, \eta \in \mathcal{S}_\mathbf{v}(\mathbb{X})$.

Now assume that $\mathbb{M}_C$ is contractive on $\mathcal{S}_\mathbf{v}(\mathbb{X})$ and let $\boldsymbol{\mu} \in \mathcal{S}_\mathbf{v}(\mathbb{X})$ be the invariant measure for $\mathbb{M}_C$. We have $|\boldsymbol{\mu}|(B) = |\mathbb{M}_C(\boldsymbol{\mu})|(B)$. Denoting $\mu = |\boldsymbol{\mu}|$, we have that μ is the attractor of the IFSP $(\mathbf{w}, \mathbf{p})$:

$$\mu(B) = \sum_i p_i \mu \left(w_i^{-1}(B) \right).$$

Heuristically, we can think of $\boldsymbol{\mu}$ as

$$d\boldsymbol{\mu}(x) = \mathbf{f}(x)\, d\mu(x),$$

where $\mathbf{f}(x)$ is the "attractor" of the "IFS"

$$\mathbf{f}(x) \longrightarrow \sum_i M_i \mathbf{f} \left(w_i^{-1}(x) \right).$$

Since $\boldsymbol{\mu}$ is absolutely continuous with respect to μ, we know that $\mathbf{f}(x)$ exists and is the Radon-Nikodym derivative of $\boldsymbol{\mu}$ with respect to μ.

The matrices M_i that correspond to the "chaos game with colour" are

$$M_i = (1 - pc_i)I + pc_i J_i,$$

where

$$J_i = \begin{pmatrix} 0 & 0 & 0 & \cdots & 0 \\ 0 & 0 & 0 & \cdots & 0 \\ \vdots & \vdots & \vdots & \ddots & \vdots \\ 1 & 1 & 1 & \cdots & 1 \\ \vdots & \vdots & \vdots & \ddots & \vdots \\ 0 & 0 & 0 & \cdots & 0 \\ 0 & 0 & 0 & \cdots & 0 \end{pmatrix}$$

and the ith row is a row of ones. It is easy to see that $\|M_i\|_1 = 1$, so that the contractivity condition becomes

$$\sum_i s_i p_i < 1.$$

For the Sierpinski gasket example with equal probabilities, $s_i = \frac{1}{2}$ and $p_i = \frac{1}{3}$, so this sum is $\frac{1}{2}$.

The connection between the colour chaos game and the IFS operator $\mathbb{M}_C$ is simple. Given a set $B \subseteq \mathbb{X}$ with $\mu(B) > 0$, the probability measure on Λ defined by

$$\frac{\boldsymbol{\mu}(B)}{\mu(B)}$$

describes the distribution of the colours on the set B. If B represents a pixel on the computer screen, then this distribution describes the percentage of time that this pixel is each colour over the run of the chaos game.

Note that $J_i \tau = e_i$ for any probability vector τ (where e_i is the ith basis vector in $\mathbb{R}^n$). So, if $pc_i = pc$ for all i and letting $\mathbf{p} = (p_1, p_2, \ldots, p_N)$, we see that

$$\left(\sum_i p_i M_i \right) \mathbf{p} = (1 - pc)\mathbf{p} + pc\mathbf{p} = \mathbf{p}$$

so $\mathbf{p}$ is the invariant distribution for $\sum_i p_i M_i$. Therefore $\boldsymbol{\mu}(\mathbb{X}) = \mathbf{p}$. The interpretation is as follows. If the probability of changing to any colour is the same for all colours, then the distribution of colour content of the entire image is just proportioned by the p_i. This is not to say that $\boldsymbol{\mu}$ will be the same regardless of the value of pc. In the limiting case where $pc = 1$, each subtile of the attractor will be coloured only by its own colour.

Notice that the distribution of colours in a region is a convex combination of the colours. Thus, this framework can also be used to produce a self-similar field where at each point we have a convex combination of objects (in this case, colours).

5.2.4 Line integrals on fractal curves

Now we apply the IFS on a vector-measures framework to the construction of the tangent field of a fractal curve and hence to the definition of the line integral of a smooth vector field over such a fractal curve.

The reader should keep the example of the von Koch curve (see Figs. 5.2 and 5.3) in mind as they read this section. This example is a good illustration of the construction of the tangent vector measure to a fractal curve. Another example is the IFS with maps

$$w_1(x, y) = \begin{pmatrix} 1/2 & -s \\ s & 1/2 \end{pmatrix} + (0, 0)$$

and

$$w_2(x, y) = \begin{pmatrix} 1/2 & s \\ -s & 1/2 \end{pmatrix} + (1/2, s).$$

The associated IFS map on vector-valued measures has $\sum_i s_i p_i = 2(1 + 4s^2)$, so it is less than 1 for $s < 1/2$. The first three iterations of the IFS operator on tangent vector measures is illustrated in Fig. 5.6, and the fractal curve is illustrated in Fig. 5.7.

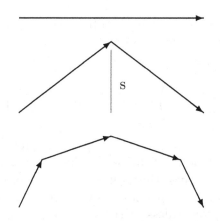

Fig. 5.6: First three iterations of IFS.

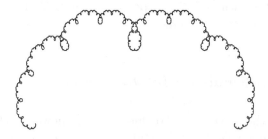

Fig. 5.7: Attractor for IFS.

Let $\mathcal{C} \subset \mathbb{R}^2$ be a continuous curve that is the attractor of an affine IFS $\{w_i\}$. For our purposes, we will let $\mathbb{X}$ be any compact subset of $\mathbb{R}^2$ that contains $\mathcal{C}$. The precise specification of $\mathbb{X}$ does not really matter as long as it is compact and contains $\mathcal{C}$. Let A and B denote the endpoints of $\mathcal{C}$, L be the line segment from A to B, and v be the displacement vector from A to B. Then the image of L under the IFS will be the first stage in the construction of $\mathcal{C}$ and will be a piecewise linear approximation to $\mathcal{C}$. Define v_i to be the displacement vector from the initial point of $w_i(L)$ to the final point of $w_i(L)$. Then $\sum_i v_i = v$ since the IFS $\{w_i\}$ generates a continuous curve. Let R_i be the rotation that takes the direction of v to the direction of v_i. Finally, let $p_i = \|v_i\|/\|v\|$, the scaling factor from L to $w_i(L)$. Then $p_i R_i(v) = v_i$, so

$$\left(\sum_i p_i R_i\right) v = \sum_i v_i = v.$$

Now, define the IFS operator $\mathbb{M}$ by taking $\alpha_i = p_i R_i$; that is,

$$\mathbb{M}\mu = \sum_i p_i R_i \mu \circ w_i^{-1}.$$

Notice that $\|\alpha_i\| = \|p_i R_i\| = p_i$ since R_i is a rotation.

Assume for the moment that $\mathbb{M}$ is a contraction on an appropriate space, and let μ be the invariant measure of $\mathbb{M}$. Then $T^n(\eta) \to \mu$ for appropriate η (a simple choice of η is $\eta = v\delta_x$, where δ_x is a point mass at some $x \in \mathbb{X}$). Suppose that we start with $\eta = v\lambda$, where λ is the normalized arclength measure on L. Then η is the tangent measure to L and $\mathbb{M}^n(\eta)$ is the tangent measure to $W^n(L)$ (the nth stage in the construction of $\mathcal{C}$). Thus, it is natural to consider μ as the tangent measure to $\mathcal{C}$.

Heuristically, one thinks of μ as $d\mu(s) = \overrightarrow{T}(s)ds$, where $\overrightarrow{T}(s)$ is the "unit tangent" vector at the point s and s is the arclength parameter. However, we hasten to say that this interpretation is not precise since neither the unit tangent nor the arclength exist, as $\mathcal{C}$ is, in general, a fractal curve with infinite length and no tangent anywhere. However, this analogy is very useful, and there are some striking similarities. Let $B \subseteq \mathcal{C}$ be a segment. Then

$$\int_B d\mu = \mu(B) = B_{final} - B_{initial},$$

the displacement vector from the initial point of the segment B to the final point of the segment B. Compare this with the case of a smooth curve where

$$\int_B \overrightarrow{T}(s) \, ds = B_{final} - B_{initial}.$$

Now we must return to the question of the contractivity of M and the existence of some $\mathcal{M}_{q,k}(X, \mathbb{R}^2)$ that is invariant with respect to M. Considering the von Koch curve example, we found that $\sum_i s_i \|R_i\| = 4/9 < 1$, so the IFS operator M should be contractive. However, if we let μ_0 be the tangent measure to the initial line segment L and μ_n be the tangent measure to the piecewise linear curve $W^n(L)$, we find that

$$\|\mu_n\| = \lfloor 4^n/3 \rfloor (2/3^n) + (1/3^n),$$

which clearly shows that $\|\mu_n\| \to \infty$. Thus, there is no $C > 0$, so if $\|\nu\| < C$, then $\|M\nu\| < C$ as well. The basic reason for this is that the von Koch curve has infinite length (also $\sum_i \|\alpha_i\| = \frac{4}{3}$, which is related to the growth rate of $\|\mu_n\|$).

This is the same situation that we saw in the case of signed measures and that prompted consideration of the "generalized measures" in Sect. 5.1.3.

5.2.5 Generalized vector measures

Since the vector measures we are considering have values in $\mathbb{R}^m$, the situation for "generalized vector measures" is a simple extension of that in Sect. 5.1.3. That is, we have

$$\mathrm{Lip}(X, \mathbb{R}^m) = \mathrm{Lip}(X, \mathbb{R}) \oplus \mathrm{Lip}(X, \mathbb{R}) \oplus \cdots \oplus \mathrm{Lip}(X, \mathbb{R}) \qquad (5.16)$$

and thus

$$\mathrm{Lip}(X, \mathbb{R}^m)^* = \mathrm{Lip}(X, \mathbb{R})^* \oplus \mathrm{Lip}(X, \mathbb{R})^* \oplus \cdots \oplus \mathrm{Lip}(X, \mathbb{R})^*. \qquad (5.17)$$

As in the scalar case, we also have

$$d_{MK}(\mu, \nu) = \sup \left\{ \int_X f \cdot d(\mu - \nu) : f \in \mathrm{Lip}_1(X, \mathbb{R}^m) \right\}$$

$$= \sup \left\{ \int_X f \cdot d(\mu - \nu) : \|f\|_{\mathrm{Lip}} \leq 1 \right\}$$

$$= \|\mu - \nu\|_{\mathrm{Lip}^*}$$

for two vector measures for which $\mu(X) = \nu(X)$.

Definition 5.31. Let $q \in \mathbb{R}^m$ be fixed and define

$$S_q(X, \mathbb{R}^m) = \{\varphi \in \mathrm{Lip}(X, \mathbb{R}^m)^* : \varphi(1) = q\}.$$

We extend the action of the IFS operator from Definition 5.24 to $\mathrm{Lip}(\mathbb{X}, \mathbb{R}^m)^*$ in the same way as we did for signed measures.

Definition 5.32. Fix $q \in \mathbb{R}^m$. Let the N maps $w_i : \mathbb{X} \to \mathbb{X}$ be Lipschitz with constant s_i and $\alpha_i \in \mathbb{R}$ satisfy $\sum_i \alpha_i q = q$. Define the IFS operator M on $\mathcal{S}_q(\mathbb{X}, \mathbb{R}^m)$ by setting for each $\varphi \in \mathcal{S}_q(\mathbb{X}, \mathbb{R}^m)$ and $f \in \mathrm{Lip}(\mathbb{X}, \mathbb{R}^m)$

$$\mathrm{M}\varphi(f) = \varphi \left(\sum_i \alpha_i^* f \circ w_i \right). \tag{5.18}$$

The proof of the next proposition is virtually identical to the proof of Proposition 5.12. We leave the obvious modifications to the reader.

Proposition 5.33. *The operator M defined as in (5.18) satisfies*

$$\|\mathrm{M}\varphi - \mathrm{M}\psi\|_{Lip^*} \leq \left(\sum_i s_i |\alpha_i^*| \right) \|\varphi - \psi\|_{Lip^*}.$$

Corollary 5.34. *If the operator M defined in (5.18) satisfies the condition $\sum_i s_i |\alpha_i^*| < 1$, then there is a unique "generalized vector measure" $\varphi \in \mathcal{S}_q(\mathbb{X}, \mathbb{R}^m)$ for which*

$$\|\mathrm{M}^n \mu - \varphi\|_{Lip^*} \to 0$$

for any $\mu \in \mathcal{M}(\mathbb{X}, \mathbb{R}^m)$ with $\mu(\mathbb{X}) = q$.

5.2.6 Green's theorem for planar domains with fractal boundaries

Suppose that we have a compact domain $D \subset \mathbb{R}^2$ whose boundary is the union of M fractal curves $\mathcal{C}_i$, each generated by an IFS. Suppose further that each curve $\mathcal{C}_i$ has no self-intersections and that $\mathcal{C}_i$ and $\mathcal{C}_j$ only intersect at a point, and only if they are adjacent on the boundary of D. Finally, we suppose that the Lebesgue measure of each curve $\mathcal{C}_i$ is zero.

Let $f : \mathbb{R}^2 \to \mathbb{R}^2$ be a smooth function. For each $\mathcal{C}_i$, we can compute

$$\int_{\mathcal{C}_i} f(x) \cdot d\mu_i(x)$$

where μ_i is the tangent vector measure on $\mathcal{C}_i$. Thus,

$$\int_{\partial D} f(x) \cdot d\mu(x) = \sum_i \int_{\mathcal{C}_i} f(x) \cdot \ d\mu_i(x),$$

where we orient the curves $\mathcal{C}_i$ in a consistent counterclockwise manner.

Let A_i and B_i denote the initial point and endpoint, respectively, for each curve $\mathcal{C}_i$, and let $v_i = \overrightarrow{A_i B_i}$. When we combine these vectors (or the line segments) together, we obtain a polygon D_1 as a first approximation to D. Applying the IFS of $\mathcal{C}_i$ to v_i yields a polygonal curve with endpoints A_i and B_i. Combining these segments together yields a polygon D_2 as a second approximation to D. The boundary of D_2, ∂D_2, is an approximation to ∂D. Iterating this procedure yields a sequence of polygons D_n that approximate D along with the boundaries ∂D_n that approximate ∂D. By construction, we know that $\partial D_n \to \partial D$ in the Hausdorff metric. Furthermore, we can use the IFS for $\mathcal{C}_i$ to obtain an IFS operator $\mathbb{M}_i$ for the tangent vector measure on $\mathcal{C}_i$. Using the $\mathbb{M}_i$ we obtain a sequence of vector measures μ_n such that μ_n is the tangent vector measure for ∂D_n.

Lemma 5.35. *Let χ_D and χ_{D_n} be the characteristic functions of D and D_n, respectively. Then $\chi_D \to \chi_{D_n}$ pointwise for almost all x.*

Proof. By assumption, $\lambda(\partial D) = 0$, so we only consider $x \notin \partial D$.

Suppose that $x \in \text{int}(D)$. Let $\varepsilon = dt(x, \partial D)$. Let N be large enough that for $n \geq N$ we have $d_{\mathbb{H}}(\partial D_n, \partial D) \leq \varepsilon/2$. Then $x \in \text{int}(D_n)$ as well, so $\chi_{D_n}(x) = \chi_D(x) = 1$.

Suppose that $x \notin D$. Again, let $\varepsilon = d(x, \partial D)$ and N be large enough that if $n \geq N$ we have $d_{\mathbb{H}}(\partial D_n, \partial D) \leq \varepsilon/2$. Then $x \notin D_n$ as well, so $\chi_{D_n}(x) = \chi_D(x) = 0$.

Thus, $\chi_{D_n} \to \chi_D$ pointwise for almost every x.   □

This convergence is illustrated in Fig. 5.8.

We are now in a position to prove Green's theorem for fractal curves.

Theorem 5.36. *(Green's theorem for fractal domains) Let $D \subset \mathbb{R}^2$ be a compact domain, with ∂D being the disjoint union of finitely many fractal curves and $\dim_H(\partial D) < 2$. If f is a smooth vector field, then*

$$\int_{\partial D} f(x) \cdot d\mu(x) = \int_D \left(\frac{\partial f_2}{\partial x} - \frac{\partial f_1}{\partial y} \right) dx dy,$$

where μ is the tangent vector measure to ∂D.

Proof. By our assumptions on f, we know that

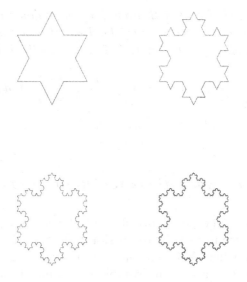

Fig. 5.8: Illustration of D_1, D_2, D_3, and D and their boundaries.

$$\chi_{D_n} \left(\frac{\partial f_2}{\partial x} - \frac{\partial f_1}{\partial y} \right)$$

is a bounded sequence of functions that converges pointwise almost everywhere. Thus, by the bounded convergence theorem,

$$\int_{D_n} \left(\frac{\partial f_2}{\partial x} - \frac{\partial f_1}{\partial y} \right) dxdy \to \int_{D} \left(\frac{\partial f_2}{\partial x} - \frac{\partial f_1}{\partial y} \right) dxdy$$

as $n \to \infty$. Furthermore, for each n, since D_n is a polygonal set, we have

$$\int_{\partial D_n} f(x) \cdot d\mu_n(x) = \int_{D_n} \left(\frac{\partial f_2}{\partial x} - \frac{\partial f_1}{\partial y} \right) dxdy,$$

where μ_n is the tangent vector measure for ∂D_n.

Finally, we know that $\mu_n \to \mu$ by construction and the contractivity of the IFS operators $\mathbb{M}_i$, and thus

$$\int_{\partial D_n} f(x) \cdot d\mu_n(x) \to \int_{\partial D} f(x) \cdot d\mu(x).$$

$\square$

The same techniques allow us to prove the corresponding divergence theorem for the normal vector measure to the boundary of D, which we state simply below.

Theorem 5.37. *(Divergence theorem for fractal domains) Let $D \subset \mathbb{R}^2$ be a compact domain with ∂D being the disjoint union of finitely many fractal curves and $\dim_H(\partial D) < 1$. If f is a smooth vector field, then*

$$\int_{\partial D} f(x) \cdot d\mu(x) = \int_D \left(\frac{\partial f_1}{\partial x} + \frac{\partial f_2}{\partial y} \right) \, dx dy,$$

where μ is the normal vector measure to ∂D.

5.2.7 Some generalizations for vector measures

We now discuss some generalizations of the ideas regarding vector measures. First, we mirror the discussion from Sect. 5.1.4. That is, we discuss the case where $\mathbb{X}$ is noncompact but we do assume that $\mathbb{X}$ is locally compact, separable, and complete as a metric space.

We use the same notation as before:

- $C_c(\mathbb{X}, \mathbb{R}^m)$ is the collection of continuous $f : \mathbb{X} \to \mathbb{R}^m$ such that there is some compact $K \subseteq \mathbb{X}$ with $f(\mathbb{X} \setminus K) = 0$.
- $C_0(\mathbb{X}, \mathbb{R}^m)$ is the collection of continuous $f : \mathbb{X} \to \mathbb{R}^m$ such that for all $\epsilon > 0$ there is some compact $K \subseteq \mathbb{X}$ with $\|f(x)\| < \epsilon$ for all $x \notin K$. The uniform closure of $C_c(\mathbb{X}, \mathbb{R}^m)$ is equal to $C_0(\mathbb{X}, \mathbb{R}^m)$.
- $C^*(\mathbb{X}, \mathbb{R}^m)$ is the collection of all bounded continuous $f : \mathbb{X} \to \mathbb{R}^m$.

Again by the Riesz representation theorem, we have that $\mathcal{M}(\mathbb{X}, \mathbb{R}^m)$ with its variation norm is the dual space to $C_0(\mathbb{X}, \mathbb{R}^m)$ equipped with the supremum norm.

Definition 5.38. Suppose $q \in \mathbb{R}^m$ and $k > 0$. Let $\mathcal{M}_{q,k}(\mathbb{X}, \mathbb{R}^m)$ be the set of all $\mu \in \mathcal{M}(\mathbb{X}, \mathbb{R}^m)$ such that

1. $\mu(\mathbb{X}) = q$;
2. $\|\mu\| = |\mu|(\mathbb{X}) \le k$;
3. $\int_{\mathbb{X}} d(x, a) \, d|\mu|(x) < \infty$ for some, and hence any, $a \in \mathbb{X}$.

We still use the same definition for d_{MK} as in the case when $\mathbb{X}$ is compact. That is,

$$d_{MK}(\mu, \nu) = \sup \left\{ \int_{\mathbb{X}} f \, d(\mu - \nu) : f \in \mathrm{Lip}_1(\mathbb{X}, \mathbb{R}^m)) \right\}.$$

Theorem 5.39. $(\mathcal{M}_{q,k}(\mathbb{X}, \mathbb{R}^m), d_{MK})$ *is a complete metric space.*

Proof. The fact that d_{MK} is finite on $\mathcal{M}_{q,k}(\mathbb{X}, \mathbb{R}^m)$ follows from the inequality $|\mu^{\mathbf{e}_i}| \leq |\mu|$ and the finiteness of d_{MK} on $\mathcal{M}_{q,k}(\mathbb{X}, \mathbb{R})$ from Theorem 5.18.

If (μ_n) is a d_{MK}-Cauchy sequence in $\mathcal{M}_{q,k}(\mathbb{X}, \mathbb{R}^m)$, then for each i we know that $(\mu_n^{\mathbf{e}_i})$ is d_{MK}-Cauchy in $\mathcal{M}_{q,k}(\mathbb{X}, \mathbb{R})$ and hence converges. Putting the components together, we know that $\mu_n \to \mu \in \mathcal{M}(\mathbb{X}, \mathbb{R}^m)$ and that $\mu(\mathbb{X}) = q$ as well. The fact that $\|\mu\| \leq k$ is again due to the fact that a weak limit cannot increase the norm and hence $\|\mu\| \leq \limsup \|\mu_n\| = k$. □

5.2.7.1 Infinite-dimensional vector measures

Now we turn to a brief discussion of vector measures that take their values in an infinite-dimensional Banach space $\mathbb{B}$. Again we point the reader interested in a more complete development of the theory of vector measures to the very good book [48]. Here we are content to give only enough details to describe our particular situation.

There are several different assumptions you can make about a vector measure. We will restrict ourselves to the situation most useful to us. Let $\mathbb{X}$ be a fixed compact (or locally compact, complete, and separable for a slightly more general situation) metric space.

By a *vector measure* μ *with values in* $\mathbb{B}$ we will mean a set function $\mu : \mathcal{B} \to \mathbb{B}$ that satisfies

1. $\mu(\emptyset) = 0$ and
2. for each disjoint sequence $E_n \in \mathcal{B}$ and each $\varphi \in \text{Ball}_{\mathbb{B}^*}$, we have $\varphi\mu(\bigcup_n E_n) = \sum_n \varphi\mu(E_n)$,

where we use $\text{Ball}_{\mathbb{B}^*}$ to denote the set $\{\varphi \in \mathbb{B}^* : \|\varphi\| \leq 1\}$.

Notice that by assumption the sum $\sum_n \varphi\mu(E_n)$ converges unconditionally (that is, all rearrangements converge and to the same value) and so does each subseries. It turns out that assumption 2 implies the stronger property that $\sum_n \mu(E_n)$ converges in the norm of $\mathbb{B}$ (this is the Orlicz-Pettis theorem).

Notice that we are also assuming that $\varphi \circ \mu$ is a signed measure for all $\varphi \in \mathbb{B}^*$. To make things simpler, we will assume that each of these signed measures has finite variation. In the language of vector measures, we assume that the *semivariation* of μ is finite. For vector measures with infinite-dimensional values, there are two important notions of variation, the *total variation* and the *semivariation*; these two notions coincide for finite-dimensional vector measures (see [48]). The semivariation is defined to be the quantity

$$\sup\{|\varphi\mu|(\mathbb{X}) : \varphi \in \text{Ball}_{\mathbb{B}^*}\},$$

while the total variation is (as usual)

$$\sup\left\{\sum_i \|\mu(E_i)\| : \text{ all finite partitions}\{E_i\} \text{ of } \mathbb{X}\right\}.$$

If μ is of bounded variation, it turns out that the total variation $|\mu|$ of μ is a finite (positive) measure. Measures of bounded semivariation have bounded ranges and thus are called *bounded vector measures*; thus all our measures are bounded in this sense. In a slight change of notation, we will denote the total variation of μ by $|\mu|$ and the semivariation of μ by $\|\mu\|$. In general, $E \to \|\mu(D)\|$ is only a monotone finitely subadditive set function on $\mathcal{B}$.

Definition 5.40. We denote by $\mathcal{M}(\mathbb{X}, \mathbb{B})$ the collection of all vector measures $\mu : \mathcal{B} \to \mathbb{B}$ that are of bounded semivariation. Furthermore, for a fixed $q \in \mathbb{B}$ and $k > 0$, we denote by $\mathcal{M}_{q,k}(\mathbb{X}, \mathbb{B})$ the subset of $\mathcal{M}(\mathbb{X}, \mathbb{B})$ that satisfies

1. $\mu(\mathbb{X}) = q$,
2. $\|\mu\|(\mathbb{X}) \leq k$, and
3. for some fixed $a \in \mathbb{X}$ we have

$$\sup\left\{\int_{\mathbb{X}} d(x, a)\, d|\varphi\mu|(x) : \varphi \in \text{Ball}_{\mathbb{B}^*}\right\} < \infty.$$

In the case where $\mathbb{X}$ is compact, the third condition is not necessary, as it follows from the first two and the boundedness of $\mathbb{X}$. Notice that the second condition implies that $|\varphi\mu|(\mathbb{X}) \leq k$ for all $\varphi \in \text{Ball}_{\mathbb{B}^*}$, and thus each $\varphi\mu \in \mathcal{M}_{\rho,k}(\mathbb{X}, \mathbb{R})$, where $\rho = \varphi(q)$.

If $\mathbb{B}$ is a Hilbert space and (x_n) is an orthonormal basis, then it is simple to construct elements of $\mathcal{M}_{q,k}(\mathbb{X}, \mathbb{B})$ for large k. Take a sequence (a_n) such that $\sum_n a_n x_n = q$ and a sequence (μ_n) of signed measures in $\mathcal{M}_{1,k}(\mathbb{X}, \mathbb{R})$. Define μ by

$$\mu(E) = \sum_n a_n x_n \mu_n(E).$$

Then $\mu(\mathbb{X}) = q$ by construction, and by the Cauchy-Bunyakovskii-Schwarz inequality we know $\|\mu\| \leq k \left(\sum_n |a_n|^2\right)^{1/2} = k\|q\|$ so $\mu \in \mathcal{M}_{q,\|q\|k}(\mathbb{X}, \mathbb{B})$.

In a Banach space, a similar construction is possible as long as a suitably nice basis (x_n) exists. In particular, what is needed is a bound on the norm of $\sum_n a_n x_n$ based on some bound on the sequence (a_n).

On $\mathcal{M}_{q,k}(\mathbb{X}, \mathbb{B})$, we use a modified Monge-Kantorovich metric defined by

$$d_{MK}(\mu, \nu) = \sup\left\{ \int_{\mathbb{X}} f\, d\varphi\mu : f \in \mathrm{Lip}_1(\mathbb{X}), \varphi \in \mathrm{Ball}_{\mathbb{B}^*} \right\}$$
$$= \sup\{d_{MK}(\varphi\mu, \varphi\nu) : \varphi \in \mathrm{Ball}_{\mathbb{B}^*}\}. \tag{5.19}$$

At this point, we mention that it is not necessary to use every $\varphi \in \mathrm{Ball}_{\mathbb{B}^*}$; it is sufficient to use a dense set of such φ, even only restricting to $\|\varphi\| = 1$. Furthermore, even though it might change the metric, it is also possible to use smaller sets of φ, say $S \subset \mathrm{Ball}_{\mathbb{B}^*}$, as long as S has enough elements to separate points in $\mathbb{B}$ (i.e., if $\varphi(x) = \varphi(y)$ for all $\varphi \in S$ implies $x = y$). One such choice of subset could be a basis of appropriate type (a Schauder basis, for instance, would work) if such a basis exists.

Theorem 5.41. *The space $\mathcal{M}_{q,k}(\mathbb{X}, \mathbb{B})$ is complete under the metric d_{MK}.*

Proof. We prove the case where $\mathbb{X}$ is compact. Just as in the proof of Theorem 5.4, we have $d_{MK}(\varphi\mu, \varphi\nu) \leq 2k\, \mathrm{diam}(\mathbb{X})$, and therefore $d_{MK}(\mu, \nu) \leq 2k\, \mathrm{diam}(\mathbb{X})$ and thus is finite.

Suppose that (μ_n) is a d_{MK}-Cauchy sequence from $\mathcal{M}_{q,k}(\mathbb{X}, \mathbb{B})$. Then, for each $\varphi \in \mathrm{Ball}_{\mathbb{B}^*}$, we know that $(\varphi\mu_n)$ is d_{MK}-Cauchy in $\mathcal{M}_{\varphi q,k}(\mathbb{X}, \mathbb{R})$ and thus converges to some $\nu_\varphi \in \mathcal{M}_{\varphi q,k}(\mathbb{X}, \mathbb{R})$. Thus, for each $\varphi \in \mathrm{Ball}_{\mathbb{B}^*}$ and each E, we have $\varphi\mu_n(E) \to \nu_\varphi(E)$. However, then this means that there is some $\nu(E) \in \mathbb{B}$ for which $\nu_\varphi(E) = \varphi\nu(E)$ and so $\nu : \mathcal{B} \to \mathbb{B}$ is well-defined and $\varphi\nu = \nu_\varphi$ for all $\varphi \in \mathrm{Ball}_{\mathbb{B}^*}$. Now, since each ν_φ is countably additive, so is $\varphi\nu$, and so $\nu \in \mathcal{M}(\mathbb{X}, \mathbb{B})$. Since $\nu_\varphi(\mathbb{X}) = \varphi q = \lim_n \varphi\mu(\mathbb{X})$, we know that $\nu(\mathbb{X}) = q$. Finally, we have $|\varphi\nu|(\mathbb{X}) = |\nu_\varphi|(\mathbb{X}) \leq k$ for all φ and thus $\nu \in \mathcal{M}_{q,k}(\mathbb{X}, \mathbb{B})$. $\quad\square$

The same types of IFS operators can be defined on $\mathcal{M}_{q,k}(\mathbb{X}, \mathbb{B})$ as in both the signed measure and finite-dimensional vector-measure cases. Similar results about contractivity hold. If $\mathbb{B} = \mathcal{P}(\Lambda)$, the space of probability measures on Λ, we can also define probability measure-valued measures and an IFS on these objects. This is an infinite-dimensional version of the IFS defined in Sect. 5.2.3 on coloured fractals.

5.3 Set-valued measures

The final class of measures we discuss is the situation of *set-valued measures* (or *multimeasures*). We will discuss two different types of multimeasures, depending on the notion of "additivity" used. The first one is the more common, where *Minkowski* (or pointwise) addition of sets is used. For this type of multimeasure, we will mostly restrict our attention to multimeasures that take values as compact and convex subsets of $\mathbb{R}^m$. The second type of multimeasure is less common, where we "add" a collection of sets by forming their union, and in this case our context will be a general complete metric space. We present results on "union-additive" multimeasures at the end of this section.

The material on Minkowski-additive measures is primarily based on the paper in progress [76], while the material on "union-additive" multimeasures is based on [106].

As before, we start with a brief review of basic definitions and properties, then define an appropriate space of measures, and finally construct IFS operations on this space. In the case of the "usual" (Minkowski additive) multimeasures, we again have the complication of noncompact spaces to deal with. The papers [6, 139] contain a wealth of basic information on multimeasures. Basic results on convex sets and support functions are drawn from many sources, but particularly from [146].

Let A_n be a sequence of subsets of $\mathbb{R}^m$. We define the *Minkowski sum* of A_n to be

$$\sum_n A_n = \left\{ \sum_n x_n : x_n \in A_n, \sum_n \|x_n\| < \infty \right\}.$$

Definition 5.42. Let $(\Omega, \mathcal{A})$ be a measure space. A *set-valued measure* or *multimeasure* μ on $(\Omega, \mathcal{A})$ is a set-function defined on the σ-algebra $\mathcal{A}$ and with values nonempty subsets of $\mathbb{R}^m$, which is additive in the Minkowski sense. That is, for all disjoint A_n, we have

$$\mu \left(\bigcup_n A_n \right) = \sum_n \mu(A_n).$$

If $0 \neq x \in \mu(\emptyset)$, then $nx \in \mu(\emptyset)$ for all $n \in \mathbb{N}$ and thus $\mu(\emptyset)$ is unbounded. However, by additivity, this means that $\mu(A)$ is unbounded for all A since $\mu(A)$ contains a translation of $\mu(\emptyset)$. Since we wish to consider only bounded multimeasures, we impose the condition that $\mu(\emptyset) = \{0\}$.

It is not generally true that if $A_1 \subseteq A_2$, then $\mu(A_1) \subseteq \mu(A_2)$. However, if $0 \in \mu(A_1)$, then it is true. In fact it is easy to see that $0 \in \mu(A)$ for all A gives that $A \subseteq B$ implies $\mu(A) \subseteq \mu(B)$, so μ is monotone. This is one type of *positivity* that can be defined for multimeasures.

For a set $A \subset \mathbb{R}^m$, we denote by $|A|$ the quantity (called the *norm* of A)

$$|A| = \sup\{|a| : a \in A\}.$$

Definition 5.43. A multimeasure μ is *bounded* if $\mu(\Omega)$ is a bounded set. An *atom* of a multimeasure is a set $A \in \mathcal{A}$ such that $\mu(A) \neq \{0\}$ but for any $E \subseteq A$ either $\mu(E) = \{0\}$ or $\mu(A \setminus E) = \{0\}$. A multimeasure with no atoms is *nonatomic*.

Notice that μ being bounded is equivalent to $\mu(A)$ being bounded for all A since $\mu(\Omega)$ contains a translation of $\mu(A)$ for any A.

Proposition 5.44. *Let μ be a bounded multimeasure.*

- *If μ is nonatomic, then $\mu(A)$ is convex for all A.*
- *If μ is nonatomic, then the* range *of μ, the set $\cup_A \mu(A)$, is convex.*
- *The set-function $\bar{\mu}$ defined by $\bar{\mu}(A) = \mathrm{cl}(\mu(A))$ is also a multimeasure.*
- *The set-function $\mathrm{co}\,\mu$ defined by $\mathrm{co}\,\mu(A) = \mathrm{co}(\mu(A))$ is also a multimeasure.*

For a multimeasure μ, the set-function $|\mu|$ is defined as

$$|\mu|(A) = \sup\left\{\sum |\mu(A_i)| : \text{ finite measurable partitions of } A\right\}.$$

It is not so hard to check that $|\mu|$ is a measure. It is called the *variation measure* of the multimeasure μ. If $|\mu|$ is a finite measure, then μ is said to be of *bounded variation*; a multimeasure of bounded variation is clearly bounded.

Definition 5.45. (Support function) Let $A \subset \mathbb{R}^m$. The *support function* of A is the function $\mathrm{supp}\,(\cdot, A) : \mathbb{R}^m \to \mathbb{R}$ given by

$$\mathrm{supp}\,(p, A) = \sup_{a \in A} p \cdot a.$$

Notice that $|\mathrm{supp}\,(\cdot, A)| \leq |A|$ for any p and, in fact, $|A| = \sup_{\|p\|=1} \mathrm{supp}\,(p, A)$. The support function $\mathrm{supp}\,(\cdot, A)$ is convex and positively homogeneous (i.e., if $\lambda \geq 0$, then $\mathrm{supp}\,(\lambda p, A) = \lambda \,\mathrm{supp}\,(p, A)$). If $A \subset \mathbb{R}^m$ is convex, then

$$A = \bigcap_{p \in \mathbb{R}^m} \{x \in \mathbb{R}^m : x \cdot p \leq \operatorname{supp}(p, A)\}.$$

This means that A can be recovered from its support function. Further note that A is bounded iff $\operatorname{supp}(p, A)$ is bounded over $\{p : \|p\| = 1\}$.

Lemma 5.46. *If s is a convex function from $\mathbb{R}^m$ to $(-\infty, \infty)$ that is positively homogeneous, then it is the support function of a certain compact and convex set A, namely*

$$A = \bigcap_{p \in \mathbb{R}^m} \{x \in \mathbb{R}^m : x \cdot p \leq s(p)\}.$$

For multimeasures, the support function plays a particularly useful role, as can be seen in the following result.

Proposition 5.47. *(Scalarization) Let μ be a multimeasure. Then, for all $p \in \mathbb{R}^m$, the set-function μ^p defined by*

$$\mu^p(A) = \operatorname{supp}(p, \mu(A))$$

is a (countably additive) signed measure. If μ is bounded, then so is μ^p for all p.

Notice that if $0 \in \mu(A)$ for all A, then $\mu^p(A) \geq 0$, so μ^p is a measure for all p. There is a converse to Proposition 5.47.

Our basic approach to working with a multimeasure is to work with the collection of its scalarizations.

Proposition 5.48. *Let ν_p, $p \in \mathbb{R}^m$ be a family of signed measures, and suppose that the function $p \mapsto \nu_p(E)$ is convex and positively homogeneous and that $|\nu_p(E)| < \infty$ for all E. Define μ by*

$$\mu(E) = \bigcap_{\|p\|=1} \{x \in \mathbb{R}^m : x \cdot p \leq \nu_p(E)\}.$$

Then μ is a multimeasure and $\mu^p = \nu_p$ for all p.

Proof. We give the idea of how to prove the additive property. For simplicity, we restrict our discussion to the case of two disjoint sets A_1, A_2. First, we comment that by Lemma 5.46 and positive homogeneity we have $\nu_p(E) = \operatorname{supp}(p, \mu(E))$ for all p and $E \in \mathcal{B}$.

Since each ν_p is a signed measure, $\nu_p(A_1 \cup A_2) = \nu_p(A_1) + \nu_p(A_2)$. For $x \in \mu(A_1)$ and $y \in \mu(A_2)$, we see that

$$(x + y) \cdot p = x \cdot p + y \cdot p \leq \nu_p(A_1) + \nu_p(A_2) = \nu_p(A_1 \cup A_2)$$

for all p and thus $x + y \in \mu(A_1 \cup A_2)$, so $\mu(A_1) + \mu(A_2) \subseteq \mu(A_1 \cup A_2)$.

For the reverse inclusion, suppose that $z \in \mu(A_1 \cup A_2)$ with $z \notin \mu(A_1) + \mu(A_2)$. Since $\mu(A_1) + \mu(A_2)$ is a compact and convex set, there is some p^* such that

$$z \cdot p^* > (x + y) \cdot p^* \quad \text{for all } x \in \mu(A_1), y \in \mu(A_2).$$

However, since $\mu(A_1), \mu(A_2)$ are compact, there are $x^* \in \mu(A_1)$ and $y^* \in \mu(A_2)$ with $x^* \cdot p^* = \mathrm{supp}\,(p^*, \mu(A_1)) = \nu_{p^*}(A_1)$ and $y^* \cdot p^* = \mathrm{supp}\,(p^*, \mu(A_2)) = \nu_{p^*}(A_2)$ such that

$$(x^* + y^*) \cdot p^* < z \cdot p^* \leq \nu_{p^*}(A_1 \cup A_2) = \nu_{p^*}(A_1) + \nu_{p^*}(A_2),$$

which is a contradiction. $\square$

5.3.1 Complete space of multimeasures

We now turn to the question of constructing a space of multimeasures that will be a complete metric space. Our metric is again a modification of the Monge-Kantorovich metric, but adapted to the set-valued situation. *We again start with the basic case of $\mathbb{X}$ being a compact metric space.*

We will let $\mathcal{K}(\mathbb{R}^m)$ denote the collection of all nonempty, compact, and convex subsets of $\mathbb{R}^m$. All multimeasures will take values in $\mathcal{K}(\mathbb{R}^m)$. This is not a strong restriction since by the comments above, if μ is bounded and nonatomic, its values are bounded and convex, so we simply take closures to obtain compact values.

Definition 5.49. Let $Q, K \in \mathcal{K}(\mathbb{R}^m)$ with $Q \subseteq K$. Let $\mathcal{M}_{Q,K}(\mathbb{X}, \mathbb{R}^m)$ denote the set of all Borel multimeasures ϕ on $\mathbb{X}$ such that $\phi(\mathbb{X}) = Q$ and $\phi(E) \subseteq K$ for all E.

If $0 \in \phi(E)$ for all E, then the sets K and Q can be the same, as, in this case, $\phi(E) \subseteq \phi(\mathbb{X})$ for all E. Notice that $\phi(E) \subseteq K$ implies that $\phi^p(E) \leq \mathrm{supp}\,(p, K)$ and thus $|\phi^p|(E) \leq |K|$ for $\|p\| = 1$, so the variation of ϕ^p is finite and uniformly bounded in $p \in \mathcal{S}^1$.

Let $\mathcal{S}^1 = \{p \in \mathbb{R}^m : \|p\| = 1\}$ denote the unit sphere in $\mathbb{R}^m$. Recalling the definition of $\mathcal{M}_{q,k}(\mathbb{X}, \mathbb{R})$ (Definition 5.3 in Sect. 5.1.1), it is important to note that $\phi \in \mathcal{M}_{Q,K}(\mathbb{X}, \mathbb{R}^m)$ and $p \in \mathcal{S}^1$ together imply that $\phi^p \in \mathcal{M}_{q,k}(\mathbb{X}, \mathbb{R})$, where

$$q = \mathrm{supp}\,(p, Q) \quad \text{and} \quad k \geq \mathrm{supp}\,(p, K).$$

Definition 5.50. Let $\phi, \psi \in \mathcal{M}_{Q,K}(\mathbb{X}, \mathbb{R}^m)$. Define

$$d_{MK}(\phi, \psi) = \sup_{p \in \mathcal{S}^1} d_{MK}(\phi^p, \psi^p)$$

$$= \sup_{p \in \mathcal{S}^1} \sup \left\{ \int_{\mathbb{X}} f \cdot (d\phi^p - d\psi^p) : f \in \mathrm{Lip}_1(\mathbb{X}) \right\}.$$

Theorem 5.51. *The space $\mathcal{M}_{Q,K}(\mathbb{X}, \mathbb{R}^m)$ is complete under the metric d_{MK}.*

Proof. Since $\|\phi^p\| \leq |K|$ for any $\phi \in \mathcal{M}_{Q,K}(\mathbb{X}, \mathbb{R}^m)$ and $p \in \mathcal{S}^1$, we have by the same inequalities as in Theorem 5.4 that for $\phi, \psi \in \mathcal{M}_{Q,K}(\mathbb{X}, \mathbb{R}^m)$

$$\int_{\mathbb{X}} f d(\phi^p - \psi^p) \leq 2|K| \operatorname{diam}(\mathbb{X})$$

uniformly in $p \in \mathcal{S}^1$, and thus $d_{MK}(\phi, \psi)$ is finite.

If $d_{MK}(\phi, \psi) = 0$, then $d_{MK}(\phi^p, \psi^p) = 0$ and thus $\phi^p = \psi^p$ for all $p \in \mathcal{S}^1$. However, as $\phi(A)$ and $\psi(A)$ are convex, this means that $\phi(A) = \psi(A)$ (since convex sets are determined by their support functions). Thus $d_{MK}(\phi, \psi) = 0$ implies that $\phi = \psi$. The converse is simple.

The other properties of a metric are easy to show, so all we have left to demonstrate is completeness.

Let ϕ_n be a Cauchy sequence in $\mathcal{M}_{Q,K}(\mathbb{X}, \mathbb{R}^m)$, so for all $\epsilon > 0$ there exists n_0 such that for $n, m \geq n_0$ we have $d_{MK}(\phi_n, \phi_m) < \epsilon$. From the definition of d_{MK}, for fixed p, the sequence of scalar signed measures ϕ_n^p is a Cauchy sequence with respect to the d_{MK} metric. Furthermore, the mass $\phi_n^p(\mathbb{X})$ of each measure ϕ_n^p is fixed if p is fixed since

$$\phi_n^p(\mathbb{X}) = \sup_{l \in \phi_n(\mathbb{X})} l \cdot p = \sup_{l \in Q} l \cdot p.$$

So, by the completeness of the space of signed measures with respect to this metric (Theorem 5.4), we know that $\phi_n^p \to \mu_p$ for some signed measure μ_p, with convergence in the d_{MK} metric. We also observe that $\mu_p(\mathbb{X}) = \sup_{l \in Q} l \cdot p$. For all $E \in \mathcal{B}$, we get that

$$|\phi_n^q(E)| = \sup_{l \in \phi_n(E)} |q \cdot l| \leq \sup_{l \in K} |q \cdot l| \leq |K|,$$

and so $\phi_n^q(E)$ is uniformly bounded. This convergence is also uniform with respect to p by the definition of a Cauchy sequence using the d_{MK} metric on $\mathcal{M}_{Q,K}(\mathbb{X}, \mathbb{R}^m)$.

Now we wish to show that $\mu_p(E)$ (as p varies over S_1) is a support function for any given $E \in \mathcal{B}$. We need only show that, as a function of p, $\mu_p(E)$ satisfies the following properties for all fixed $E \in \mathcal{B}$:

- $p \to \mu_p(E)$ is convex and
- $p \to \mu_p(E)$ is positively homogeneous.

The functions $p \to \phi_n^p(E)$ (being support functions) are obviously convex and positively homogeneous (as functions of p for any fixed n and E). So we get for all $\alpha \geq 0$, for $p, p_1, p_2 \in \mathbb{R}^m$, and $E \in \mathcal{B}$,

$$\phi_n^{\alpha p}(E) - \alpha\phi_n^p(E) = 0$$

and

$$\phi_n^{p_1}(E) + \phi_n^{p_2}(E) - \phi_n^{p_1+p_2}(E) \geq 0.$$

Since these properties are satisfied for all E, taking the limit we get that $p \to \mu_p(E)$ is subadditive and positively homogeneous. This implies that $p \to \mu_p(E)$ is convex.

Using similar arguments, one can prove that $|\mu_p(E)| \leq |K|$ for all $E \in \mathcal{B}$. The function $\mu_p(E)$ is continuous in p for any fixed $E \in \mathcal{B}$, being convex and everywhere finite. We define the set-valued function ϕ by

$$\phi(E) = \bigcap_{q \in \mathcal{S}^1} \{x \in \mathbb{R}^m : x \cdot q \leq \mu_q(E)\}.$$

We next observe that

$$Q \subseteq \phi(\mathbb{X}) = \bigcap_{p \in \mathcal{S}^1} \{x : x \cdot p \leq \mu_p(\mathbb{X})\} = \bigcap_{p \in \mathcal{S}^1} \{x : x \cdot p \leq \sup_{l \in Q} l \cdot p\}.$$

If there exists $x^* \in \phi(\mathbb{X})$ and $x^* \notin Q$, then, using a standard separation argument in $\mathbb{R}^m$, we see there exists p^* such that $p^* \cdot x^* > p^* \cdot l$ for all $l \in Q$, and taking the maximum, we get $p^* \cdot x^* > \sup_{l \in Q} p^* \cdot l$, which is absurd. So $Q = \phi(\mathbb{X})$. On the other hand, ϕ is a multimeasure and $\operatorname{supp}(p, \phi) = \mu_p$, that is, $\phi_n \to \phi$ in the d_{MK} metric. $\square$

It is worth pointing out the fact that the same proof can show that $\mathcal{M}_{Q,k}(\mathbb{X}, \mathbb{R}^m)$ is complete, where this time the constraints on the multimeasures μ are $\mu(\mathbb{X}) = Q$ and $|\mu(E)| \leq k$ for all E. Clearly, $\mathcal{M}_{Q,K}(\mathbb{X}, \mathbb{R}^m) \subset \mathcal{M}_{Q,|K|}(\mathbb{X}, \mathbb{R}^m)$.

Finally, if ϕ_n is a sequence of *positive* multimeasures and $\phi_n \to \phi$ in the metric d_{MK}, then ϕ is positive as well. To see, this just notice that ϕ is positive iff ϕ^q is a nonnegative measure for all q and that weak convergence of measures preserves positivity. Thus, if $0 \in Q$, the subset of $\mathcal{M}_{Q,K}(\mathbb{X}, \mathbb{R}^m)$ consisting of positive multimeasures is a closed subspace and hence also complete.

5.3.2 *IFS operators on multimeasures*

We now turn to the task of defining appropriate IFS operators on the complete metric space $\mathcal{M}_{Q,K}(\mathbb{X}, \mathbb{R}^m)$ defined in the previous section. This task turns out to be rather more delicate than the similar task for either signed measures or vector-valued measures. The problem is defining an operator that maps some $\mathcal{M}_{Q,K}(\mathbb{X}, \mathbb{R}^m)$ into itself, so that there is a possibility of having a fixed point; the conditions for this are even more restrictive than the corresponding conditions for either signed or vector-valued measures. As the values of the measures become more complicated, the self-mapping conditions become more complicated as well. This is a problem similar to the case of vector-valued measures and the tangent measure to the von Koch curve, where the natural IFS operator increases the variation at each iterate. In that case, one solution (given in Sect. 5.2.5) was to view these "measures" via their action on the space of Lipschitz functions; recall that the same type of solution was also used for signed measures (see Sect. 5.1.3).

However, first we present the basic and more restrictive framework, so we work in the space $\mathcal{M}_{Q,K}(\mathbb{X}, \mathbb{R}^m)$, where $\mathbb{X}$ is compact and $Q \subseteq K \in \mathcal{K}(\mathbb{R}^m)$. Take $w_i : \mathbb{X} \to \mathbb{X}$, for $i = 1, 2, \ldots, N$ to be Lipschitz with Lipschitz constants c_i. Furthermore, take $T_i : \mathbb{R}^m \to \mathbb{R}^m$ to be linear and assume that

1. $\sum_i T_i Q = Q$ (total mass preservation) and
2. $\sum_{i \in S} T_i K \subset K$ for all $S \subseteq \{1, 2, \ldots, N\}$ (boundedness preservation).

We define the IFS operator $M : \mathcal{M}_{Q,K}(\mathbb{X}, \mathbb{R}^m) \to \mathcal{M}_{Q,K}(\mathbb{X}, \mathbb{R}^m)$ as

$$M\phi(B) = \sum_i T_i(\phi(w_i^{-1}(B))). \tag{5.20}$$

A simple argument will show that if $\phi \in \mathcal{M}_{Q,K}(\mathbb{X}, \mathbb{R}^m)$, then so is $M\phi$. To see the countable additivity of $M\phi$, it is useful to note that each T_i is continuous with respect to the Hausdorff metric on $\mathcal{K}(\mathbb{R}^m)$ since T_i is linear and thus Lipschitz.

These restrictions on the T_i can be thought of in two ways. First, one can fix Q and K and ask for all operators of the form (5.20) that will map $\mathcal{M}_{Q,K}(\mathbb{X}, \mathbb{R}^m)$ to itself. This naturally leads to the restrictions on T, that is, the conditions that $\sum_i T_i Q = Q$ and $\sum_{i \in S} T_i K \subset K$ for all $S \subset \{1, 2, \ldots, N\}$. If $0 \in K$, then the choice $T_i = p_i I$, where $\{p_i\}$ are a set of probabilities, always works. Of course, this simple structure might not result in very interesting behaviour for the IFS operator.

The other view is where we fix T_i and ask for some choice of K and Q such that the conditions are satisfied. This one is perhaps a bit more natural in the context of IFS theory. That is, we start with an operator of interest and then we look for some context (space) in which we can study the convergence properties of this operator. Notice that if the conditions are satisfied for K and Q, then because T_i are linear these conditions are also satisfied for λK and λQ for any $\lambda \in \mathbb{R}$. Thus a choice of w_i and T_i can define an IFS operator on $\mathcal{M}_{Q,K}(\mathbb{X}, \mathbb{R}^m)$ for many different choices of Q and K, and thus the fixed point of such an operator will depend on which space $\mathcal{M}_{Q,K}(\mathbb{X}, \mathbb{R}^m)$ we are interested in, in particular on the choice of K. This is similar to the standard case of IFS with probabilities where normally we consider probability measures but we are also free to study positive measures with any given (fixed) mass.

Notice that if $0 \in \phi(E)$ for all E, then $0 \in M\phi(E)$ for all E, as each T_i is linear. Thus, if ϕ is *positive*, then so is $M\phi$.

Theorem 5.52. *For the IFS operator as defined, above we have*

$$d_{MK}(M\phi_1, M\phi_2) \le \left(\sum_i c_i \|T_i\| \right) d_{MK}(\phi_1, \phi_2). \qquad (5.21)$$

Proof. First we note that for linear T and convex A we have

$$\sup_{q \in \mathcal{S}^1} \operatorname{supp}(q, TA) = \sup_{x \in TA} \sup_{q \in \mathcal{S}^1} q \cdot x = \sup_{y \in A} \|Ty\|$$
$$\le \|T\| \sup_{y \in A} \|y\| = \|T\| \sup_{q \in \mathcal{S}^1} \sup_{y \in A} q \cdot y$$
$$= \|T\| \sup_{q \in \mathcal{S}^1} \operatorname{supp}(q, A).$$

Now, for a given Lipschitz function f, we have

$$\sup_{q \in S_1} \int_{\mathbb{X}} f(x)\, d\left[\operatorname{supp}(q, M\phi_1(x)) - \operatorname{supp}(q, M\phi_2(x))\right]$$

$$= \sup_{q \in S_1} \int_{\mathbb{X}} f(x)\, d\left[\operatorname{supp}\left(q, \sum_i T_i\phi_1(w_i^{-1}(x))\right)\right.$$

$$\left. - \operatorname{supp}\left(q, \sum_i T_i\phi_2(w_i^{-1}(x))\right)\right]$$

$$= \sup_{q \in S_1} \int_{\mathbb{X}} f(x) \sum_i d\left[\operatorname{supp}(T_i^* q, \phi_1(w_i^{-1}(x))) - \operatorname{supp}(T_i^* q, \phi_2(w_i^{-1}(x)))\right]$$

$$\le \sup_{q \in S_1} \int_{\mathbb{X}} \left\{ \sum_i \|T_i\| f(w_i(y)) \right\} d\left[\operatorname{supp}(q, \phi_1(y)) - \operatorname{supp}(q, \phi_2(y))\right].$$

The function $\hat{f} = \sum_i \|T_i\| f \circ w_i$ has Lipschitz factor at most $\sum_i c_i \|T_i\|$. Thus, when we take the supremum over all functions with Lipschitz factor 1, we get

$$d_{MK}(M\phi_1, M\phi_2) \leq \left(\sum_i c_i \|T_i\| \right) d_{MK}(\phi_1, \phi_2),$$

as desired. $\square$

We say that M is *average contractive* if $\sum_i c_i \|T_i\| < 1$. As an immediate corollary, we get the following.

Corollary 5.53. *Suppose that the operator M is average contractive on $\mathcal{M}_{Q,K}(\mathbb{X}, \mathbb{R}^m)$. Then there is a unique invariant multimeasure $\phi \in \mathcal{M}_{Q,K}(\mathbb{X}, \mathbb{R}^m)$ with $M\phi = \phi$.*

We now give a few examples of this. We thank John Hutchinson for the examples of zonotopes.

Example 5.54. Our first example is a very simple type where the multimeasure is a product of a scalar (probability) measure μ and a compact and convex set Q. Let $Q \in \mathcal{K}(\mathbb{R}^m)$ be a *balanced* set (that is, for all $|\lambda| \leq 1$ we have $\lambda Q \subset Q$). Let $\mathbb{X} = [0,1]$ and $w_i(x) = x/2 + i/2$ for $i = 0, 1$. Furthermore, let $p_0 \in (0,1)$ and $p_1 = 1 - p_0$. Define $T_i = p_i I$. Then the invariant multimeasure for the IFS operator

$$M\phi(B) = T_0\phi(w_0^{-1}(B)) + T_1\phi(w_1^{-1}(B))$$

satisfies $\phi(B) = Q\mu(B)$, where μ is the invariant probability measure for the IFS with probabilities $\{w_0, w_1, p_0, p_1\}$ on $\mathbb{X}$. $\square$

Example 5.55. Take $\mathbb{X} = [0,1]$ and $w_i(x) = x/3 + 2i/3$ for $i = 0, 1$. Furthermore, let

$$T_0(x,y) = \begin{pmatrix} \alpha & 0 \\ 0 & 1-\alpha \end{pmatrix} \begin{pmatrix} x \\ y \end{pmatrix}, \quad T_1(x,y) = \begin{pmatrix} 1-\alpha & 0 \\ 0 & \alpha \end{pmatrix} \begin{pmatrix} x \\ y \end{pmatrix},$$

with $1/2 < \alpha < 1$. Finally, take $Q = K = [0,1]^2$. Clearly $T_0(Q) + T_1(Q) = Q$. The invariant multimeasure ϕ is supported on the standard $1/3$-Cantor subset of $[0,1]$, and the values are rectangles that are more "vertical" to the left and more "horizontal" to the right. $\square$

Figure 5.9 illustrates two multimeasure attractors of IFS Markov operators. The first image in the figure, the "circular" one, is as in Example 5.54 where $Q \subset \mathbb{R}^2$ is the unit disk, $p_0 = 0.3$, and $p_1 = 0.7$. The second image, the "rectangular" one, is as in Example 5.55, with $\alpha = 0.3$. However, for both images we use the maps $w_i(x) = x/2 + i/2$, for $i = 0, 1$. This differs from Example 5.55 since we obtain a measure with full support, rather than being supported on a Cantor set. The image we show is easier to see than one supported on a Cantor set.

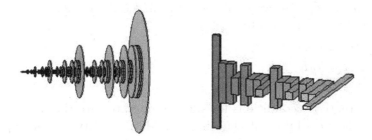

Fig. 5.9: Circular and rectangular multimeasures.

Example 5.56. A set $Q \subset \mathbb{R}^m$ is a *zonotope* if $Q = l_1 + l_2 + \cdots + l_p$, where $l_i \subset \mathbb{R}^m$ are compact line segments. Many natural convex sets are zonotopes or can be approximated by zonotopes; see [27, 87]. By translating, we can assume that l_i has its midpoint at the origin, so that $l_i \subset Q$ and $Q = -Q$. Let $Q = l_1 + l_2 + \cdots + l_p \subset \mathbb{R}^m$ be a zonotope as above, and let $P_i : \mathbb{R}^m \to \mathbb{R}^m$ be the orthogonal projection onto the subspace spanned by l_i. Furthermore, let $\alpha_i = |l_i|/|P_iQ|$ where $|\cdot|$ denotes the diameter, and let $T_i = \alpha_i P_i$. Note that $\alpha_i < 1$. Then $l_i = T_iQ$, so $\sum_i T_iQ = Q$. Let $K = Q$.

Take any IFS maps $w_i : X \to X$ for $i = 1, 2, \ldots, N$ with contraction factors c_i, and take $\beta_{i,j} \in [0, 1]$ with $\sum_j \beta_{i,j} = 1$. Define $T_{i,j} = \beta_{i,j} T_i$ such that $T_i = \sum_j T_{i,j}$. Notice that $Q = \sum_{i,j} T_{i,j}Q$. Finally, we define M on $\mathcal{M}_{Q,Q}(X, \mathbb{R}^m)$ by

$$M\phi(B) = \sum_{i,j} T_{i,j}\phi(w_j^{-1}(B)).$$

By a simple calculation, we see that M is average contractive if $\sum_i c_i \alpha_i < 1$. □

5.3.2.1 IFS with condensation

We can also define a type of IFS with condensation on multimeasures. We briefly indicate a possible construction. In addition to the w_i and T_i from (5.20), we also have an additional multimeasure ψ and an additional linear map T_0. We define the operator as

$$M\phi(B) = T_0(\psi(B)) + \sum_i T_i\phi(w_i^{-1}(B)). \qquad (5.22)$$

If $\phi \in \mathcal{M}_{Q,K}(\mathbb{X}, \mathbb{R}^m)$, we need to ensure $M\phi \in \mathcal{M}_{Q,K}(\mathbb{X}, \mathbb{R}^m)$ as well. The conditions for this are:

1. $\sum_i T_i Q + T_0 \psi(\mathbb{X}) = Q$.
2. $\sum_i T_i K + T_0 K \subseteq K$.
3. $\psi(\mathbb{X}) \subseteq K$.

The contractivity conditions for such an operator are the same as for (5.20). The possibility of choosing ψ adds flexibility to the type of invariant multimeasures one can construct with this type of operator.

As is usually the case with IFS operators with condensation, it is sometimes useful to think of the operator M as

$$M\phi = A\phi + \hat{\psi},$$

where A is the "linear" part of M and $\hat{\psi} = T_0\psi$. Then, assuming the operator is contractive, the fixed point ϕ of M can be thought of as

$$\phi = \hat{\psi} + A\hat{\psi} + A^2\hat{\psi} + A^3\hat{\psi} + \cdots = (I - A)^{-1}\hat{\psi},$$

where the sum is thought of in the appropriate fashion.

5.3.3 Generalizations for spaces of multimeasures

In this section, we briefly give some results and definitions for the situation where the space $\mathbb{X}$ is not compact and/or the operator M does not preserve boundedness. In the latter case, we again consider a wider class of "multimeasures" as bounded linear operators on the space of Lipschitz functions, as in Sects. 5.1.3 and 5.2.5. Again our basic approach will be to work with the scalarizations of the multimeasure.

We know that $\mathrm{Lip}(\mathbb{X}, \mathbb{R})^*$ is complete, being the dual space of a Banach space. Consider the space

$$\mathcal{L}(\mathcal{S}^1, \mathrm{Lip}(\mathbb{X}, \mathbb{R})^*) = \{\Phi : \mathcal{S}^1 \to \mathrm{Lip}(\mathbb{X}, \mathbb{R})^* : \|\Phi^q\|_{\mathrm{Lip}^*} < \infty, \forall q \in \mathcal{S}^1\}.$$

On this space, we place the norm

$$\|\Phi\|_{\text{Lip}^*} := \sup_{q \in \mathcal{S}^1} \|\Phi^q\|_{\text{Lip}^*}. \qquad (5.23)$$

Clearly $\mathcal{L}(\mathcal{S}^1, \text{Lip}(\mathbb{X}, \mathbb{R})^*)$ is complete under this norm. Each multi-measure $\phi \in \mathcal{M}_{Q,K}(\mathbb{X}, \mathbb{R}^m)$ naturally induces an element of $\Phi \in \mathcal{L}(\mathcal{S}^1, \text{Lip}(\mathbb{X}, \mathbb{R})^*)$ via

$$\Phi^q(f) = \int_{\mathbb{X}} f(x) \, d\phi^q(x).$$

However, clearly the map $\Xi : \mathcal{M}_{Q,K}(\mathbb{X}, \mathbb{R}^m) \to \mathcal{L}(\mathcal{S}^1, \text{Lip}(\mathbb{X}, \mathbb{R})^*)$ is not surjective (though it is continuous, linear, and injective). Notice that the image depends on both Q and K. A necessary condition for an element $\Phi \in \mathcal{L}(\mathcal{S}^1, \text{Lip}(\mathbb{X}, \mathbb{R})^*)$ to be associated with a multimeasure is given by the conditions in Lemma 5.46. That is, fix $f \in \text{Lip}(\mathbb{X}, \mathbb{R})$. Then the function $\mathbb{R}^m \to \mathbb{R}$ given by

$$q \mapsto \Phi(f)^{\hat{q}}, \quad \hat{q} = q/\|q\|, \qquad (5.24)$$

should be convex and positively homogeneous. Notice that if $\Phi_n \in \mathcal{L}(\mathcal{S}^1, \text{Lip}(\mathbb{X}, \mathbb{R})^*)$ is a sequence that satisfies (5.24) and if Φ_n converges to Φ in $\| \cdot \|_{\text{Lip}^*}$, then Φ will also satisfy (5.24).

Finally, as usual we need a mass preservation condition, so for a fixed set $Q \in \mathcal{K}(\mathbb{X})$ we define

$$\mathcal{L}_Q(\mathbb{X}, \mathbb{R}^m) = \{\Phi \in \mathcal{L}(\mathcal{S}^1, \text{Lip}(\mathbb{X}, \mathbb{R})^*) : \Phi^q(1) = \text{supp}\,(q, Q), \forall q \in \mathcal{S}^1\}.$$

Now let $w_i : \mathbb{X} \to \mathbb{X}$ and $T_i : R^m \to \mathbb{R}^m$ be as in the definition of the IFS operator in (5.20). Take $\Phi \in \mathcal{L}(\mathcal{S}^1, \text{Lip}(\mathbb{X}, \mathbb{R})^*)$, and define $\mathbb{M}\Phi$ as

$$\mathbb{M}\Phi = \sum_i \Phi^{T_i^* q}(f^q \circ w_i),$$

where $f^q : \mathbb{X} \to \mathbb{R}$ is Lipschitz for each $q \in \mathcal{S}^1$. Then $\mathbb{M}\Phi \in \mathcal{L}(\mathcal{S}^1, \text{Lip}(\mathbb{X}, \mathbb{R})^*)$ as well (as is straightforward to show). Furthermore, as in the case of (5.20), if $\Phi \in \mathcal{L}_Q(\mathbb{X}, \mathbb{R}^m)$, then so is $\mathbb{M}\Phi$.

Theorem 5.57. *The operator $\mathbb{M}$ as defined above satisfies*

$$\|\mathbb{M}\Phi - \mathbb{M}\Psi\|_{Lip^*} \leq \left(\sum c_i \|T_i^*\|\right) \|\Phi - \Psi\|_{Lip^*}.$$

Proof. The scheme of the proof is the same as in the other simpler cases. First we observe that if $\Phi, \Psi \in \mathcal{L}_Q(\mathbb{X}, \mathbb{R}^m)$, then $\Phi^q(c) = \Psi^q(c)$ and thus

$$\begin{aligned}
\|\Phi - \Psi\|_{\mathrm{Lip}^*} &= \sup_{q \in \mathcal{S}^1} \sup\{\Phi^q(f) - \Psi^q(f) : \|f\|_{\mathrm{Lip}} \leq 1\} \\
&= \sup_{q \in \mathcal{S}^1} \sup\{\Phi^q(f) - \Psi^q(f) : f \in \mathrm{Lip}_1(\mathbb{X})\}.
\end{aligned}$$

We next observe that $f \in \mathrm{Lip}_1(\mathbb{X})$. Then the Lipschitz constant of $\hat{f} = \sum_i \|T_i^*\| f \circ w_i$ is at most $\sum_i \|T_i^*\| c_i$.   □

It is also possible to deal with noncompact $\mathbb{X}$ by adding a type of first-moment condition on the multimeasures. That is, as before, we have $Q \subset K \in \mathcal{K}(\mathbb{R}^m)$ and we define $\mathcal{M}_{Q,K}(\mathbb{X}, \mathbb{R}^m)$ to be the set of all multimeasures on $\mathbb{X}$ such that

1. $\phi(\mathbb{X}) = Q$;
2. $\phi(E) \subseteq K$ for all E;
3. there is some set $D \in \mathcal{K}(\mathbb{R}^m)$ and an $a \in \mathbb{X}$ such that $\int_X f(x) d\phi(x) \subseteq D$ for all $f \in \mathrm{Lip}_1(\mathbb{X})$ with $f(a) = 0$.

The last condition is the first-moment type condition. This translates into a type of uniform tightness condition for the collection of scalarizations ϕ^q. Following the same general pattern as before, it is possible to prove completeness. We can also combine the construction for "generalized multimeasures" (as elements of $\mathcal{L}_Q(\mathbb{X}, \mathbb{R}^m)$) with a first-moment condition to obtain a general construction on noncompact $\mathbb{X}$. The details are left to the references.

5.3.4 Union-additive multimeasures

The second type of set-valued measures we consider are those that are additive in a different way. Instead of using a pointwise (Minkowski) sum of a collection of sets, we use the union of the collection. This simple modification results in a considerable change in the theory of these measures, simplifying much of the analysis and vastly generalizing the range of behaviour. However, the resulting union-additive multimeasures are quite different from the usual measures and are not really properly thought of as direct generalizations of scalar measures.

All our results on union-additive multimeasures are based on the paper [106].

Let Ω be a complete metric space and $\mathcal{B}$ be its Borel σ-algebra. Additionally, let $\mathbb{X}$ be another complete metric space with $\mathbb{H}(\mathbb{X})$ (again) the collection of all nonempty compact subsets of $\mathbb{X}$ under the Hausdorff metric $d_{\mathbb{H}}$.

For a set $B \subseteq \mathbb{X}$, recall our notation of ϵ-*dilation* of B as $B_\epsilon = \{x \in \mathbb{X} : d(x, B) < \epsilon\}$. One way of characterizing the Hausdorff distance is

$$d_{\mathbb{H}}(A, B) < \epsilon \text{ iff } A \subseteq B_\epsilon \text{ and } B \subseteq A_\epsilon.$$

We begin with a simple proposition.

Proposition 5.58. *Let $A_n \subseteq \mathbb{X}$ be compact, and suppose that $S = \bigcup_{n=1}^{\infty} A_n$ is compact. Let $S_n = \bigcup_{i=1}^{n} A_i$. Then $S_n \to S$ in the Hausdorff metric.*

Proof. Let $\epsilon > 0$ be fixed. Obviously, for all n,

$$S_n \subseteq S \subseteq S_\epsilon.$$

On the other hand, the collection $\{(S_n)_\epsilon\}_{n \in \mathbb{N}}$ is an open cover for the compact set S, so there is some n_0 such that

$$S \subseteq (S_n)_\epsilon$$

for all $n \geq n_0$. $\qquad\square$

A union-additive multimeasure is a set-valued function $\phi : (\Omega, \mathcal{B}) \to \mathbb{H}(\mathbb{X})$, with the one exception that we require that $\phi(\emptyset) = \emptyset$. Thus, $\phi(B) \subseteq \mathbb{X}$ is compact and nonempty for all nonempty B. For additivity, let $A_i \in \mathcal{B}$ be pairwise disjoint. We require that any union-additive multimeasure ϕ satisfy

$$\phi\left(\bigcup_{i=1}^{\infty} A_i\right) = \lim_n \bigcup_{i=1}^{n} \phi(A_i) = \overline{\bigcup_{i=1}^{\infty} \phi(A_i)}, \qquad (5.25)$$

where we take the limit in the Hausdorff metric. The Hausdorff limit is necessary since in general a countable union of compact sets need not be compact. Notice that since $\phi(\Omega) \in \mathbb{H}(\mathbb{X})$ and $\phi(A_i) \subseteq \phi(\Omega)$ we have $\bigcup_i \phi(A_i) \subseteq \phi(\Omega)$, and thus the closure above will result in a compact set. This and Proposition 5.58 explain why the closure is the Hausdorff limit.

One interesting difference between union-additive multimeasures and the usual measures is the fact that it is equivalent to asking that countable union-additivity (5.25) hold for any sets A_i, not necessarily only those pairwise disjoint. For the case of the usual measures, additivity is only guaranteed to hold for disjoint sets.

There is another useful way to look at union-additive multimeasures. The collection $\mathcal{B}$ of Borel subsets of Ω ordered under inclusion is a lattice, as is $\mathbb{H}(\mathbb{X})$. A *join homomorphism* from $\mathcal{B}$ to $\mathbb{H}(\mathbb{X})$ is an order-preserving map $\phi : \mathcal{B} \to \mathbb{H}(\mathbb{X})$ for that $\phi(A \cup B) = \phi(A) \cup \phi(B)$. This is the same condition as that for a finitely union-additive multimeasure. Thus, a countably union-additive multimeasure is simply a join homomorphism which preserves countable joins.

We now define the collection of multimeasures of interest.

Definition 5.59. Denote by $UA(\Omega, \mathbb{X})$ the collection of countably union-additive multimeasures with $\phi(\emptyset) = \emptyset$ and $\phi(A) \in \mathbb{H}(\mathbb{X})$ for all nonempty $A \in \mathcal{B}$.

Constructing union-additive multimeasures is simple. One general method goes as follows. Let $f : \Omega \to \mathbb{X}$ be any function with $\overline{f(\Omega)} \subseteq \mathbb{X}$ compact, and define $\phi(A) = \overline{f(A)}$. Then $\phi(\emptyset) = \emptyset$ and, for any set $A \subseteq \Omega$, we have $\phi(A) \in \mathbb{H}(\mathbb{X})$. Furthermore, $\overline{f(A \cup B)} = \overline{f(A)} \cup \overline{f(B)}$ and so, by Proposition 5.58,

$$\phi\left(\bigcup_{i=1}^{\infty} A_i\right) = \overline{f\left(\bigcup_{i=1}^{\infty} A_i\right)} = \lim_{N} \bigcup_{i=1}^{N} \overline{f(A_i)} = \lim_{N} \bigcup_{i=1}^{N} \phi(A_i) = \overline{\bigcup_{i=1}^{\infty} \phi(A_i)},$$

and thus ϕ is countably union-additive. One can even use a multifunction $f : \Omega \rightrightarrows \mathbb{X}$ and define

$$\phi(A) = \overline{\bigcup_{a \in A} f(a)}.$$

Not every $\phi \in UA(\Omega, \mathbb{X})$ is of this form, however. As an example, consider $\mathbb{X} = \Omega = \mathbb{R}$ with $\phi(\emptyset) = \emptyset$, and $\phi(F) = \{1\}$ for any countable set F, and $\phi(A) = [0, 1]$ for any uncountable Borel set A. In Sect. 5.3.5, we provide another general construction method using the theory of IFSs.

Having specified our class of multimeasures, we now define a metric under which $UA(\Omega, \mathbb{X})$ will be complete. We define

$$\hat{d}_{\mathbb{H}}(\phi_1, \phi_2) = \sup_{\emptyset \neq A \in \mathcal{B}} d_{\mathbb{H}}(\phi_1(A), \phi_2(A)) \tag{5.26}$$

(thus, $\hat{d}_{\mathbb{H}}$ is a metric of uniform convergence for measures in $UA(\Omega, \mathbb{X})$).

Theorem 5.60. *The space* $(UA(\Omega, \mathbb{X}), \hat{d}_{\mathbb{H}})$ *is a complete metric space.*

Proof. Let ϕ_n be a Cauchy sequence of measures in $(UA(\Omega, \mathbb{X}), \hat{d}_{\mathbb{H}})$. We define the set-valued function ϕ by taking $\phi(\emptyset) = \emptyset$ and, for each nonempty $A \in \mathcal{B}$, $\phi(A) = \lim_n \phi_n(A)$ in the Hausdorff metric on $\mathbb{X}$. Since $\mathbb{H}(\mathbb{X})$ is complete under the Hausdorff metric, this limit exists.

We must show that ϕ is an element of $UA(\Omega, \mathbb{X})$. Clearly, for each nonempty A we have that $\phi(A)$ is nonempty and compact since each $\phi_n(A)$ is nonempty and compact and the Hausdorff limit preserves these properties.

Let $A_i \in \mathcal{B}$ be pairwise disjoint and nonempty. From the hypotheses, we have that for any $\epsilon > 0$ there exists $m_0(\epsilon)$ such that $d_{\mathbb{H}}(\phi_m(A), \phi(A)) < \epsilon/2$ for all $m \geq m_0$. But this means that $\phi_m(A) \subseteq (\phi(A))_{\epsilon/2}$ and $\phi(A) \subseteq (\phi_m(A))_{\epsilon/2}$. Because of the supremum in the definition of $d_{\mathbb{H}}$, these inclusions can be true for all A_i with the same m_0. Then

$$\overline{\bigcup_{i=1}^{\infty} \phi_m(A_i)} \subseteq \overline{\bigcup_{i=1}^{\infty} (\phi(A_i))_{\epsilon/2}} \subseteq \left(\overline{\bigcup_{i=1}^{\infty} (\phi(A_i))}\right)_{\epsilon}$$

and

$$\overline{\bigcup_{i=1}^{\infty} \phi(A_i)} \subseteq \overline{\bigcup_{i=1}^{\infty} (\phi_m(A_i))_{\epsilon/2}} \subseteq \left(\overline{\bigcup_{i=1}^{\infty} (\phi_m(A_i))}\right)_{\epsilon}.$$

This means that $d_{\mathbb{H}}(\overline{\bigcup_{i=1}^{\infty} \phi(A_i)}, \overline{\bigcup_{i=1}^{\infty} \phi_m(A_i)}) < \epsilon$ and thus

$$\lim_n \lim_m \bigcup_{i=1}^{n} \phi_m(A_i) = \lim_m \lim_n \bigcup_{i=1}^{n} \phi_m(A_i).$$

But this gives

$$\phi\left(\bigcup_{i=1}^{\infty} A_i\right) = \lim_m \phi_m\left(\bigcup_{i=1}^{\infty} A_i\right) = \lim_m \lim_n \phi_m\left(\bigcup_{i=1}^{n} A_i\right)$$

$$= \lim_m \lim_n \bigcup_{i=1}^{n} \phi_m(A_i) = \lim_n \lim_m \bigcup_{i=1}^{n} \phi_m(A_i)$$

$$= \lim_n \bigcup_{i=1}^{n} \lim_m \phi_m(A_i) = \lim_n \bigcup_{i=1}^{n} \phi(A_i),$$

which is what we wished to prove. $\square$

5.3.5 IFS on union-additive multimeasures

It is simpler to define an IFS operator on union-additive multimeasures than on regular (Minkowski) additive multimeasures. The construction is more "geometrical" and the restrictions are less rigid.

As usual, let $w_i : \Omega \to \Omega$ for $i = 1, 2, \ldots, N$. However, in this case, we only require that the w_i map Borel sets to Borel sets (so continuous is more than enough). Furthermore, let $T_i : \mathbb{H}(\mathbb{X}) \to \mathbb{H}(\mathbb{X})$ be contractions with contractivity factor k_i and such that

$$T_i \left(\overline{\bigcup_{n=1}^{\infty} A_n} \right) = \overline{\bigcup_{n=1}^{\infty} T_i(A_n)}$$

for disjoint A_n. This condition on T_i is easily met; for example, if $T_i(A) = t_i(A)$ for some contractive $t_i : \mathbb{X} \to \mathbb{X}$. We are not requiring any of the w_i to be contractive.

We define two different types of operators $M : UA(\Omega, \mathbb{X}) \to UA(\Omega, \mathbb{X})$. For the first one, we assume that $\bigcup_i w_i(\Omega) = \Omega$, so that Ω is the attractor of the IFS $\{w_i\}$. In particular, for any nonempty $S \in \mathcal{B}$, we have $w_i^{-1}(S) \neq \emptyset$ for at least one i. In this case, we define

$$M_1\phi(B) = \bigcup_{w_i^{-1}(B) \neq \emptyset} T_i(\phi(w_i^{-1}(B))), \quad \emptyset \neq B \in \mathcal{B}. \qquad (5.27)$$

For the second type, we make no additional assumption on the maps w_i but in addition we take some fixed $\psi \in \mathbb{H}(\mathbb{X})$ and define

$$M_2\phi(B) = \psi(B) \cup \bigcup_{w_i^{-1}(B) \neq \emptyset} T_i(\phi(w_i^{-1}(B))), \quad \emptyset \neq B \in \mathcal{B}. \qquad (5.28)$$

In both cases, we see that since this is a finite union we have no need to take a Hausdorff limit of the union. Also notice we can simplify the formulas if we extend each T_i by defining $T_i(\emptyset) = \emptyset$, and then, for example, we can write (5.27) as

$$M_1\phi(B) = \bigcup_i T_i(\phi(w_i^{-1}(B))).$$

Clearly $M_j\phi(\emptyset) = \emptyset$ for either Markov operator. Take $\emptyset \neq S \in \mathcal{B}$. Then, in (5.27) we have $w_i^{-1}(S) \neq \emptyset$ for at least one i and thus $M_1\phi(S) \neq \emptyset$. For the M_2 as defined in (5.28), we have $\psi(S) \neq \emptyset$, so again $M_2\phi(S) \neq \emptyset$. The fact that $M_j\phi$ is countably union-additive follows from the assumptions on T_i. Thus, in either case, $M_j : UA(\Omega, \mathbb{X}) \to UA(\Omega, \mathbb{X})$.

Theorem 5.61. *Let $k = \max_i k_i$ be the maximum of the contractivities of the operators T_i. Then*

$$\hat{d}_{\mathbb{H}}(M_j\phi_1, M_j\phi_2) \leq k\hat{d}_{\mathbb{H}}(\phi_1, \phi_2).$$

Proof. Let $A \in \mathcal{B}$. Then, starting with M_1, we have

$$
\begin{aligned}
d_{\mathbb{H}}(M_1\phi_1(A), M_1\phi_2(A)) &= d_{\mathbb{H}}\left(\bigcup_i T_i(\phi_1(w_i^{-1}(A))), \bigcup_i T_i(\phi_2(w_i^{-1}(A)))\right) \\
&\leq \max_i d_{\mathbb{H}}(T_i(\phi_1(w_i^{-1}(A))), T_i(\phi_2(w_i^{-1}(A)))) \\
&\leq k \max_i d_{\mathbb{H}}(\phi_1(w_i^{-1}(A)), \phi_2(w_i^{-1}(A)))) \\
&\leq k\, \hat{d}_{\mathbb{H}}(\phi_1, \phi_2)
\end{aligned}
$$

and thus

$$\hat{d}_{\mathbb{H}}(M_1\phi_1, M_1\phi_2) \leq k\hat{d}_{\mathbb{H}}(\phi_1, \phi_2),$$

as desired.

For M_2, we notice that

$$d_{\mathbb{H}}(A \cup S, B \cup S) \leq \max\{d_{\mathbb{H}}(A, B), d_{\mathbb{H}}(S, S)\} = d_{\mathbb{H}}(A, B)$$

for any $A, B, S \in \mathbb{H}(\mathbb{X})$, and thus the contractivity factor for M_2 is similarly obtained. □

Example 5.62. Our first example is one where the values of the fractal multimeasure are subsets of some fractal. Let $\Omega = [0, 1]$ and $\mathbb{X} = [0, 1] \times [0, 1]$. We let $w_i : [0, 1] \to [0, 1]$ be defined by $w_i(x) = x/3 + i/3$ for $i = 0, 1, 2$ and $T_i : \mathbb{X} \to \mathbb{X}$ be defined as

$$
\begin{aligned}
T_0(x, y) &= (x/2, y/2), & T_1(x, y) &= (x/2 + 1/2, y/2), \\
T_2(x, y) &= (x/2, y/2 + 1/2).
\end{aligned}
$$

Since $[0, 1] = w_0([0, 1]) \cup w_1([0, 1]) \cup w_2([0, 1])$, the maps w_i satisfy the conditions necessary to define M_1 using (5.27).

The attractor for the IFS $\{w_0, w_1, w_2\}$ is $[0, 1]$ while the attractor for the IFS $\{T_0, T_1, T_2\}$ is the classical Sierpinski gasket, G.

Using these to define a Markov operator, we get the fixed-point multimeasure ϕ. This multimeasure has the property that, for all Borel $S \subseteq [0, 1]$,

$$\phi(S) = \{x \in G : \text{the address of } x \text{ (represented in ternary)} \in S\}.$$

□

Example 5.63. For our next example, we take $\Omega = \mathbb{X} = [0, 1]$, $w_0(x) = x/3$, $w_1(x) = x/3 + 2/3$ (so that the attractor of the IFS $\{w_0, w_1\}$ is the classical Cantor set), and $T_0(x) = px$ and $T_1(x) = (1-p)x + p$ (so that the attractor of the IFS $\{T_0, T_1\}$ is $[0, 1]$) for some $0 < p < 1$.

As the attractor of $\{w_0, w_1\}$ is the Cantor set, we must use (5.28) to define the Markov operator. We are free to choose any $\psi \in UA(\Omega, \mathbb{X})$ for this, and for each different choice we are likely to get a different fixed point. A simple choice is to select some fixed $S_0 \in \mathbb{H}(\mathbb{X})$ and define $\psi(B) = S_0$ for all nonempty B. For a specific choice, let $S_0 = \{0\}$.

In this case, the fixed-point multimeasure ϕ has the property that $\phi(S) \subseteq [0, 1]$, and the Lebesgue measure λ of $\phi(S)$ could be thought of as some sort of "probability" of the set S. For instance, $\lambda(\phi([0, 1/3])) = \lambda([0, p]) = p$, $\lambda(\phi([1/3, 2/3])) = \lambda(\emptyset) = 0$, and $\lambda(\phi([2/3, 1])) = \lambda([p, 1]) = 1 - p$.

Furthermore, $\{0\} \subseteq \phi(B)$ for any nonempty B, and $\phi(B) = \{0\}$ if the intersection of B with the Cantor set is empty.   □

5.3.5.1 IFS operators with infinitely many maps

This framework easily allows one to deal with IFS operators with infinitely many maps w_i and T_i. The setup is similar to that in Sect. 2.6.4. We have Ω a complete metric space and Λ and $\mathbb{X}$ compact metric spaces. The space Λ is used as the "indexing" space for the infinite IFS operator. We suppose that $w : \Lambda \times \Omega \to \Omega$ with w_λ^{-1} mapping Borel sets to Borel sets for all λ. Finally, take $T : \Lambda \times \mathbb{H}(\mathbb{X}) \to \mathbb{H}(\mathbb{X})$ with $T_\lambda : \mathbb{H}(\mathbb{X}) \to \mathbb{H}(\mathbb{X})$ Lipschitz for each λ. For convenience, as before, we extend the definition of T by defining $T_\lambda(\emptyset) = \emptyset$ for all λ. Finally, we also need the condition that for each λ we have

$$T_\lambda \left(\bigcup_n A_n \right) = \bigcup_n T_\lambda(A_n)$$

for all sequences of disjoint A_n (where the implied limit in the union is in the Hausdorff metric on $\mathbb{H}(\mathbb{X})$).

Using these data, we define two operators $M_j : UA(\Omega, \mathbb{X})$, in a fashion analogous to (5.27) and (5.28), by

$$M_1(\phi)(B) = \bigcup_\lambda T_\lambda(\phi(w_\lambda^{-1}(B))) \qquad (5.29)$$

and

$$M_2(\phi)(B) = \psi(B) \cup \bigcup_\lambda T_\lambda(\phi(w_\lambda^{-1}(B))), \tag{5.30}$$

where in the union over Λ the limit is taken in the Hausdorff metric on $\mathbb{H}(\mathbb{X})$.

We use the same metric, $\hat{d}_\mathbb{H}$, and by an argument similar to Theorems 5.60 and 5.61 we get the following result.

Theorem 5.64. *Suppose that $0 < k < 1$ is a uniform upper bound for the Lipschitz constant of T_λ over $\lambda \in \Lambda$. Then $M_j : UA(\Omega, \mathbb{X}) \to UA(\Omega, \mathbb{X})$ and*

$$\hat{d}_\mathbb{H}(M_j\phi_1, M_j\phi_2) \le k\hat{d}_\mathbb{H}(\phi_1, \phi_2).$$

5.3.6 Generalities on union-additive multimeasures

In the most basic sense, an IFS with probabilities performs a recursive partitioning of probability (total mass of 1) over some recursively defined set. So, an ultimate generalization of an IFS with probability would be to define some set of operators that perform a recursive partition of some underlying structure. We briefly outline this abstract view and give some examples of general constructions along this line.

Let Ω and $\mathbb{X}$ be sets. Our "multimeasures" will take as input subsets in some algebra $\mathcal{A}$ of subsets of Ω (an algebra, not necessarily a σ-algebra). The output of the multimeasure will be subsets of $\mathbb{X}$, that will represent parts of a recursively defined partition of $\mathbb{X}$.

So, let $w_i : \Omega \to \Omega$ be (somewhat arbitrary) functions and $T_i : \mathrm{Pow}(\mathbb{X}) \to \mathrm{Pow}(\mathbb{X})$ be functions that satisfy

1. $\bigcup_i T_i(S) = S$ for all $S \in \mathcal{A}$,
2. $T_i(S) \cap T_j(S) = \emptyset$ for all $S \in \mathcal{A}$, $i \ne j$, and
3. $T_i(A \cup B) = T_i(A) \cup T_i(B)$ for all $A, B \in \mathcal{A}$ and i.

Clearly, properties 1 and 2 ensure that $\{T_i(S)\}$ is a partition of S for any appropriate S. Property 3 will ensure that if ϕ is union-additive, then so is $M\phi$.

Define our "operator" M by

$$M\phi(B) = \bigcup_{i=1}^{N} T_i(\phi(w_i^{-1}(B))). \tag{5.31}$$

Then any fixed point of M is a self-similar recursively defined partition-valued multimeasure. In particular, once we have specified the value of $\phi(\Omega)$, (5.31) will recursively define the values of the fixed point ϕ on sets of the form

$$w_{i_1} \circ w_{i_2} \circ \ldots \circ w_{i_k}(\Omega), \text{ some } i_1, i_2, \ldots, i_k \in \{1, 2, \ldots, N\}, \quad (5.32)$$

and thus any fixed point of M will be uniquely defined on the algebra of sets generated by sets of this form. However, in general there is no reason to believe that a fixed point ϕ is countably union-additive.

We will say that an algebra $\mathcal{A}$ of subsets of Ω is $\{w_i\}$-*compatible* if for all $A \in \mathcal{A}$ and all sequences $i_1, i_2, \ldots, i_k \in \{1, 2, \ldots, N\}$ we have

$$w_{i_1} \circ w_{i_2} \circ \cdots \circ w_{i_k}(A) \in \mathcal{A}.$$

Choosing some $\{w_i\}$-compatible algebra $\mathcal{A}$, define $UFA(\Omega, \mathbb{X})$ as the collection of all finitely union-additive multimeasures from $(\Omega, \mathcal{A})$ to $\mathrm{Pow}(\mathbb{X})$. With this notation, it is clear that M as defined in equation (5.31) defines a linear operator $M : UFA(\Omega, \mathbb{X}) \to UFA(\Omega, \mathbb{X})$.

We end our discussion of multimeasures with two examples that are variations on this type of generalization.

Example 5.65. Let $\Omega = \mathbb{X} = [0, 1]$, $w_i(x) = x/2 + i/2$, and $T_i(x) = x/2 + i/2$ for $i = 0, 1$. Then the fixed-point multimeasure ϕ satisfies $\phi(S) = S$ for any $S \subseteq [0, 1]$. This is clearly a rather silly example. It can be made slightly more interesting by defining $T_0(x) = x/2 + 1/2$ and $T_1(x) = x/2$. To describe the invariant multimeasure ϕ, we need to define an auxiliary function. Let $\tau : [0, 1] \to [0, 1]$ be defined by the ith binary digit of $\tau(x)$ as 1 iff the ith binary digit of x is 0 (and similarly for 0 and 1). That is, τ "flips" all the binary digits of x. Using τ, we see that $\phi(S) = \tau(S)$. Actually, these situations fall into the framework of (5.27) with each T_i contractive with contractivity factor $1/2$, and thus the fixed point is unique and is a countably union-additive multimeasure.

Another slightly different example is $\Omega = [0, 1]$, with the same w_i but $\mathbb{X} = \mathbb{N} \cup \{0\}$, the set of whole numbers. Then we define $T_0(x) = 2x$ and $T_1(x) = 2x + 1$ (such that $T_0(\mathbb{X}) = 2\mathbb{X}$ and $T_1(\mathbb{X}) = 2\mathbb{X} + 1$, the sets of even and odd numbers, respectively). The operator M in this case first partitions $\mathbb{X}$ into even and odd, then into the modular four classes, then modular eight, modular sixteen, and so on.

Interestingly, this example can also be put into the framework of the Markov operator (5.27). To do this, we recall some facts about 2-adic numbers (see [145]). We place the metric d_2 on $\mathbb{X}$ by setting

$$d_2(n, m) = 2^{-k} \text{ whenever } n - m = a2^k \text{ with } \gcd(a, 2) = 1.$$

Using this metric, $d_2(T_i(n), T_i(m)) = (1/2)d_2(n, m)$, so the two maps T_i are contractive. However, $\mathbb{X}$ is not complete under this metric. The completion of $\mathbb{X}$ is contained in the set of *2-adic integers* (call this $\mathbb{Z}_2$), which can be viewed as formal infinite sums of the form $\sum_{n=1}^{\infty} a_n 2^n$. It turns out that $\mathbb{Z}_2$ is not only complete but also compact.

The two maps T_i can be extended naturally to $\mathbb{Z}_2$ and $T_0(\mathbb{Z}_2)$ and $T_1(\mathbb{Z}_2)$ form a partition of $\mathbb{Z}_2$ just as $T_0(\mathbb{X})$ and $T_1(\mathbb{X})$ formed a partition of $\mathbb{X}$. Thus, the unique fixed point of the operator M is a countably union-additive multimeasure with values in $\mathbb{H}(\mathbb{Z}_2)$. $\qquad\square$

Example 5.66. The "partition" can also be in the sense of a direct sum in the context of a vector space. Let Ω be a complete metric space and $\mathcal{H}$ be a vector space. We let $w_i : \Omega \to \Omega$ be contractive and $T_i : \mathcal{H} \to \mathcal{H}$ be linear and injective and satisfy $\sum_i T_i = I$ (the identity) and $T_i(\mathcal{H}) \cap T_j(\mathcal{H}) = \{0\}$. These conditions ensure that the T_i "partition" the space $\mathcal{H}$ into a direct sum of the subspaces $T_i(\mathcal{H})$.

Our multimeasures will be defined on the Borel σ-algebra of Ω and take values as linear subspaces of $\mathcal{H}$. We define the IFS "operator" M by

$$M\phi(B) = \sum_i T_i(\phi(w_i^{-1}(B))).$$

As above, for each n, we get a direct-sum decomposition of $\mathcal{H}$ as

$$\mathcal{H} = \sum T_{i_1} \circ T_{i_2} \circ \cdots \circ T_{i_n}(\mathcal{H}),$$

where the sum is over all sequences i_j. Furthermore, the multimeasure ϕ maps the Borel set

$$B = w_{i_1} \circ w_{i_2} \circ \cdots \circ w_{i_n}(\Omega)$$

to the subspace

$$T_{i_1} \circ T_{i_2} \circ \cdots \circ T_{i_n}(\mathcal{H}).$$

As a more concrete example of this situation, we let $w_i(x) = x/2 + i/2$ for $i = 0, 1$, and $\mathcal{H} = L^2[0, 1]$ and $T_i(f) = f \circ w_i^{-1}$ (so that $T_0(f)$ is supported on $[0, 1/2]$ and $T_1(f)$ is supported on $[1/2, 1]$). The fixed point of M is then the multimeasure

$$\phi(S) = \{f \in L^2[0, 1] : f \text{ is supported on } S\}.$$

Another (and more interesting) example of this construction is where we again have $w_i(x) = x/2 + i/2$ but have $\mathcal{H} = L^2(\mathbb{R})$, and T_0 as the "high-pass filter," and T_1 the "low-pass filter" from a two-scale MRA (multiresolution analysis) associated with a wavelet basis. These T_0, T_1 satisfy the recursive partitioning conditions, so the invariant multimeasure ϕ is a subspace-valued measure that is compatible with the wavelet basis. $\qquad\square$

5.3.7 Extension of finitely union-additive multimeasures

One might wish to use the standard procedures to extend a finitely union-additive multimeasure to a countably union-additive multimeasure. Unfortunately, this does not work in general. Given $\phi \in UFA(\Omega, \mathbb{X})$, it is certainly possible to define an "outer measure" ϕ^* on $\mathrm{Pow}(\Omega)$ by

$$\phi^*(S) = \bigcap \left\{ \bigcup_{i=1}^{\infty} \phi(A_i) : S \subseteq \bigcup_{i=1}^{\infty} A_i, A_i \in \mathcal{A} \right\},$$

where $S \subseteq \Omega$ (the intersection acts like taking the infimum). The following properties are easy to verify:

1. $\phi^*(\emptyset) = \emptyset$.
2. $A \in \mathcal{A}$ implies that $\phi^*(A) \subseteq \phi(A)$.
3. $A \subseteq B$ implies that $\phi^*(A) \subseteq \phi^*(B)$.
4. $\bigcup_{n=1}^{\infty} \phi^*(B_n) \subseteq \phi^*(\bigcup_{n=1}^{\infty} B_n)$.

Notice that the last property has the opposite inclusion of the usual countable additivity.

However, ϕ^* is not necessarily an extension of ϕ. To see this, let $\Omega = \mathbb{X} = \mathbb{N}$, and let $\mathcal{A}$ be the algebra generated by the finite subsets of Ω (so $\mathcal{A}$ consists of finite and cofinite subsets of $\mathbb{N}$). We define $\phi(\emptyset) = \emptyset$ and $\phi(F) = \{1\}$ for any finite set F and $\phi(\mathbb{N}) = \phi(C) = \{0, 1\}$ for any cofinite set C. Then it is easy to see that ϕ^* will satisfy $\phi^*(\emptyset) = \emptyset$ and $\phi^*(F) = \{1\}$ for any finite set F and $\phi^*(I) = \{1\}$ for any infinite set I. Thus, ϕ^* does not agree with ϕ, even on the algebra $\mathcal{A}$.

Chapter 6
The Chaos Game

We saw the chaos game in Chapter 2, where it was introduced first as a means of generating an image of the attractor of an IFS in $\mathbb{R}^2$. In this chapter, we will see several other things one can do with the chaos game. First we will modify the chaos game to obtain a way of generating approximations of the invariant function for an IFSM (see Chapter 3 for the basic properties and results about an IFS on functions). Our modification is inspired by work of Berger [21, 22, 23], who constructed a chaos game for generating the graph of a wavelet. Our next topic will extend the work of Berger and will generate both a wavelet analysis and a wavelet synthesis. That is, using a chaos game, we can compute the coefficients, in a wavelet expansion of a given L^2 function f and, conversely, given these coefficients we can use a chaos game to generate approximations to the function f. Finally we also extend the chaos game to the set-valued setting, constructing a chaos game for generating an approximation to multifunctions as well as for multimeasures.

The chaos game is an example of the iteration of random functions, where you have a dynamical system whose dynamics is generated by more than one function. There is a substantial body of research in this area, as it has strong connections with the theory of stochastic processes, in particular Markovian processes. The survey paper by Diaconis and Freedman [47] is a good overview of this area and a good place to see some further applications.

6.1 Chaos game for IFSM

The simplest way of thinking of using a chaos game for an IFS on functions would be to think about the function in terms of its graph and then use a "geometric" IFS in one higher dimension to render an image of the graph of the function (this is the approach used in fractal interpolation functions; see [15]). This only works, however, if the IFS that generates the function is nonoverlapping. Otherwise it is not clear how to combine the different parts of the IFS. Instead of this geometric view, we use the IFSM operator more directly and generate a type of piecewise approximation to the function, but this piecewise approximation is a function, not a geometric set.

We will present two different versions of the chaos game for an IFS on functions. The first one requires a rather strong condition on the IFS maps $\{w_i\}$: we require them to be an IFS partition of the domain $\mathbb{X}$. The second requires a strong condition on the grey-level maps $\{\phi_i\}$: we require them to be affine. The material in this section is based on [62].

The basic theory and notation for an IFS on functions (IFSM) is presented in Sect. 3.2.

6.1.1 Chaos game for nonoverlapping IFSM

We start out by assuming that $\mathbb{X}$ is a set and that the IFS geometric maps $\{w_i\}$ form an IFS partition of $\mathbb{X}$. That is, we assume that $\mathbb{X} = \cup_i w_i(\mathbb{X})$ and $w_i(\mathbb{X}) \cap w_j(\mathbb{X}) = \emptyset$ if $i \neq j$. We are not making any assumptions on the set $\mathbb{X}$ or any contractivity assumptions on w_i. However, we do assume that each $\phi_i : \mathbb{R} \to \mathbb{R}$ is contractive. This is the setup from Sect. 3.2.1. Let $f : \mathbb{X} \to \mathbb{R}$ be the fixed point of the resulting IFSM operator. Recall from (3.2) that

$$f(x) = \lim_n \phi_{\sigma_1} \circ \phi_{\sigma_2} \circ \cdots \circ \phi_{\sigma_n}(t)$$

for any t, where σ is the code that corresponds to the point $x \in \mathbb{X}$.

To get our approximation to f, we choose some partition $\{B_i\}$ of $\mathbb{X}$. Associated with each B_i is a "cumulative sum" variable S_i, which will initially be set to 0. The only requirement we impose on the B_i is that $a^{-1}(B_i) \subset \Sigma$ be closed for each i. Since Σ is compact, this means that each $a^{-1}(B_i)$ is in fact compact, so they are strictly separated by a positive distance. Choose $\{p_i\}$ to be any set of probabilities (later we will specialize and specify each p_i). The collection of p_i induces the product measure P on Σ, which in turn induces a measure μ on $\mathbb{X}$, the push-forward of P on Σ via the address map $a : \Sigma \to \mathbb{X}$.

Our first algorithm is described as follows:

1. Initialize $x_0 = a(1, 1, 1, 1, \ldots)$, the image of the point $(1, 1, 1, \ldots) \in \Sigma$ under a.
2. Initialize u_0 to be the fixed point of ϕ_1. Alternatively, u_0 can be set to be $\phi_1^{\circ m}(1)$ for some large value of m, so that u_0 is sufficiently close to the fixed point of ϕ_1.
3. Initialize the sum $S_{j_0} = u_0$, where $x_0 \in B_{j_0}$.
4. Choose a pair $(w_{\sigma_0}, \phi_{\sigma_0})$, $\sigma_0 \in \{1, 2, \ldots, N\}$, according to the probabilities p_i.
5. Set $x_1 = w_{\sigma_0}(x_0)$ and $u_1 = \phi_{\sigma_0}(u_0)$.
6. Increment the sum S_{j_1} by u_1, where $x_1 \in B_{j_1}$.
7. Continue in this way by returning to step 4 above,

$$x_{n+1} = w_{\sigma_n}(x_n), \quad u_{n+1} = \phi_{\sigma_n}(u_n), \quad \sigma_n \in \{1, 2, 3, \ldots, \mathbb{N}\},$$

where σ_n is chosen according to the probabilities p_i and then by updating the appropriate $S_{j_{n+1}}$.

Proposition 6.1. *For each $k \in \{1, 2, \ldots, K\}$,*

$$\frac{S_k}{n} \to \int_{B_k} f(x) \, d\mu(x) \quad as \quad n \to \infty, \tag{6.1}$$

where μ is the push-forward measure under the address map $a : \Sigma \to \mathbb{X}$ of the product measure P on Σ. Thus, $\frac{S_k}{n\mu(B_k)}$ converges to the μ-average value of f over B_k.

Proof. From the two assumptions that the maps ϕ_i are contractive and that the sets $w_i(\mathbb{X})$ are nonoverlapping, it follows that $u_n \approx f(x_n)$ and, in fact, that $|u_n - f(x_n)| \to 0$. Let I_k denote the characteristic function of B_k. Then, at the nth stage of this chaos game, we have

$$\frac{S_k}{n} \approx \frac{1}{n} \sum_{m \leq n} I_k(x_m) f(x_m)$$

and thus by the ergodic theorem (see [53]) the result follows. $\qquad\square$

Proposition 6.2. *Let $\mathcal{P}_n$ be a nested sequence of finite partitions of $\mathbb{X}$ with $a^{-1}(S) \subset \Sigma$ compact for all $S \in \mathcal{P}_n$ and with*

$$\{x\} = \bigcap \{S : x \in S, S \in \mathcal{P}_n\} \tag{6.2}$$

for all $x \in \mathbb{X}$. Let $\bar{f}_n$ be the μ-average value function of f associated with $\mathcal{P}_n$. Then $\bar{f}_n$ converges pointwise μ almost everywhere to f.

Proof. We first note that $\bar{f}_n$ is the μ-conditional expectation of f given $\mathcal{P}_n$. Thus $\bar{f}_n$ forms a martingale sequence. Since f is bounded and μ is a probability measure, this martingale is $\mathcal{L}^1$ bounded and thus we get the pointwise convergence from the martingale convergence theorem (see [160]). □

Notice that the condition specified in (6.2) is thought of as requiring that the partitions $\mathcal{P}_n$ limit to a partition consisting of only singletons. This is a control of the "size" of the elements of the partition $\mathcal{P}_n$.

We can specialize these results a little bit to the case of $\mathbb{X} \subset \mathbb{R}^m$. In this case, we can hope that μ will be the Lebesgue measure, which we will denote by λ. The way to ensure this is to set $p_i = \lambda(w_i(\mathbb{X}))/\lambda(\mathbb{X})$. It is easy to check that with this choice of p_i we will have $\mu = \lambda$, and then the functions $\bar{f}_n$ will be the usual average value function of f over the partition $\mathcal{P}_n$ and we will again obtain pointwise convergence of $\bar{f}_n$ to f.

Thus with our chaos game we can obtain an approximation to any accuracy (in the $\mathcal{L}^p$ sense) by using a sufficiently fine partition of $\mathbb{X}$. Notice that in this context we can slightly relax the nonoverlapping condition to $\mu(w_i(\mathbb{X}) \cap w_j(\mathbb{X})) = 0$, so that the images are measure disjoint.

6.1.1.1 Limits to this chaos game

It would be natural to try to use some variation of this basic algorithm for an IFSM whose maps overlap; that is, with $w_i(\mathbb{X}) \cap w_j(\mathbb{X}) \neq \emptyset$ for some i and j. However, in general this will not work, as the following simple example shows. The basic problem is that if there is overlapping, then in step 2 of the algorithm above it is not guaranteed that u_n would be close to $f(x_n)$ since $f(x_n)$ might be a sum of various contributions and only one of these will be reflected in the update to u_{n-1}.

Example 6.3. Let $\mathbb{X} = [0, 1]$ and $w_1(x) = w_2(x) = x/2$ and $w_3(x) = x/2 + 1/2$. Furthermore, let $\phi_1(t) = \phi_2(t) = 1/2$ and $\phi_3(t) = 1$. Then it is easy to see that the fixed-point function of this IFSM is $f(x) = 1$ for all $x \in [0, 1]$. However, anytime either of the first two maps are chosen we have that $u_n = \phi_i(u_{n-1}) = 1/2$. If we let $B_1 = [0, 1/2]$ and $B_2 = [1/2, 1]$, then we have

$$\int_{B_1} f(x) \, d\mu(x) = \mu(B_1) = p_1 + p_2.$$

However, for our accumulation sum S_1, we will have

$$\frac{S_1}{n} \to \frac{p_1}{2} + \frac{p_2}{2} = \frac{p_1 + p_2}{2}$$

by the law of large numbers. The problem is, as mentioned above, that there are two ways of reaching B_1 and the value of u_n does not reflect the true value of f on B_1, only one of these contributions.   □

What this example indicates is that we cannot hope to use this basic chaos game for an overlapping IFSM.

6.1.2 Chaos game for overlapping IFSM

If we have an overlapping IFSM, the chaos game from the preceding section will not work to generate approximations to the invariant function. In this section, we give a modification that will work for some overlapping IFSM operators. However, there is a rather severe restriction on the IFSM grey-level maps, the ϕ_i.

We first present these results in the simple case where $\mathbb{X} = [0, 1]$. This specific restriction is not completely necessary. However, as we will see, something like it is necessary for our results.

Let $\{w_i, \phi_i\}$ be the maps from an N-map IFSM of the form

$$w_i(x) = s_i x + a_i, \quad \phi_i(t) = \alpha_i t + \beta_i, \quad 1 \le i \le N.$$

We do not assume that $w_i(\mathbb{X}) \cap w_i(\mathbb{X}) = \emptyset$. The associated IFSM transform then has the form

$$T(g)(x) = \sum_i \alpha_i g\left(\frac{x - a_i}{s_i}\right) \chi_{w_i(\mathbb{X})}(x) + \sum_i \beta_i \chi_{w_i(\mathbb{X})}(x),$$

where χ_S is the characteristic function of the set S. We will collect the terms in the second sum together into a function $\beta : \mathbb{X} \to \mathbb{R}$ defined by

$$\beta(x) = \sum_i \beta_i \chi_{w_i(\mathbb{X})}(x).$$

This function β is a "condensation function" for the IFSM T. We assume that $\alpha_i, \beta_i \ge 0$ and that

$$\sum_i |s_i| \alpha_i < 1,$$

so that T is contractive in $\mathcal{L}^1(\mathbb{X})$. Let $f \in \mathcal{L}^1(\mathbb{X})$ be the fixed point of T. Notice that because of our assumptions on α_i and β_i we have that

$f(x) \geq 0$ and $\beta(x) \geq 0$ for all x. From the fact that $T(f) = f$, we get that

$$\langle f \rangle = \int_{\mathbb{X}} f(x) \, dx = \frac{\sum_i |s_i|\beta_i}{1 - \sum_i |s_i|\alpha_i}. \tag{6.3}$$

We can also easily compute that

$$\langle \beta \rangle = \int_{\mathbb{X}} \beta(x) \, dx = \sum_i |s_i|\beta_i.$$

Now, for any Borel subset $S \subset \mathbb{X}$, we have

$$\int_S f(x) \, dx = \sum_i \alpha_i |s_i| \int_{w_i^{-1}(S)} f(x) \, dx + \int_S \beta(x) \, dx. \tag{6.4}$$

The strategy of our algorithm for the IFSM T is to think of the right-hand side of (6.4) as defining an IFS on probabilities, but with a condensation measure, and then using the usual chaos game for such probability measures to generate an approximation to f. To do this, we need to normalize the terms in (6.4). Thus, define the normalized functions $\bar{f}(x) = f(x)/\langle f \rangle$ and $\bar{\beta}(x) = \beta(x)/\langle \beta \rangle$. Dividing both sides of (6.4) by $\langle f \rangle$ and rearranging, we recast (6.4) as an IFSP operator by defining

$$M\nu(S) = \sum_i \alpha_i |s_i| \nu(w^{-1}(S)) + \left[1 - \sum_i \alpha_i |s_i|\right] \theta(S), \tag{6.5}$$

where S is a Borel set and the probability measure θ is defined by $\theta(S) = \int_S \bar{\beta}(x) \, dx$. Let μ be the invariant probability measure of the IFSP Markov operator M. Then we will have

$$\mu(B) = \int_B \bar{f}(x) \, dx,$$

and so running a chaos game on M should give us an approximation to $\bar{f}$ and thus to f. Motivated by (6.5), we set $p_i = \alpha_i |s_i|$ for $i = 1, 2, \ldots, N$ and $p_0 = 1 - \sum_i \alpha_i |s_i|$, so that $p_0 + p_1 + \cdots + p_N = 1$.

Again, the starting point is a partition $\{B_i\}$ of $\mathbb{X}$, this time into Borel subsets. The sum variables S_i are again initialized to be zero. However, this time S_i gives information only about visitation to the set B_i. The algorithm is:

1. Initialize x_0 to be the fixed point of w_1 for convenience.
2. Set $S_{j_0} = 1$, where $x_0 \in B_{j_0}$.
3. Choose $\sigma_1 \in \{0, 1, 2, \ldots, N\}$ according to the probabilities p_i. If

   a. $\sigma_1 \geq 1$, then define $x_1 = w_{\sigma_1}(x_0)$, set j_1 by $x_1 \in B_{j_1}$, and
     increment S_{j_1}.

   b. $\sigma_1 = 0$, then choose x_1 according to the distribution with $\bar{\beta}(x)$
     as its density. Set j_1 by $x_1 \in B_{j_1}$, and increment S_{j_1}.

4. Continue in this way, choosing the next σ_n according to the probabilities p_i and either (a) setting $x_n = w_{\sigma_n}(x_{n-1})$ or (b) sampling
  from the distribution with density $\bar{\beta}$. Then update the appropriate
  S_{j_n}.

At the nth stage, the approximation to f on B_k yielded by the algorithm above is given by

$$f_{avg}(B_k) \approx \frac{1}{n} \left(\frac{S_k}{\lambda(B_k)} \right) \left(\frac{\sum_i \beta_i |s_i|}{1 - \sum_i \alpha_i |s_i|} \right). \qquad (6.6)$$

Proposition 6.4. *The approximation given in (6.6) converges to the
average value of $\bar{f}$ over the set B_k as $n \to \infty$.*

Proof. By Proposition 6 in [23], we have $S_k/n \to \mu(B_k)$, where μ is
the invariant measure of the IFSP with condensation defined in (6.5).
We write the IFSP Markov operator as

$$M\nu(B) = A\nu(B) + p_0\theta(B),$$

where A is the linear part of M. Thus,

$$\mu = \sum_n A^n(\theta).$$

This shows that μ is absolutely continuous with respect to the Lebesgue
measure since each term $A^n(\theta)$ is absolutely continuous and the series converges absolutely. Since μ is invariant with respect to M, its
Radon-Nikodym derivative (i.e., its density) must also be invariant
with respect to M. By scaling, we obtain the desired result.   □

   Again, with a sequence of partitions $\mathcal{P}_n$ that converge in the appropriate sense (condition 6.2) is satisfied), we again obtain that f is
the limit of the sequence of approximations. The proof of the following
result is the same as that of Proposition 6.2, so we leave it out.

Proposition 6.5. *Let $\mathcal{P}_n$ be a nested sequence of finite partitions of
$\mathbb{X}$ into Borel sets with the condition (6.2) satisfied. Let $\bar{f}_n$ be the μ-
average value function of f associated with $\mathcal{P}_n$. Then $\bar{f}_n$ converges
pointwise μ almost everywhere to f.*

The connection between the IFSM and the IFSP with condensation is only possible because the IFSM operator is affine. It is possible to generalize a little bit, however.

First, the terms β_i in the maps ϕ_i can be made to be functions of x. This will make the function $\beta(x)$ more complicated than piecewise constant. It is only necessary for $\beta \in \mathcal{L}^1(\mathbb{X})$, so it can be thought of as the density of a measure θ. The nonnegativity condition can also be relaxed.

Suppose that $\beta \in \mathcal{L}^1(\mathbb{X})$ but is bounded below on X. Let $\gamma \in \mathcal{L}^1(\mathbb{X})$ be a nonnegative function with $\beta(x) + \gamma(x) \geq 0$ for all x (it is possible simply to choose $\gamma(x)$ to be a constant). Define the IFSM operator T', where

$$T'(g) = A(g) + (\beta + \gamma), \quad g \in \mathcal{L}^1(\mathbb{X}), \tag{6.7}$$

where we have $A(g) = \sum_i \alpha_i |s_i| g \circ w_i^{-1}$ as before. Then T' is contractive with fixed point $f' \in \mathcal{L}^1(\mathbb{X})$ given by

$$f' = \sum_n A^n(\beta + \gamma) = \sum_n A^n(\beta) + \sum_n A^n(\gamma) = f + h,$$

where f is as before and h is the fixed point for the IFSM operator $T_\gamma(g) = A(g) + \gamma$. Therefore, $f = f' - h$. We now run two separate chaos games (as above), one for T' and one for T_γ, since they both have nonnegative condensation functions. We keep two accumulation sum variables for each B_i and at the end we can compute f by subtracting the appropriate scaled version of these sums. Notice that in choosing the σ_n we can use the same σ_n for both of these separate chaos games, since there is no requirement that the two stochastic sequences be independent.

Finally, we could also generalize to spaces other than $\mathbb{X} = [0, 1]$. What is necessary is to be able to make the link between the two quantities

$$\int_S g(w_i^{-1}(x)) \, d\eta(x)$$

and

$$\int_S g(y) \, d\eta(w_i(y)).$$

That is, we need some relationship between $d\eta$ and $d(\eta \circ w_i)$. In the case above, we see that $d(w_i(x)) = s_i dx$ because of the form of the maps w_i and because we are using the Lebesgue measure. This property is key in being able to interpret the IFSM operator as an operator acting on densities of measures and thus inducing an IFSP operator. As in Sect. 3.2.2, two broad classes of examples of this type of link are where either

μ is the d-dimensional Hausdorff measure and each ϕ_λ is a similarity (in which case $d\mu(w_i(x)) = c_\lambda^d d\mu(x))$ or $\mathbb{X}$ is the attractor of the IFS $\{w_\lambda\}$ and μ is the attractor of the IFSP $\{w_\lambda, dP(\lambda)\}$ (in which case $d\mu(w_\lambda(x)) \leq dP(\lambda)d\mu(x)$).

6.2 Chaos game for wavelets

In this section, we present chaos game algorithms for wavelets. First we discuss an algorithm due to M. Berger [21, 22, 23], that will generate an approximation to the graph of a wavelet function. This algorithm is similar in spirit to our chaos game for an IFSM presented in the previous section. After discussing Berger's algorithm, we present an algorithm that will "mix" chaos games for several wavelets (all of which are translations and dilations of a single "mother" wavelet) in order to generate a chaos game approximation for a generic function in $\mathcal{L}^2$. Finally, we present a version of the chaos game that will compute the coefficients in a wavelet expansion. The results and discussion of this section primarily come from [133].

Some basic discussion concerning wavelets was presented in Sect. 3.3.1, so we refer the reader to that section for basic notions and notation. The one additional fact we will need is that the mother wavelet function ψ is a finite linear combination of dilated and translated copies of the scaling function; that is,

$$\psi(x) = \sum_k (-1)^k h_{N-k}\phi(2x - k). \tag{6.8}$$

For the convenience of the reader, we recall that the dilation equation is given as

$$\phi(x) = \sum_n h_n\phi(2x - n). \tag{6.9}$$

If we assume that the only nonzero coefficients in the dilation equation, the h_i, are $h_0, h_1, \ldots, h_N$, then both ϕ and ψ are supported on the interval $[0, N] \subset \mathbb{R}$.

Define the mappings

$$w_{i,j}(x) = \frac{x}{2^i} + \frac{j}{2^i}, \quad i \geq 0, j \in \mathbb{Z}. \tag{6.10}$$

Then we have that the wavelet $\psi_{i,j}$ is given by $\psi_{i,j}(x) = 2^{i/2}\psi \circ w_{i,j}^{-1}$. That is, these mappings "map" the mother wavelet onto the various $\psi_{i,j}$.

6.2.1 Rendering a compactly supported scaling function

In general, the scaling function does not have a closed analytic form. The typical way to obtain values of ϕ at a point x involves using the "cascade algorithm" (see Sect. 7.2 in [158]). This recursive algorithm starts with the values of ϕ at integer points and then uses the dilation equation (6.9) to generate the values of ϕ at the half-integer points, then the quarter-integer points, and so on. The values of ϕ at the integer points are also obtained from the dilation equation.

We present a method for approximating the scaling function ϕ by using a chaos game. As mentioned above, this algorithm was conceived by Berger [21, 22]. The idea is to properly view the dilation equation as an IFSM.

Let us suppose we have chosen a sequence h_n with $h_n = 0$ for $n \notin \{0, 1, 2, \ldots, N\}$ and such that ϕ generates a multiresolution analysis on $\mathcal{L}^2(\mathbb{R})$. Daubechies and Lagarias [45] and Micchelli and Prautzsch [136] independently noticed that one could vectorize the dilation equation.

Define $V_\phi : [0, 1] \to \mathbb{R}^N$ as

$$V_\phi(x) = \begin{pmatrix} \phi(x) \\ \phi(x+1) \\ \vdots \\ \phi(x+N-1) \end{pmatrix} \tag{6.11}$$

Define the $N \times N$ matrices T_0 and T_1 by $(T_0)_{i,j} = h_{2i-j-1}$ and $(T_1)_{i,j} = h_{2i-j}$; that is,

$$T_0 = \begin{pmatrix} h_0 & 0 & 0 & \cdots & 0 & 0 \\ h_2 & h_1 & h_0 & \cdots & 0 & 0 \\ \vdots & \vdots & \vdots & \ddots & \vdots & \vdots \\ 0 & 0 & 0 & \cdots & h_N & h_{N-1} \end{pmatrix}$$

and

$$T_1 = \begin{pmatrix} h_1 & h_0 & 0 & \cdots & 0 & 0 \\ h_3 & h_2 & h_1 & \cdots & 0 & 0 \\ \vdots & \vdots & \vdots & \ddots & \vdots & \vdots \\ 0 & 0 & 0 & \cdots & 0 & h_N \end{pmatrix}.$$

Let $\tau : [0, 1] \to [0, 1]$ be defined as $\tau(x) = 2x \bmod 1$. Then the dilation equation becomes (assuming that ϕ is continuous)

$$V_\phi(x) = T_\omega V_\phi(\tau x), \tag{6.12}$$

where ω is the first digit in the binary expansion of x. Notice that (6.12) is an example of a *vector IFSM*.

As an example, we compute the transformation T_0 corresponding to the Daubechies-4 wavelet. In this case, only h_0, h_1, h_2, and h_3 are nonzero, so ϕ is supported on the interval $[0, 3]$. Thus

$$V_\phi(x) = \begin{pmatrix} \phi(x) \\ \phi(x+1) \\ \phi(x+2) \end{pmatrix}.$$

From the dilation equation, we have the equations

$$\begin{aligned}
\phi(x) &= h_0\phi(2x) &&+ h_1\phi(2x-1) + h_2\phi(2x-2) + h_3\phi(2x-3), \\
\phi(x+1) &= h_0\phi(2x+2) + h_1\phi(2x+1) + h_2\phi(2x) &&+ h_3\phi(2x-1), \\
\phi(x+2) &= h_0\phi(2x+4) + h_1\phi(2x+3) + h_2\phi(2x+2) &&+ h_3\phi(2x+1).
\end{aligned}$$

After rejecting the contributions from outside the support of ϕ, it follows that (for $0 < x < 1/2$)

$$\begin{pmatrix} \phi(x) \\ \phi(x+1) \\ \phi(x+2) \end{pmatrix} = \begin{pmatrix} h_0 & 0 & 0 \\ h_2 & h_1 & h_0 \\ 0 & h_3 & h_2 \end{pmatrix} \begin{pmatrix} \phi(2x) \\ \phi(2x+1) \\ \phi(2x+2) \end{pmatrix}.$$

Clearly the matrix is T_0. The computations for $x > 1/2$ are similar.

In the papers [21, 22], Berger first noticed that this IFS approach to the dilation equation naturally leads to the following chaos game, which approximates the graph of the scaling function ϕ:

1. Initialize $x = 0$ and the vector $V_\phi(x)$ to be the fixed point of T_0.
2. Pick $\alpha \in \{0, 1\}$ with equal probability.
3. $x \mapsto x/2 + \alpha/2$.
4. $V_\phi(x) \mapsto T_\alpha V_\phi(x)$.
5. Plot the points $(x, \phi(x)), (x + 1, \phi(x + 1)), \ldots, (x + N - 1, \phi(x + N - 1))$.
6. If the number of iterations is less than maximum, go to step 2.

The basic reason that this simple algorithm works is that the vectorized IFS is nonoverlapping; that is, $\lambda(w_0([0, 1]) \bigcap w_1([0, 1])) = 0$. Also notice that this algorithm treats ϕ as a geometric set in that it generates an image of the graph of ϕ. This is substantially different from our chaos game for IFSM, which generates an averaged version of the attractor of the IFSM.

6.2.2 Modified chaos game algorithm for wavelet generation

We now present a slight modification to the algorithm in the previous section in order to generate the wavelet ψ. However, the algorithm in this section is more in the spirit of the algorithm from Sect. 6.1.1, where we generate a piecewise constant approximation to ψ.

Let $\{B_i\}$ be a partition of $[0, N] = \mathbb{X}$ into Borel sets and $\{S_i\}$ be a set of accumulation variables. Initialize $x_0 = 0$, and let $y \in \mathbb{R}^N$ be the normalized fixed point of T_0 (eigenvector for eigenvalue 1). Initialize all the S_i to be zero.

1. Choose $\alpha \in \{0, 1\}$ with equal probability.
2. $x \mapsto x/2 + \alpha/2$.
3. $y \mapsto T_\alpha(y)$.
4. for each $k = 0, 1, \ldots, 2N - 1$,

   - if $x/2 + k/2 \in B_m$, then update

$$S_m \mathrel{+}= \sum_{i=\max(0,k-N+1)}^{\min(N,k)} (-1)^i h_{N-i} y_{k-i}.$$

5. If the number of iterations is less than some maximum, go to step 1.

The approximation to ψ on B_m is

$$\frac{S_m}{2 \times \lambda(B_m) \times n}. \tag{6.13}$$

The basic idea of the algorithm is to use a slight modification of the previous chaos game (for ψ) and take the appropriate dilation and linear combinations (from equation (6.8)) in order to form ψ. The extra factor of 2 in the denominator of (6.13) is there because ψ is defined in terms of $\phi(2x - i)$ rather than $\psi(x)$.

Another way to view step 4 of the algorithm is as follows. We have the values $y_j = \phi(x + j)$ for $j = 0, 1, \ldots, N - 1$ and the coefficients h_i for $i = 0, 1, \ldots, N$. For $z = (x + i + j)/2$, we know that $\psi(z)$ depends on the term $(-1)^i h_{N-i} y_j$. So, for each i, j such that $(x+i+j)/2 \in B_m$, we update

$$S_m \mathrel{+}= (-1)^i h_{N-i} y_j.$$

In our algorithm in step 4, we simply group all those i, j so that $i + j = k$ and loop through k.

Proposition 6.6. *For each* m,

$$\frac{S_m}{2 \times n \times \lambda(B_m)} \to \frac{1}{\lambda(B_m)} \int_{B_m} \psi(x) \, dx \quad as \quad n \to \infty.$$

Proof. The process is a nonoverlapping vector-valued IFSM on $[0, 1]$. We map the vector-valued function $g : [0, 1] \to \mathbb{R}^N$ to the function $\hat{g} : [0, N] \to \mathbb{R}$ by

$$\hat{g}(x) = \sum_{i=1}^{N} \chi_{[i-1,i]} g(x)_i.$$

Through this mapping, the process is equivalent to a nonoverlapping IFSM on $[0, N]$. Now, at each stage, y represents

$$(\phi(x), \phi(x + 1), \ldots, \phi(x + N - 1)).$$

To make the notation simpler, we just set $\phi(x + i) = 0$ if $i < 0$ or $i > N - 1$ in the formulas below. So, letting x_n be the orbit generated by the chaos game, we have

$$\frac{S_m}{n \times 2} = \frac{1}{2 \times n} \sum_{n} \sum_{k=0}^{2N-1} \sum_{i=0}^{N} (-1)^i h_{N-i} \phi(x_n + k - i) \chi_{B_m} \left((x_n + k)/2 \right)$$

$$\longrightarrow 1/2 \int_0^1 \sum_{k=0}^{2N-1} \sum_{i=0}^{N} (-1)^i h_{N-i} \phi(x + k - i) \chi_{B_m} \left((x + k)/2 \right) \, dx$$

$$= \sum_{k=0}^{2N-1} \int_{k/2}^{(k+1)/2} \sum_{i=0}^{N} h_{N-i} \phi(2y - i) \chi_{B_m}(y) \, dy$$

$$= \int_0^N \sum_i h_{N-i} \phi(2y - i) \chi_{B_m}(y) \, dy$$

$$= \int_{B_m} \psi(y) \, dy.$$

Thus, we have the desired result. □

As we saw previously in Propositions 6.2 and 6.5, we obtain convergence if we have a sequence of partitions that refines in an appropriate way. The following result is similar.

Proposition 6.7. *Let* $\mathcal{P}_n$ *be a sequence of Borel partitions of* $[0, N]$ *that satisfy condition (6.2). Let* $\bar{\psi}_n$ *be the average value functions of* ψ *associated with* $\mathcal{P}_n$. *Then* $\bar{\psi}_n$ *converges to* ψ *pointwise almost every-where.*

If instead of ψ we want to generate the wavelet $\psi_{i,j}$, the only change necessary to the algorithm above is to replace step 4 with

- if $w_{i,j}(x/2 + k/2) \in B_m$, then update

$$S_m \mathrel{+}= 2^{i/2} \sum_{l=\max(0,k-N+1)}^{\min(N,k)} (-1)^l h_{N-l} y_{k-l},$$

and then the approximation to $\psi_{i,j}$ on B_m is

$$\frac{S_m}{2 \times n \times \lambda(w_{i,j}^{-1}(B_m))} = \frac{S_m}{n \times \lambda(B_m) \times 2^{i+1}}.$$

The scaling factor $2^{i/2}$ is necessary since $\psi_{i,j}(x) = 2^{i/2}\psi(2^i x - j)$. With the modified step 4, we have

$$\frac{S_m}{2 \times n \times \lambda(w_{i,j}^{-1}(B_m))} \to 1/\lambda(B_m) \int_{B_m} \psi_{i,j}(x) \, dx$$

since

$$1/\lambda(B_m) \int_{B_m} \psi_{i,j}(x) \, dx = \frac{1}{2^i \lambda(B_m)} \int_{w_{i,j}^{-1}(B_m)} 2^{i/2}\psi(x) \, dx.$$

6.2.3 Chaos game for wavelet analysis

Given an f in L^2, we can represent

$$f = \sum c_{i,j}\psi_{i,j}$$

since the wavelets $\psi_{i,j}$ form an orthonormal basis. This is the wavelet analysis of f.

Suppose we modify the algorithm of the previous section by replacing step 4 with

$$S_m \mathrel{+}= f(x/2 + k/2) \times \sum_{i=\max(0,k-N+1)}^{\min(N,k)} (-1)^i h_{N-i} y_{k-i}.$$

Then, by arguments similar to those in the proof of Proposition 6.6, we would have that

$$\frac{S_m}{2 \times n \times \lambda(B_j)} \to \frac{1}{\lambda(B_j)} \int_{B_j} f(x)\psi(x) \, dx \quad \text{as} \quad n \to \infty. \quad (6.14)$$

Now, in wavelet analysis, we wish to compute integrals of the form

$$c_{i,j} = \int f(x)\psi_{i,j}(x)\ dx.$$

We can also modify the algorithm of the previous section in such a way as to do this in parallel for many different pairs (i, j). We use as motivation (6.14).

Suppose that we want to calculate $c_{i,j}$ for $(i, j) \in \Lambda$, some collection of coefficients. For each $\sigma = (i, j) \in \Lambda$, let S_σ be an accumulation variable, initialized to zero, and define $\mathrm{scale}(\sigma) = i$.

Our modified algorithm is as follows:

1. Initialize $x = 0$ and $y \in \mathbb{R}^N$ to be the fixed point of T_0.
2. Choose $\alpha \in \{0, 1\}$ with equal probability.
3. $x \mapsto x/2 + \alpha/2$.
4. $y \mapsto T_\alpha(y)$.
5. For each $\sigma \in \Lambda$, we do

   - for each $k = 0, 1, \ldots 2N - 1$, update

$$S_\sigma\ += \ f(w_\sigma(x/2 + k/2)) \times \sum_{i=\max(0,k-N+1)}^{\min(N,k)} (-1)^i h_{N-i} y_{k-i}.$$

6. If the number of iterations is less than maximum, go to step 2.

The approximation to c_σ is

$$\frac{S_\sigma}{2 \times 2^{\mathrm{scale}(\sigma)} \times n}. \tag{6.15}$$

Proposition 6.8. *For each $\sigma \in \Lambda$,*

$$\frac{S_\sigma}{2 \times 2^{\mathrm{scale}(\sigma)} \times n} \to \int f(x)\psi_\sigma(x)\ dx \quad as \quad n \to \infty.$$

Proof. The proof is similar to the proof of Proposition 6.6. In this case, we are evaluating the function $g(x) = \psi_\sigma(x)f(x)$ along the trajectory. Thus,

$$\frac{S_\sigma}{2 \times 2^{\mathrm{scale}(\sigma)} \times n} \to \int \psi_\sigma(x)f(x)\ dx,$$

which is what we wished to show.   □

What if we do not have the true function f but only some approximation to f? The typical case would be where we have a piecewise constant approximation to f (as in the case of a function that we only sample at certain values or a function that is the result of a chaos game approximation). Call this approximation $\hat{f}$. Then what we can compute is

$$\hat{c}_{i,j} = \int \hat{f}(x)\psi_{i,j}(x)\ dx.$$

As long as we know that $\hat{f}$ is close to f, then we know that $\hat{c}_{i,j}$ is close to $c_{i,j}$.

A special case of potential interest is when the function f is also the attractor for a related IFS. In this case, we can run two chaos games simultaneously to obtain a wavelet decomposition of f.

6.2.4 Chaos game for wavelet synthesis

Let

$$f = \sum_{i,j} c_{i,j}\psi_{i,j}$$

be the decomposition of a function in a wavelet basis. We wish to recover the function f from the coefficients $c_{i,j}$. For the moment, we assume that there are only finitely many nonzero coefficients and let $\Lambda = \{(i,j) | c_{i,j} \neq 0\}$.

Since Λ is finite, the support of f is compact, say the interval $[A, B]$.

Let $E = \sum_{i,j} |c_{i,j}|$ and $p_{i,j} = |c_{i,j}|/E$. Using the notation from previous sections, we have the following algorithm.

Initialize $x = 0$ and $y \in \mathbb{R}^N$ to be the fixed point of T_0. Let B_j be a partition of $[A, B]$ into Borel sets with associated accumulation variables S_j initialized to be 0.

1. Choose $\alpha \in \{0, 1\}$ with equal probability.
2. $x \mapsto x/2 + \alpha/2$.
3. $y \mapsto T_\alpha(y)$.
4. Choose $(i, j) \in \Lambda$ with probability $p_{i,j}$.
5. For $k = 0, 1, \ldots, 2N - 1$,

   - if $w_{i,j}(x/2 + k/2) \in B_m$, then update

$$S_m \mathrel{+}= SGN(c_{i,j})\frac{2^{i/2}}{2^i}\left(\sum_{l=\max(0,k-N+1)}^{\min(N,k)} (-1)^l h_{N-l} y_{k-l}\right).$$

6. If the number of iterations is less than maximum, go to step 1.

The approximation to f on B_m is

$$\frac{E \times S_m}{2 \times \lambda(B_m) \times n}. \tag{6.16}$$

Proposition 6.9. *For each* m

$$\frac{S_m}{2 \times n \times \lambda(B_m)} \to \frac{1}{E \times \lambda(B_m)} \int_{B_m} f(x)\, dx \quad \text{as} \quad n \to \infty.$$

Proof. For each $\sigma \in \Lambda$, let S_m^σ be an accumulation variable corresponding to B_m. Suppose we modify the algorithm above so that when we are in state σ, we update only S_m^σ. Then, by Proposition 6.6, we know that

$$\frac{S_m^\sigma}{2 \times n \times \lambda(B_m)} \to SGN(c_{i,j}) p_\sigma / \lambda(B_m) \int_{B_m} \psi_\sigma(x)\, dx.$$

Now,

$$S_m = \sum_{\sigma \in \Lambda} S_m^\sigma,$$

so

$$\frac{S_m}{2 \times n \times \lambda(B_m)} = \sum_{\sigma \in \Lambda} (S_m^\sigma 2 \times n \times \lambda(B_m))$$

$$\longrightarrow \sum_{\sigma \in \Lambda} SGN(c_{i,j}) p_\sigma / \lambda(B_m) \int_{B_m} \psi_\sigma(x)\, dx$$

and

$$\sum_{\sigma \in \Lambda} SGN(c_{i,j}) p_\sigma / \lambda(B_m) \int_{B_m} \psi_\sigma(x)\, dx = \frac{1}{\lambda(B_m) \times E} \int_{B_m} f(x)\, dx,$$

and thus we have completed the proof. $\square$

As in the case of the chaos game for ψ, we have that if we refine the partition B_m, we get a refined estimate of ψ. For completeness, we record this fact in the next proposition.

Proposition 6.10. *Let* $\mathcal{P}_n$ *be a nested sequence of Borel partitions of* $[A, B]$ *whose "sizes" go to zero as* $n \to \infty$. *Let* $\bar{f}_n$ *be the average value function of* f *associated with* $\mathcal{P}_n$. *Then* $\bar{f}_n$ *converges to* f *pointwise almost everywhere.*

6.2.5 Some extensions

6.2.5.1 Arbitrary function in $L^2(\mathbb{R})$

Extending the wavelet synthesis algorithm to generate approximations to arbitrary functions $f \in L^2(\mathbb{R})$ is simple. The idea is that there is some finite subset $\mathcal{F}$ of wavelet coefficients such that the truncation of f to these basis elements, f_{tr}, satisfies $\|f - f_{tr}\|_2 < \epsilon$. Notice that f_{tr} is compactly supported since ψ is compactly supported and f_{tr} is also a finite linear combination of $\psi_{i,j}$ for $(i, j) \in \mathcal{F}$. Now, if we choose a sufficiently fine partition, $\mathcal{P}$, we can make the average value function $\bar{f}$ associated with it satisfy $\|f_{tr} - \bar{f}\|_2 < \epsilon$. Finally, by choosing a sufficiently large number of iterations, we can make the chaos game approximation f_{cg} to f_{tr} satisfy $\|f_{cg} - \bar{f}\|_2 < \epsilon$.

In this way, we can use a chaos game to approximate any $\mathcal{L}^2(\mathbb{R})$ function to any degree of accuracy.

Proposition 6.11. *Let* $f \in L^2(\mathbb{R})$ *and* $\epsilon > 0$. *Then there exists a chaos game defined by the algorithm in Sect. 6.2.4 that produces a function* f_{cg} *that almost surely satisfies*

$$\|f - f_{cg}\|_{L^2} < \epsilon.$$

6.2.5.2 General dilations and higher dimensions

The results can be extended to more general dilation equations and more general wavelets.

For dilation factor N (instead of 2), we simply get maps $T_0, \ldots, T_{N-1}$ corresponding to the N-map vector IFSM

$$\mathcal{T}(f) = \sum_{i=0}^{N-1} T_i f(Nx - i),$$

with f being the vectorized version of the scaling function as usual. Then the chaos game for the scaling function is a simple modification of our chaos game from Sect. 6.2.1. Once we have a chaos game for the scaling function, the chaos games for the wavelet and for wavelet analysis and wavelet synthesis follow.

In order to extend our results to multidimensional scaling functions (that is, $\phi : \mathbb{R}^n \to \mathbb{R}$), wavelets, and wavelet expansions, we need a formulation of the vector IFS in this more general context.

We briefly indicate the theory here. For a more complete discussion see the references [24, 34, 35, 113, 169].

Given a *dilation matrix* $A \in M_n(\mathbb{Z})$ and a set of coefficients c_n, we have the dilation equation

$$\phi(x) = \sum_{n \in X} c_n \phi(Ax - n), \qquad (6.17)$$

where $X \subset \mathbb{Z}^n$ is a finite set. The scaling function will be the solution to this equation.

For $\Omega \subset \mathbb{Z}^n$, we let K_Ω be the attractor of the IFS $W_\Omega = \{A^{-1}x + A^{-1}d \ : \ d \in \Omega\}$. Then, if ϕ is a compactly supported solution to (6.17), the support of ϕ is a subset of K_X.

In one dimension, the vectorized function V_ϕ is based covering the support of the scaling function with copies of $[0, 1]$. In $\mathbb{R}^n$, we need to base the vectorized function on a self-affine tiling of $\mathbb{R}^n$. Under certain conditions (see [169]), there is a *digit set* $\mathcal{D} \subset \mathbb{Z}^n$ (necessarily a set of complete representatives of $\mathbb{Z}^n / A\mathbb{Z}^n$) such that $K_\mathcal{D}$ is a tile for $\mathbb{R}^n$ (i.e. that $\mathbb{Z}^n + K_\mathcal{D}$ tiles $\mathbb{R}^n$).

So, we convert the dilation equation (6.17) into the vectorized equation

$$V_\phi(x) = T_\alpha V_\phi(\tau x). \qquad (6.18)$$

Here $V_\phi : K_\mathcal{D} \to \mathbb{R}^S$, where $S \subset \mathbb{Z}^n$ is a "covering set" (see [113], p. 85), and for each $\alpha \in \mathcal{D}$ we have that T_α is an $|S| \times |S|$ matrix defined by $(T_\alpha)_{m,n} = c_{\alpha + Am - n}$ for $m, n \in S$ and $\tau : K_\mathcal{D} \to K_\mathcal{D}$ is the shift map (see [113], p. 83).

This gives a nonoverlapping vector IFS, so we can use the chaos game to generate its attractor. The following algorithm will accomplish this (assuming that the dilation equation has a solution).

Given $c_n \ : \ n \in X \subset \mathbb{Z}^n$, the dilation matrix A, $\mathcal{D} \subset \mathbb{Z}^n$ a digit set for a tiling based on A, and $S \subset \mathbb{Z}^n$ a covering set, let B_n be a partition of compact sets such that $K_X \subset \bigcup B_i$. For each B_i, let S_i be an accumulation variable initialized to zero.

1. Choose $d \in \mathcal{D}$, and initialize x as the fixed point of the map $x \to A^{-1}x + A^{-1}d$. Initialize $y \in \mathbb{R}^S$ as the fixed point of T_d.
2. Choose $\alpha \in \mathcal{D}$ randomly (with equal probability).
3. Update $y = T_\alpha y$ and $x = A^{-1}x + A^{-1}\alpha$.
4. For each $m \in S$, if $x + m \in B_i$, update

$$S_i \ += \ y_m.$$

5. If the number of iterations is less than the maximum, go to step 2.

The approximation to ϕ on B_i is

$$\frac{S_i}{\lambda(B_i) \times n,} \tag{6.19}$$

and the proof of the convergence to the average value of ϕ on B_i is the same as in the one-dimensional case.

It is relatively straightforward to modify this basic algorithm to obtain a chaos game algorithm for generating the mother wavelet, doing wavelet analysis, or doing wavelet synthesis.

6.3 Chaos game for multifunctions and multimeasures

We now turn to a more abstract framework, where we consider set-valued IFS operators. Furthermore, our viewpoint has changed in that we are interested in using the random iteration to compute integrals in the classical sense of the ergodic theorem. We first start with a slight extension of the classical chaos game result.

6.3.1 Chaos game for fractal measures with fractal densities

In this section, we consider a chaos game for a measure of the form

$$\nu(B) = \int_B \psi(x) \, d\mu(x),$$

where μ is the attractor of the IFS with probabilities $\{w_i, p_i\}$ on the compact space $\mathbb{X}$ and ψ is the attractor of the IFS on maps $\{w_i, \phi_i\}$. We assume that $w_i(\mathbb{X}) \cap w_j(\mathbb{X}) = \emptyset$ for $i \neq j$ and that each $\phi_i : \mathbb{R} \to \mathbb{R}$ is contractive. Thus there is some $M > 0$ such that $|\psi(x)| \leq M$ for all $x \in \mathbb{X}$.

Recall that the probabilities p_i induce the product measure, P, on the code space $\Sigma = \{1, 2, \ldots, N\}^{\mathbb{N}}$.

Theorem 6.12. *Suppose that $f : \mathbb{X} \to \mathbb{R}$ is bounded and continuous, $x_0 \in \mathbb{X}$, and $y_0 \in \mathbb{R}$. Let $\sigma_n \in \{1, 2, \ldots, N\}$ be iid and chosen according to the probabilities $\{p_i\}$. Let $x_{n+1} = w_{\sigma_n}(x_n)$ and $y_{n+1} = \phi_{\sigma_n}(y_n)$. Then, for all x_0 and P a.e. σ, we have that*

$$\lim_n 1/n \sum_{i \leq n} y_i f(x_i) = \int_X f(x) d\nu(x) = \int_X f(x) \psi(x) d\mu(x).$$

Proof. Let $\kappa < 1$ be the maximum of the contraction factors of the ϕ_i. Then, for any $y_0, z_0 \in \mathbb{R}$ and any $\sigma \in \Sigma^\infty$, we have

$$|\phi_{\hat{\sigma}^n}(y_0) - \phi_{\hat{\sigma}^n}(z_0)| \leq \kappa^n |y_0 - z_0| \to 0.$$

Thus, we may assume that $y_0 = \psi(x_0)$ with no loss of generality. Notice that then

$$y_{n+1} = \phi_{\sigma_n}(y_n) = \phi_{\sigma_n}(\psi(x_n)) = \psi(x_{n+1}).$$

Let $\Omega = \mathbb{X} \times \mathbb{R}$ with the metric $d((x_1, y_1), (x_2, y_2)) = d(x_1, x_2) + |y_1 - y_2|$ and $F_i : \Omega \to \Omega$ be defined as $F_i(x, y) = (w_i(x), \phi_i(y))$. Then each F_i is a contraction on the compact space Ω. Let $g : \Omega \to \mathbb{R}$ be defined as

$$g(x, y) = \begin{cases} yf(x) & \text{if } |y| \leq 2M, \\ -2Mf(x) & \text{if } y < -2M, \\ 2Mf(x) & \text{if } y > 2M. \end{cases}$$

Finally, let $z_0 = (x_0, y_0) \in \Omega$ and $z_{n+1} = F_{\sigma_n}(z_n) = (x_{n+1}, y_{n+1})$.

If we denote by $A \subset \mathbb{X}$ the attractor of the IFS $\{w_i\}$, it is easy to check that the attractor of the IFS $\{F_i\}$ on Ω is the set

$$\Delta = \{(x, \psi(x)) : x \in A\}$$

and that the invariant measure θ of the IFS with probabilities $\{F_i, p_i\}$ is supported on Δ and projects onto the measure μ on $\mathbb{X}$. Notice that $g(x, y) = yf(x)$ for any $(x, y) \in \Delta$.

Thus, by Theorem 3 (iv) in [54],

$$1/n \sum_{i \leq n} y_i f(x_i) = 1/n \sum_{i \leq n} g(x_i, y_i) = 1/n \sum_{i \leq n} g(z_n)$$

$$\to \int_{\mathbb{X}} g(z) \, d\theta(z) = \int_{\mathbb{X}} f(x) \psi(x) \, d\mu(x),$$

as was desired. $\square$

The advantage of having a fractal density can be seen from the fact that one need not know the value of the density at any point – the IFSM maps ϕ_i will compute these automatically. However, if one has a (finite) measure ν of the form

$$\nu(B) = \int_B \theta(x) \, d\mu(x),$$

where θ is any bounded function in $L^1(\mu)$, it is simple to see that

$$\frac{1}{n} \sum_{i \leq n} f(x_i)\theta(x_i) \to \int_{\mathbb{X}} f(x)\theta(x) \, d\mu(x) = \int_{\mathbb{X}} f(x) \, d\nu(x)$$

for all bounded continuous f.

6.3.2 Chaos game for multifunctions

Now we turn to the situation of an ergodic theorem for integrals of the form

$$\int_{\mathbb{X}} F(x) \, d\mu,$$

where $F : \mathbb{X} \rightrightarrows \mathbb{R}^d$ is a set-valued function and μ is the invariant measure of an IFSP. Appendix C has the basic background information on notions from set-valued analysis, but we repeat some of them here for the convenience of the reader.

We will require the multifunction F to be continuous. We say that F is *continuous* at $x \in \mathbb{X}$ if for every $\epsilon > 0$ there is a $\delta > 0$ such that whenever $d(x,y) < \delta$ we have $d_{\mathbb{H}}(F(x), F(y)) < \epsilon$, where $d_{\mathbb{H}}$ is the Hausdorff distance. We say that F is *bounded* if there is a compact set $K \subset \mathbb{R}^d$ such that $F(x) \subset K$ for all $x \in \mathbb{X}$.

Of primary importance is the *support function* of a compact and convex set K: the function $\operatorname{supp}(\cdot, K) : \mathbb{R}^d \to \mathbb{R}$ defined by

$$\operatorname{supp}(q, K) = \sup_{k \in K} q \cdot k.$$

The support function is clearly defined by its values on $\mathcal{S}^1 = \{y \in R^d : \|y\| = 1\}$ since $\operatorname{supp}(\lambda q, K) = \lambda \operatorname{supp}(q, K)$ for $\lambda \geq 0$.

The set K may be recovered from its support function as

$$K = \bigcap_{q \in \mathcal{S}^1} \{z \in \mathbb{R}^d : z \cdot q \leq z \cdot \operatorname{supp}(q, K)\}.$$

Recall that we denote by $\mathcal{K}(\mathbb{R}^d)$ the collection of nonempty compact and convex subsets of $\mathbb{R}^d$. The *norm* of a set $K \in \mathcal{K}(\mathbb{R}^d)$ is defined as $\|K\| = \sup\{|x| : x \in K\}$.

Proposition 6.13. *Suppose that F is a bounded continuous multifunction, $x_0 \in \mathbb{X}$ and $x_{n+1} = w_{\sigma_n}(x_n)$, with $\sigma_n \in \{1, 2, \ldots, N\}$ being iid* *and chosen according to $\{p_i\}$. Fix $q \in \mathbb{R}^d$. Then, for all x_0 and P a.e. σ, we have that*

$$\lim_n \operatorname{supp} \left(q, (1/n) \sum_{i \le n} F(x_i) \right) \to \operatorname{supp} \left(q, \int_{\mathbb{X}} F(x) \, d\mu(x) \right).$$

Proof. First, we note that by properties of the support function $\operatorname{supp}(q, K)$ of a compact and convex set K we have

$$\operatorname{supp} \left(q, 1/n \sum_{i \le n} F(x_i) \right) = 1/n \sum_{i \le n} \operatorname{supp}(q, F(x_i))$$

and

$$\operatorname{supp} \left(q, \int_{\mathbb{X}} F(x) \, d\mu(x) \right) = \int_{\mathbb{X}} \operatorname{supp}(q, F(x)) \, d\mu(x).$$

Furthermore, since $F(x)$ is continuous and bounded, so is $x \mapsto \operatorname{supp}(p, F(x))$ for all p. Thus, by Elton's ergodic theorem for IFSP [53], we have

$$\lim_n \operatorname{supp} \left(q, 1/n \sum_{i \le n} F(x_i) \right) = \lim_n 1/n \sum_{i \le n} \operatorname{supp}(q, F(x_i))$$

$$= \int_{\mathbb{X}} \operatorname{supp}(q, F(x)) \, d\mu(x)$$

$$= \operatorname{supp} \left(q, \int_{\mathbb{X}} F(x) \, d\mu(x) \right)$$

for P a.e. σ, as desired. $\qquad\square$

We would like to infer from this that the sequence of compact and convex sets $1/n \sum_{i \le n} F(x_i)$ converges to the compact and convex set $\int_{\mathbb{X}} F(x) \, d\mu(x)$. The issue is that for each q we have convergence only for $\sigma \in A_q \subset \Sigma^\infty$, where $P(A_q) = 1$. But then it is conceivable that $P(\bigcap_q A_q) < 1$. We prove next that this cannot happen.

Theorem 6.14. *Suppose that F is a bounded continuous multifunction, $x_0 \in \mathbb{X}$ and $x_{n+1} = w_{\sigma_n}(x_n)$, with $\sigma_n \in \{1, 2, \ldots, N\}$ being iid and chosen according to $\{p_i\}$. Then, for all x_0 and P a.e. σ, we have that*

$$\lim_n \operatorname{supp} \left(q, (1/n) \sum_{i \le n} F(x_i) \right) \to \operatorname{supp} \left(q, \int_{\mathbb{X}} F(x) \, d\mu(x) \right)$$

uniformly over all $q \in \mathcal{S}^1$.

Proof. Let $\{q_m\} \subset \mathcal{S}^1$ be a countable dense subset. Then, for each q_m, we have a subset $A_m \subset \Sigma^\infty$ with $P(A_m) = 1$ and such that for all $\sigma \in A_m$

$$\lim_n \operatorname{supp}\left(q_m, (1/n) \sum_{i \leq n} F(x_i)\right) \to \operatorname{supp}\left(q_m, \int_X F(x) \, d\mu(x)\right).$$

Let $A = \bigcap_m A_m$. Then $P(A) = 1$ with the same convergence for all $\sigma \in A$ and all q_m.

We assumed that there is some compact and convex $K \subset \mathbb{R}^d$ with $F(x) \subset K$ for all x. Let $M > 0$ be such that $\operatorname{supp}(q, K) \leq M$ for all $q \in \mathcal{S}^1$. Since K is convex, we have

$$1/n \sum_{i \leq n} F(x_i) \subset K$$

for all n and thus

$$\left| \operatorname{supp}\left(q, 1/n \sum_{i \leq n} F(x_i)\right) \right| \leq M$$

for all n as well. Similarly, we have

$$\int_X F(x) \, d\mu(x) \subset K \quad \text{and} \quad \left| \operatorname{supp}\left(q, \int_X F(x) \, d\mu(x)\right) \right| \leq M.$$

But then, for all q and all n, the functions

$$q \mapsto \operatorname{supp}\left(q, 1/n \sum_{i \leq n} F(x_i)\right)$$

and

$$q \mapsto \operatorname{supp}\left(q, \int_X F(x) \, d\mu(x)\right)$$

are Lipschitz with factor at most M (these functions are convex and bounded by M; see [146]).

Let $\epsilon > 0$ be given and $\delta = \epsilon/(3M)$, and choose $q_{m_1}, q_{m_2}, \ldots, q_{m_l}$ to form a δ-cover of the compact set $\mathcal{S}^1$. Furthermore, choose N large enough that $n \geq N$ implies that

$$\left| \operatorname{supp}\left(q_{m_i}, 1/n \sum_{i \leq n} F(x_i)\right) - \operatorname{supp}\left(q_{m_i}, \int_X F(x) \, d\mu(x)\right) \right| < \epsilon/3$$

for $i = 1, 2, \ldots, l$. Then, for any $n \geq N$ and any $q \in \mathcal{S}^1$, we have that

$$\left| \operatorname{supp} \left(q, 1/n \sum_{i \leq n} F(x_i) \right) - \operatorname{supp} \left(q, \int_{\mathbb{X}} F(x)\, d\mu(x) \right) \right| \leq$$

$$\left| \operatorname{supp} \left(q, 1/n \sum_{i \leq n} F(x_i) \right) - \operatorname{supp} \left(q_{m_i}, 1/n \sum_{i \leq n} F(x_i) \right) \right| +$$

$$\left| \operatorname{supp} \left(q_{m_i}, 1/n \sum_{i \leq n} F(x_i) \right) - \operatorname{supp} \left(q_{m_i}, \int_{\mathbb{X}} F(x)\, d\mu(x) \right) \right| +$$

$$\left| \operatorname{supp} \left(q_{m_i}, \int_{\mathbb{X}} F(x)\, d\mu(x) \right) - \operatorname{supp} \left(q, \int_{\mathbb{X}} F(x)\, d\mu(x) \right) \right|$$

$$< M \frac{\epsilon}{3M} + \frac{\epsilon}{3} + M \frac{\epsilon}{3M} = \epsilon,$$

as desired. $\qquad\square$

Corollary 6.15. *Under the conditions of the theorem, we have that*

$$1/n \sum_{i \leq n} F(x_i) \to \int_{\mathbb{X}} F(x)\, d\mu(x)$$

in the Hausdorff metric.

Proof. Just use the fact that for two compact and convex sets $A, B \subset \mathbb{R}^d$ we have

$$d_{\mathbb{H}}(A, B) = \sup_{q \in \mathcal{S}^1} |\operatorname{supp}(q, A) - \operatorname{supp}(q, B)|.$$

$\qquad\square$

The ideas used in the proof of Theorems 6.12 and 6.14 and Proposition 6.13 can be combined to produce a chaos game for integrals of the form

$$\phi(B) = \int_B f(x)\Psi(x)\, d\mu(x), \tag{6.20}$$

where $f : \mathbb{X} \to \mathbb{R}$ is continuous, $\Psi : \mathbb{X} \rightrightarrows \mathbb{R}^d$ is the multifunction attractor of the IFS $\{w_i, \Phi_i\}$, and μ is the invariant probability measure for the IFSP $\{w_i, p_i\}$. We assume that there is some compact set $K \in \mathcal{K}(\mathbb{R}^d)$ such that $\Psi(x) \subset K$ for all $x \in \mathbb{X}$ (this will always be the case if each $\Phi_i : \mathcal{K}(\mathbb{R}^d) \to \mathcal{K}(\mathbb{R}^d)$ is a contraction). With no loss of generality, we assume that $K = B_R(0)$ for some $R > 0$.

The algorithm starts with $x_0 \in \mathbb{X}$ and $S_0 \in \mathcal{K}(\mathbb{R}^d)$. We choose σ_n as before and let $x_{n+1} = w_{\sigma_n}(x_n)$ and $S_{n+1} = \Phi_{\sigma_n}(S_n)$. We show that

$$1/n \sum_{i \le n} f(x_i) S_i \to \int_{\mathbb{X}} f(x) \Psi(x) \, d\mu(x).$$

We first prove convergence for a continuous and bounded function $f : \mathbb{X} \to [0, \infty)$.

Proposition 6.16. *Suppose that $f : \mathbb{X} \to [0, \infty)$ is bounded and continuous, $x_0 \in \mathbb{X}$ and $S_0 \in \mathcal{K}(\mathbb{R}^d)$. Let $\sigma_n \in \{1, 2, \ldots, N\}$ be iid and chosen according to the probabilities $\{p_i\}$. Let $x_{n+1} = w_{\sigma_n}(x_n)$ and $S_{n+1} = \Phi_{\sigma_n}(S_n)$. Then, for all x_0 and P a.e. σ, we have that*

$$\lim_n 1/n \sum_{i \le n} S_i f(x_i) = \int_{\mathbb{X}} f(x) \Psi(x) d\mu(x),$$

with convergence in the Hausdorff metric.

Proof. We let $\Omega = \mathbb{X} \times \mathcal{K}(\mathbb{R}^d)$ with the metric $d((x_1, D_1), (x_2, D_2)) = d(x_1, x_2) + d(D_1, D_2)$, so that Ω is a compact space. We define the contractions $\Lambda_i : \Omega \to \Omega$ by $\Lambda_i(x, D) = (w_i(x), \Phi_i(D))$. Letting $A \subset \mathbb{X}$ be the attractor of the IFS $\{w_i\}$, the attractor of the IFS $\{\Lambda_i\}$ on Ω is the set

$$\Delta = \{(x, \Psi(x)) : x \in A\}$$

with invariant measure θ of the IFSP $\{\Lambda_i, p_i\}$ supported on Δ and projecting onto the measure μ on $\mathbb{X}$. For each $q \in \mathcal{S}^1$, we define the continuous and bounded function $g_q : \Omega \to \mathbb{R}$ by

$$g_q(x, D) = \begin{cases} f(x) \cdot \text{supp}(q, D) & \text{if } D \subset 2K, \\ 2f(x) \cdot \text{supp}(q, K) & \text{otherwise.} \end{cases}$$

Notice that for any $(x, D) \in \Delta$, $g_q(x, D) = f(x) \cdot \text{supp}(q, D)$ since $D \subset K$.

Then, as before, we get

$$\text{supp}\left(q, 1/n \sum_{i \le n} f(x_i) S_i\right) = 1/n \sum_{i \le n} f(x_i) \text{supp}(q, S_i)$$

$$= 1/n \sum_{i \le n} f(x_i) \text{supp}(q, \Psi(x_i)) = 1/n \sum_{i \le n} g_q(x_i, \Psi(x_i))$$

$$\to \int_{\Omega} g_q(\omega) \, d\theta(\omega) = \int_{\mathbb{X}} f(x) \cdot \text{supp}(q, \Psi(x)) \, d\mu(x)$$

$$= \mathrm{supp}\left(q, \int_{\mathbb{X}} f(x) \cdot \Psi(x) \, d\mu(x)\right).$$

Using the same argument as in the proof of Theorem 6.14, we get that

$$1/n \sum_{i \leq n} f(x_i) S_i \to \int_{\mathbb{X}} f(x) \cdot \Psi(x) \, d\mu(x)$$

with convergence in the Hausdorff metric. $\square$

Theorem 6.17. *Suppose that* $f : \mathbb{X} \to \mathbb{R}$ *is bounded and continuous,* $x_0 \in \mathbb{X}$, *and* $S_0 \in \mathcal{K}(\mathbb{R}^d)$. *Let* $\sigma_n \in \{1, 2, \ldots, N\}$ *be iid and chosen according to the probabilities* $\{p_i\}$. *Let* $x_{n+1} = w_{\sigma_n}(x_n)$ *and* $S_{n+1} = \Phi_{\sigma_n}(S_n)$. *Then, for all* x_0 *and* P *a.e.* σ, *we have that*

$$\lim_n 1/n \sum_{i \leq n} S_i f(x_i) = \int_{\mathbb{X}} f^+(x)\Psi(x) d\mu(x) - \int_{\mathbb{X}} f^-(x)\Psi(x) d\mu(x),$$

with convergence in the Hausdorff metric.

Proof. We simply modify the preceding proof, separating the sum into the two parts corresponding to when $f(x_i) < 0$ and when $f(x_i) \geq 0$.
                                                                        $\square$

6.3.3 Chaos game for multimeasures

The chaos game will always yield an integral of some function with respect to the invariant measure of some underlying IFSP. Thus, in order to construct a chaos game for more general measures Φ (such as vector-valued or set-valued measures, see [76, 135]), it is necessary to express Φ as some density F times an underlying probability measure μ that is the invariant measure of the IFSP. Now, clearly a generic multimeasure will not have this form. So, in this section we will describe an approximation procedure that will allow one, in principle at least, to use the chaos game to compute integrals with respect to a very large class of multimeasures.

A *multimeasure* will be a countably additive set function Φ defined on the Borel sets of $\mathbb{X}$ and such that $\Phi(B) \in \mathcal{K}(\mathbb{R}^d)$ for all Borel sets B. The integral of a real-valued function f with respect to a multimeasure Φ is defined in a way similar to the integral of a multifunction (see [9] for details). The *variation* $\|\Phi\|$ of Φ is defined by

$$\|\Phi\|(A) = \sup \sum_i \|\Phi(A_i)\|,$$

where the supremum is over all finite measurable partitions $\{A_i\}$ of A. The set function $\|\Phi\|$ is a positive measure.

We start with a rather standard result in IFS theory, with a sketch of the proof.

Lemma 6.18. *Let ν be an arbitrary probability measure on $\mathbb{X}$. Then there is a sequence of probability measures μ_n with $\mu_n \Rightarrow \nu$ weakly and such that each μ_n is the invariant measure of an IFSP on $\mathbb{X}$.*

Proof. (We give only a sketch of the proof.) For each n, choose a compact set K_n such that $\nu(K_n) > 1 - 1/n$. Then choose a $1/n$-net $\{x_i^n\}$ of K_n (that is, each point of K_n is within $1/n$ of some x_i^n). For this particular n, choose an IFS such that the ith map is of the form $w_i^n(z) = x_i^n$ and set the probability $p_i^n \approx \nu(\{z : d(z, x_i^n) < 1/n\})$ (to be more precise, form a partition of K_n of sets of diameter less than $1/n$). Then the attractor μ_n of this IFS has the form

$$\mu_n = \sum_i p_i^n \delta_{x_i^j}$$

and we have $\mu_n \Rightarrow \nu$. □

Lemma 6.19. *Let μ_m be a sequence of probability measures on $\mathbb{X}$ that converge weakly to the probability measure μ. Then, for any continuous and bounded multifunction F on $\mathbb{X}$, we have*

$$\int_{\mathbb{X}} F(x) \, d\mu_m(x) \to \int_{\mathbb{X}} F(x) \, d\mu(x)$$

with convergence in the Hausdorff metric.

Proof. The same technique used in the proof of Theorem 6.14 will work here. □

Take Φ to be a multimeasure of bounded variation. Further assume that there is some fixed compact and convex set K such that $\Phi(B) \subset K$ for all Borel sets B. Then the Radon-Nikodym theorem (see [6, 74]) implies that there is some multifunction F such that

$$\Phi(B) = \int_B F(x) \, d\nu(x),$$

where

$$\nu(B) = \left(\frac{\|\Phi\|}{\|\Phi\|(\mathbb{X})} \right)(B)$$

is a probability measure. Furthermore, since Φ is bounded, so is F. Let μ_m be a sequence of IFSP invariant measures that converge weakly to ν (as in Lemma 6.18).

Now, choose any continuous bounded $f : \mathbb{X} \to [0, \infty)$. Then, since F is bounded, it is easy to show that $f(x)F(x)$ is a bounded continuous multifunction. Thus, by Proposition 6.16, for each m we can construct a chaos game sequence x_i^m such that

$$\frac{1}{n} \sum_{i \leq n} f(x_i^m)F(x_i^m) \to \int_{\mathbb{X}} f(x)F(x) \, d\mu_m(x),$$

with convergence in the Hausdorff metric as $n \to \infty$. However, then, by Lemma 6.19, we have

$$\int_{\mathbb{X}} f(x)F(x) \, d\mu_m(x) \to \int_{\mathbb{X}} f(x)F(x) d\nu(x) = \int_{\mathbb{X}} f(x) \, d\Phi(x).$$

Clearly, the case of a signed measure ν is a special case of the construction above.

Chapter 7
Inverse Problems and Fractal-Based Methods

In this chapter, we consider an assortment of inverse problems for differential and integral equations, all of which can be treated within the framework of Banach's fixed point theorem and the collage theorem. As always, the essence of the method is the approximation of elements of a complete metric space by fixed points of contractive operators on that space.

The general inverse problem in applications asks us to use observational data to estimate parameters in and, perhaps, the functional form of the governing model of the phenomenon under study. The parameter estimation literature is rich in papers featuring *ad hoc* methods to answer these questions by minimizing the approximation error. Many established methods involve an iterative process of guessing parameter values, computing a numerical solution, measuring an error, refining the guess, and then repeating the process. The convergence of such schemes depends very much on making a good initial guess.

For each of the settings we consider, we introduce an appropriate complete metric space and the appropriate contractive map so that, using the collage theorem, we can switch from the minimization of the true error to the minimization of the associated collage distance. This minimization is computationally cheap, with the collage distance often being quadratic in the parameters. Based on the suboptimality theorem, Theorem 2.14, we know that the solution corresponding to the minimal-collage parameters can lie close to the optimal solution, often making further optimization unnecessary. In the case where further optimization is undertaken, the minimal-collage parameters provide a good initial guess for other methods.

For each of the settings we consider, our approach to establishing an inverse problem solution framework generally begins with an analysis of classical existence-uniqueness theory. The goal is to lift out of this theory an underlying complete metric space and the contractive map for the problem at hand.

7.1 Ordinary differential equations

We consider the following inverse problems for ODEs:

> Given a target solution curve $x(t)$ (perhaps the interpolation of data points) for $t \in I$, where I is some interval centered at x_0, and an $\epsilon > 0$, find a vector field $f(x,t)$ (subject to appropriate conditions) such that the (unique) solution $y(t)$ to the IVP $\dot{x}(t) = f(x,t)$, $x(t_0) = x_0$ satisfies $\|x - y\| < \epsilon$.

In many cases, differentiation of $x(t)$ followed by some manipulation will lead to the ODE it satisfies. However, we consider more general cases for which an exact solution is improbable, such as

1. when $x(t)$ is not given in closed form but rather in the form of data points (x_i, t_i) that may then be interpolated by a smooth function,
2. when it is desired to restrict $f(x,t)$ to a specific class of functions (e.g., polynomial in x and/or t, possibly only of first or second degree).

These situations occur in real-world applications such as biomathematics (population, disease), chemistry (reaction kinetics), and physics (damped, forced anharmonic oscillators). Many of the models used in these areas are formulated in terms of systems of polynomial ODEs. The reader is referred to [101, 102, 91, 90, 100, 93] for more discussion and a large number of examples.

We need to review the existence and uniqueness theorem for solutions of ODEs in terms of contraction mappings. For a more detailed treatment, the reader is referred to [40]. Consider the initial value problem

$$\frac{dx}{dt} = f(x,t), \qquad x(t_0) = x_0, \tag{7.1}$$

where $x : \mathbb{R} \mapsto \mathbb{R}$ is differentiable and, for the moment, $f : \mathbb{R} \times \mathbb{R} \mapsto \mathbb{R}$ is continuous. The extension to systems of ODEs, where $x : \mathbb{R} \mapsto \mathbb{R}^n$ and $f : \mathbb{R}^n \times \mathbb{R} \mapsto \mathbb{R}^n$, is straightforward. A solution to this problem satisfies the equivalent integral equation

$$x(t) = x_0 + \int_{t_0}^{t} f(x(s), s) \, ds. \tag{7.2}$$

The Picard operator T associated with (7.1) is defined as

$$v(t) = (Tu)(t) = x_0 + \int_{t_0}^{t} f(u(s), s) \, ds. \qquad (7.3)$$

Let $I = [t_0 - a, t_0 + a]$ for $a > 0$ and $C(I)$ be the Banach space of continuous functions $x(t)$ on I with norm $\|x\|_\infty = \max_{t \in I} |x(t)|$. Then $T : C(I) \mapsto C(I)$. Furthermore, a solution $x(t)$ to (7.1) is a fixed point of T. In what follows, without loss of generality, we let $x_0 = t_0 = 0$ so that $I = [-a, a]$. Nonzero values may be accommodated by appropriate shiftings and scalings. Define $D = \{(x, t) \mid |x| \le b, \ |t| \le a\}$ and $\bar{C}(I) = \{u \in C(I) \mid \|u\|_\infty \le b\}$. Also assume that

1. $\displaystyle \max_{(x,t) \in D} |f(x, t)| < \frac{b}{a}$ and
2. $f(x, t)$ satisfies the following Lipschitz condition on D

$$|f(x_1, t) - f(x_2, t)| \le K|x_1 - x_2|, \quad \forall (x_i, t) \in D,$$

such that $c = Ka < 1$.

Define a metric on $\bar{C}(I)$ in the usual way:

$$\begin{aligned} d_\infty(u, v) &= \|u - v\|_\infty \\ &= \sup_{t \in I} |u(t) - v(t)|, \quad \forall u, v \in \bar{C}(I). \end{aligned} \qquad (7.4)$$

Then $(\bar{C}(I), d_\infty)$ is a complete metric space. By construction, T maps $\bar{C}(I)$ to itself and

$$d_\infty(Tu, Tv) \le c\, d_\infty(u, v), \quad \forall u, v \in \bar{C}(I). \qquad (7.5)$$

The contractivity of T implies the existence of a unique element $\bar{x} \in \bar{C}(I)$ such that $\bar{x} = T\bar{x}$. $\bar{x}$ is the unique solution to the IVP in (7.1).

Now let $S = [-\delta, \delta]$ for some $0 < \delta \ll 1$, and redefine D as $D = \{(x, t) \mid |x| \le b + \delta, \ |t| \le a\}$. Let $\mathcal{F}(D)$ denote the set of all functions $f(x, t)$ on D that satisfy properties 1 and 2 above and define

$$\|f_1 - f_2\|_{\mathcal{F}(D)} = \max_{(x,t) \in D} |f_1(x, t) - f_2(x, t)|. \qquad (7.6)$$

We let $\Pi(I)$ denote the set of all Picard operators $T : C(I) \mapsto C(I)$ having the form in (7.3) where $x_0 \in S$ and $f \in \mathcal{F}(D)$. Equation (7.5) is satisfied by all $T \in \Pi(I)$. The continuity theorem for fixed points can be recast in our setting.

Proposition 7.1. *Let $T_1, T_2 \in \Pi(I)$ as defined above (i.e., for $u \in \bar{C}(I)$,*

$$(T_i u)(t) = x_i + \int_0^t f_i(u(s), s) \, ds, \quad t \in I, \quad i = 1, 2 \Big). \tag{7.7}$$

Also let $\bar{u}_i(t)$ denote the fixed point functions of the T_i. Then

$$d_\infty(\bar{u}_1, \bar{u}_2) \leq \frac{1}{1-c} \big[|x_1 - x_2| + a \|f_1 - f_2\|_{\mathcal{F}(D)} \big], \tag{7.8}$$

where $c = Ka < 1$.

Intuitively, closeness of the vector fields f_1, f_2 and initial conditions x_1, x_2 ensures closeness of the solutions $\bar{x}_1(t), \bar{x}_2(t)$ to the corresponding IVPs. Theorem 7.1 can be viewed as a recasting of the classical results on continuous dependence (see, for example, [40]) in terms of Picard contractive maps. From a computational point of view, however, it is not convenient to work with the d_∞ metric. It will be more convenient to work with the $\mathcal{L}^2$ metric. Note that

1. $\bar{C}(I) \subset C(I) \subset \mathcal{L}^2(I)$ and
2. $T : \bar{C}(I) \mapsto \bar{C}(I)$.

The following result establishes that the operator $T \in \Pi(I)$ is also contractive in d_2, the $\mathcal{L}^2$ metric

Proposition 7.2. *Let $T \in \Pi(I)$. Then*

$$d_2(Tu, Tv) \leq \frac{c}{\sqrt{2}} d_2(u, v), \quad \forall u, v \in \bar{C}(I), \tag{7.9}$$

where $c = Ka < 1$ and $d_2(u, v) = \|u - v\|_2$.

Proof. For $u, v \in \bar{C}(I) \subset \mathcal{L}^2(I)$,

$$\begin{aligned}
\|Tu - Tv\|_2^2 &= \int_I \left[\int_0^t (f(u(s), s) - f(v(s), s)) \, ds \right]^2 dt \\
&\leq \int_I \left[\int_0^t |f(u(s), s) - f(v(s), s)| \, ds \right]^2 dt \\
&\leq K^2 \int_I \left[\int_0^t |u(s) - v(s)| \, ds \right]^2 dt. \tag{7.10}
\end{aligned}$$

For $t > 0$, and from the Cauchy-Schwarz inequality,

$$\int_0^t |u(s) - v(s)|\, ds \le \left[\int_0^t ds\right]^{\frac{1}{2}} \left[\int_0^t |u(s) - v(s)|^2\, ds\right]^{\frac{1}{2}}$$

$$\le t^{\frac{1}{2}} \left[\int_0^t |u(s) - v(s)|^2\, ds\right]^{\frac{1}{2}}.$$

A similar result follows for $t < 0$. Note that

$$\int_0^a \left[\int_0^t |u(s) - v(s)|\, ds\right]^2 dt \le \int_0^a \int_0^t t|u(s) - v(s)|^2\, ds\, dt$$

$$= \int_0^a \int_s^a t|u(s) - v(s)|^2\, dt\, ds$$

$$= \frac{1}{2} \int_0^a (a^2 - s^2)|u(s) - v(s)|^2\, ds$$

$$\le \frac{a^2}{2} \int_0^a |u(s) - v(s)|^2\, ds.$$

A similar result is obtained for the integral over $[-a, 0]$. Substitution into (7.10) completes the proof. $\square$

Note that the space $\bar{C}(I)$ is not complete with respect to the $\mathcal{L}^2$ metric d_2. This is not a problem, however, since the fixed point $\bar{x}(t)$ of T lies in $\bar{C}(I)$. The Picard operator $T \in \Pi(I)$ is also contractive in the $\mathcal{L}^1$ metric. The following result is obtained by changing the order of integration as above.

Proposition 7.3. *Let $T \in \Pi(I)$. Then*

$$d_1(Tu, Tv) \le c d_1(u, v), \quad \forall u, v \in \bar{C}(I) \ (\subset \mathcal{L}^1(I)), \tag{7.11}$$

where $c = Ka < 1$ and $d_1(u, v) = \|u - v\|_1$.

Note that we can also work with the "Bielecki norms" [25] defined by

$$\|x\|_{\infty,\lambda} = \sup_{t \in I} e^{-\lambda K t}|x(t)|$$

and

$$\|x\|_{2,\lambda} = \left(\int_I \left(e^{\lambda K t}x(t)\right)^2 dt\right)^{\frac{1}{2}}.$$

The weighting factor λ allows us to establish an existence result regardless of a using the fact that such a weighted norm is equivalent to its associated standard norm,

$$\|x\|_{\cdot,\lambda} \le \|x\|_{\cdot} \le e^{\lambda K \delta} \|x\|_{\cdot,\lambda}. \tag{7.12}$$

However, using the weighted $\mathcal{L}^2$ norm in practice will in general lead to computational complications. The most convenient norm to work with is the standard $\mathcal{L}^2$ norm, even though contractivity in this norm depends on a. From (7.12), we see that if the $\mathcal{L}^2$ collage distance $d_2(x, Tx) < \epsilon$, then the associated weighted collage distance $d_{2,\lambda}(x, Tx) < \epsilon$.

7.1.1 Inverse problem for ODEs

The collage theorem provides a systematic method for solving our inverse problem. Find a vector field $f(x,t)$ with associated Picard operator T such that the collage distance is as small as desired. In practical applications, we consider the $\mathcal{L}^2$ collage distance $\|x - Tx\|_2$. The minimization of the squared $\mathcal{L}^2$ collage distance conveniently becomes a least-squares problem in the parameters that define f and hence T. In principle, the use of polynomial approximations to vector fields is sufficient for a formal solution to the following inverse problem for ODEs.

Theorem 7.4. *Suppose that $x(t)$ is a solution to the initial value problem $\dot{x} = f(x)$, $x(0) = 0$, for $t \in I = [-a, a]$ and $f \in \mathcal{F}(D)$. Then, given an $\epsilon > 0$, there exists an interval $\bar{I} \subseteq I$ and a Picard operator $T_\epsilon \in \Pi(\bar{I})$ such that $\|x - T_\epsilon x\|_\infty < \epsilon$, where the norm is computed over $\bar{I}$.*

Proof. From the Weierstrass approximation theorem, for any $\eta > 0$ there exists a polynomial $P_N(x)$ such that $\|P_N - f(x)\|_\infty < \eta$. Define a subinterval $\bar{I} \subseteq I$, $\bar{I} = [-\bar{a}, \bar{a}]$, such that $c_N = K_N \bar{a} < \frac{1}{2}$, where K_N is the Lipschitz constant of P_N on $\bar{I}$. (The value $\frac{1}{2}$ above is chosen without loss of generality.) Now let T_N be the Picard operator associated with $P_N(x)$ and $x_0 = 0$. By construction, T_N is contractive on $\bar{C}(\bar{I})$ with contraction factor c_N. Let $\bar{x}_N \in C(\bar{I})$ denote the fixed point of T_N. From Proposition 7.1,

$$\|x - \bar{x}_N\|_\infty \leq \frac{\bar{a}\eta}{1 - c_N} < 2\bar{a}\eta.$$

Then

$$\begin{aligned}
\|x - T_N x\|_\infty &\leq \|x - \bar{x}_N\|_\infty + \|\bar{x}_N - T_N x\|_\infty \\
&= \|x - \bar{x}_N\|_\infty + \|T_N \bar{x}_N - T_N x\|_\infty \\
&\leq (1 + c_N)\|x - \bar{x}_N\|_\infty \\
&< 2\|x - \bar{x}_N\|_\infty.
\end{aligned}$$

Since $|\bar{a}| < |a|$, given an $\epsilon > 0$ there exists an N sufficiently large that $4\bar{a}\eta < \epsilon$, yielding the result $\|x - T_N x\|_\infty < \epsilon$. We may simply rename T_N as T_ϵ, acknowledging the dependence of N on ϵ, to obtain the desired result. $\square$

We now have, in terms of the collage theorem, the basis for a systematic algorithm to provide polynomial approximations to an unknown vector field $f(x,t)$ that will admit a solution $x(t)$ as closely as desired. For the moment, we consider only the one-dimensional case (i.e., a target solution $x(t) \in \mathbb{R}$ and *autonomous* vector fields that are polynomial in x,

$$f(x) = \sum_{n=0}^{N} \lambda_n x^n, \tag{7.13}$$

for some $N > 0$). Without loss of generality, we let $t_0 = 0$, but, for reasons to be made clear below, we leave x_0 as a variable. Then

$$(Tx)(t) = x_0 + \int_0^t \left[\sum_{k=0}^{N} \lambda_k (x(s))^k \right] ds. \tag{7.14}$$

The squared $\mathcal{L}^2$ collage distance is given by

$$\Delta^2 = \int_I [x(t) - (Tx)(t)]^2 \, dt$$

$$= \int_I \left[x(t) - x_0 - \sum_{k=0}^{N} \lambda_k g_k(t) \right]^2 dt, \tag{7.15}$$

where

$$g_k(t) = \int_0^t (x(s))^k \, ds, \quad k = 0, 1, \ldots. \tag{7.16}$$

(Clearly, $g_0(t) = t$.) Δ^2 is a quadratic form in the parameters λ_k, $0 \leq k \leq N$, as well as x_0.

We now minimize Δ^2 with respect to the variational parameters λ_k, $0 \leq k \leq N$, and possibly x_0 as well. In other words, we may not necessarily impose the condition that $x_0 = x(0)$. (One justification is that there may be errors associated with the data $x(t)$.) The stationarity conditions $\frac{\partial \Delta^2}{\partial x_0} = 0$ and $\frac{\partial \Delta^2}{\partial \lambda_k} = 0$ yield the set of linear equations

$$\begin{bmatrix} 1 & \langle g_0 \rangle & \langle g_1 \rangle & \cdots & \langle g_N \rangle \\ \langle g_0 \rangle & \langle g_0 g_0 \rangle & \langle g_0 g_1 \rangle & \cdots & \langle g_0 g_N \rangle \\ \langle g_1 \rangle & \langle g_1 g_0 \rangle & \langle g_1 g_1 \rangle & \cdots & \langle g_1 g_N \rangle \\ \cdots & \cdots & \cdots & \cdots & \cdots \\ \langle g_N \rangle & \langle g_N g_0 \rangle & \langle g_N g_1 \rangle & \cdots & \langle g_N g_N \rangle \end{bmatrix} \begin{bmatrix} x_0 \\ \lambda_0 \\ \lambda_1 \\ \cdots \\ \lambda_N \end{bmatrix} = \begin{bmatrix} \langle x \rangle \\ \langle x g_0 \rangle \\ \langle x g_1 \rangle \\ \cdots \\ \langle x g_N \rangle \end{bmatrix}, \tag{7.17}$$

where $\langle f \rangle = \int_I f(t)\,dt$. If x_0 is not considered a variational parameter, then the first row and column of the matrix above (as well as the first element of each column vector) are simply removed. It is not in general guaranteed that the matrix of this linear system is nonsingular. For example, if $x(t) = C$, then $g_k(t) = C^k t$, and for $i > 1$ the ith column of the matrix in (7.17) is $C^{i-2}t$ times the first column; the matrix has rank two. In such situations, however, the collage distance can trivially be made equal to zero.

From Theorem 7.4, the collage distance Δ in (7.17) may be made as small as desired by making N sufficiently large. These ideas extend naturally to higher dimensions. Let $x = (x_1, \ldots, x_n)$ and $f = (f_1, \ldots, f_n)$. The ith component of the $\mathcal{L}^2$ collage distance is

$$\Delta_i^2 = \int_I \left[x_i(t) - x_i(0) - \int_0^t f_i(x_1(s), \ldots, x_n(s))\,ds \right]^2 dt.$$

Presupposing a particular form for the vector field components f_i determines the set of variational parameters; $x_i(0)$ may be included. Imposing the usual stationarity conditions will yield a system of equations for these parameters. If f_i is assumed to be polynomial in x_j, $j = 1, \ldots, n$, the process yields a linear system similar to (7.17).

7.1.2 Practical Considerations and examples

In practical applications, the target solution $x(t)$ may not be known in exact or closed form but rather in the form of data points (e.g. $(x_i, t_i) = x(t_i)$, $1 \leq i \leq n$ in one dimension, $(x_i, y_i) = (x(t_i), y(t_i))$ in two dimensions). One may perform some kind of smooth interpolation or optimal fitting of the data points to produce an approximate target solution, which shall be denoted as $\tilde{x}(t)$. Here, we consider approximations having the form

$$\tilde{x}(t) = \sum_{l=0}^{p} a_l \phi_l(t),$$

where the $\{\phi_l\}$ comprise a suitable basis. As we show below, the most convenient form of approximation is the best $\mathcal{L}^2$ or "least-squares" polynomial approximation (i.e. $\phi_l(t) = t^l$).

Given a target solution $x(t)$, its approximation $\tilde{x}(t)$, and a Picard operator T, the collage distance satisfies the inequality

$$\|x - Tx\| \le \|x - \tilde{x}\| + \|\tilde{x} - T\tilde{x}\| + \|T\tilde{x} - Tx\|$$
$$\le (1 + c)\|x - \tilde{x}\| + \|\tilde{x} - T\tilde{x}\|. \tag{7.18}$$

(The norm and corresponding metric are unspecified for the moment.)
Let us define the following:

1. $\delta_1 = \|x - \tilde{x}\|$, the error in approximation of $x(t)$ by $\tilde{x}$;
2. $\delta_2 = \|\tilde{x} - T\tilde{x}\|$, the collage distance of $\tilde{x}$.

Each of these terms is independent of the other. Once a satisfactory approximation $\tilde{x}(t)$ is constructed, we then apply the algorithm of Sect. 7.1.1 to it, seeking to find an optimal Picard operator for which the collage distance δ_2 is sufficiently small. The condition $(1 + c)\delta_1 + \delta_2 < \epsilon$ guarantees that the "true" collage distance satisfies $\|x - Tx\| < \epsilon$.

Also, in practical settings, data will be gathered on $[0, \tau]$, where $\tau \gg a$, the width of the interval on which T is contractive. Thus, our setup requires that the minimization of Δ_i^2 be constrained to ensure that the resulting Picard operator is contractive on the entire interval $[0, \tau]$. To avoid this issue, we subdivide the interval

$$I = [0, \tau] = \bigcup_{m=1}^{r} I_m = \bigcup_{m=1}^{r} [t_{m-1}, t_m], \text{ where } t_0 = 0 \text{ and } t_r = \tau.$$

The Picard operator T_m on I_m is contractive if we choose I_m small enough. Now, we note that

$$\sum_{m=1}^{r} \min_{\lambda_i, x_0} \int_{I_m} (x_i(t) - T_m x_i(t))^2 \, dt \le \min_{\lambda_i, x_0} \int_{I} (x_i(t) - T x_i(t))^2 \, dt \tag{7.19}$$

and choose to work with the right hand side.

We first apply the collage theorem to cases where the exact solution $x(t)$ is known in closed form. This permits all integrals in (7.19) to be calculated exactly.

Example 7.5. Let $x(t) = Ae^{Bt} + C$ be the target solution, where $A, B, C \in \mathbb{R}$. This is the exact solution of the linear ODE

$$\frac{dx}{dt} = -BC + Bx$$

for which A plays the role of the arbitrary constant. If we choose $N = 1$ in (7.13 (i.e. a linear vector field), then the solution of the linear system in (7.17) is given by

$$x_0 = A + C, \quad \lambda_0 = -BC, \quad \lambda_1 = B,$$

which agrees with the ODE above. (This solution is independent of the choice of the interval I as well as N.)

Example 7.6. Let $x(t) = t^2$ be the target solution on the half-interval $I = [0, 1]$. If we choose $N = 1$ and x_0 variable, the solution to the 3×3 system in (7.17) defines the IVP

$$\frac{dx}{dt} = \frac{5}{12} + \frac{35}{18}x, \quad x(0) = -\frac{1}{27},$$

with corresponding (minimized) collage distance $\|x - Tx\|_2 = 0.0124$. The solution to this IVP is

$$\bar{x}(t) = \frac{67}{378}e^{\frac{35}{18}t} - \frac{3}{14}.$$

Note that $\bar{x}(0) \neq x(0)$. The $\mathcal{L}^2$ distance between the two functions is $\|x - \bar{x}\|_2 = 0.0123$.

If, however, we impose the condition that $x_0 = x(0) = 0$, then the solution to the 2×2 system in (7.17) yields the IVP

$$\frac{dx}{dt} = \frac{5}{12} + \frac{35}{16}x, \quad x(0) = 0,$$

with corresponding collage distance $\|x - Tx\|_2 = 0.0186$ and solution

$$\bar{x}(t) = \frac{1}{7}e^{\frac{35}{16}t} - \frac{1}{7}.$$

As expected, the distance $\|x - \bar{x}\|_2 = 0.0463$ is larger than in the previous case where x_0 was not constrained.

Setting $N = 2$ (i.e. allowing f to be quadratic), and solving 7.17) leads to IVPs with smaller collage distances, as one would expect. Table 7.1 summarizes some results for this example. As expected, the error $\|x - \bar{x}\|_2$ decreases as N increases.

Example 7.7. Consider $x(t) = \frac{-125}{8}t^4 + \frac{1225}{36}t^3 - \frac{625}{24}t^2 + \frac{25}{3}t$, with $I = [0, 1]$, and repeat the calculations of the previous example. This quartic has a local maximum at $t = \frac{1}{3}$ and $t = \frac{4}{5}$ and a local minimum at $t = \frac{1}{2}$. The coefficients were chosen to scale $x(t)$ for graphing on $[0, 1]^2$. The results are summarized in Table 7.2; in the quadratic f case, decimal coefficients are presented here to avoid writing the cumbersome rational expressions.

In this example, the two measures of distance over $[0,1]$, $\|x - Tx\|_2$ and $\|x - \bar{x}\|_2$, appear to face impassable lower bounds. Increasing N

Table 7.1: Inverse problem results for Example 7.6, $x(t) = t^2$.

	f	x_0	$\|x - Tx\|_2$	$\|x - \bar{x}\|_2$
f linear, x_0 constrained	$\frac{5}{16} + \frac{35}{16}x$	0	0.0186	0.0463
f linear, x_0 variable	$\frac{5}{12} + \frac{35}{18}x$	$-\frac{1}{27}$	0.0124	0.0123
f quadratic, x_0 constrained	$\frac{105}{512} + \frac{945}{256}x - \frac{1155}{512}x^2$	0	0.0070	0.0300
f quadratic, x_0 variable	$\frac{35}{128} + \frac{105}{32}x - \frac{231}{128}x^2$	$-\frac{1}{60}$	0.0047	0.0049

Table 7.2: Inverse problem results for Example 7.7, $x(t) = \frac{-125}{8}t^4 + \frac{1225}{36}t^3 - \frac{625}{24}t^2 + \frac{25}{3}t$.

	f	IC	$\|x - Tx\|_2$	$\|x - \bar{x}\|_2$
f linear, x_0 constrained	$\frac{158165}{15844} - \frac{40887}{3961}x$	0	0.0608	0.0504
f linear, x_0 variable	$\frac{192815}{18444} - \frac{16632}{1537}x$	$-\frac{8425}{221328}$	0.0604	0.0497
f quadratic, x_0 constrained	$9.5235 - 8.6938x - 1.2020x^2$	0	0.0607	0.0497
f quadratic, x_0 variable	$11.0343 - 12.4563x + 1.0744x^2$	-0.0518	0.0603	0.0501

(increasing the degree of f) does not shrink either distance to zero, at least for moderate values of N. Graphically, all four cases look similar. Figure 7.1 presents two graphs to illustrate the two distance measures for the case where f is quadratic and x_0 is variable. It is the nonmonotonicity of $x(t)$ that causes difficulty.

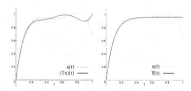

Fig. 7.1: Graphical results for Example 7.7, $x(t) = \frac{-125}{8}t^4 + \frac{1225}{36}t^3 - \frac{625}{24}t^2 + \frac{25}{3}t$.

Given the parametric representation of a curve $x = x(t)$, $y = y(t)$, $t \geq 0$, we look for a two-dimensional system of ODEs of the form

$$\dot{x}(t) = f(x, y), \ x(0) = x_0, \tag{7.20}$$
$$\dot{y}(t) = g(x, y), \ y(0) = y_0, \tag{7.21}$$

with conditions on f and g to be specified below.

Example 7.8. Applying this method to the parabola $x(t) = t$ and $y(t) = t^2$ with f and g restricted to be linear functions of x and y, we obtain, as expected, the system

$$\dot{x}(t) = 1, \ x(0) = 0,$$
$$\dot{y}(t) = 2x, \ y(0) = 0.$$

When f and g are allowed to be at most quadratic in x and y, we obtain the system

$$\dot{x}(t) = 1 + c_1(x^2 - y), \ x(0) = 0,$$
$$\dot{y}(t) = 2x + c_2(x^2 - y), \ y(0) = 0,$$

where c_1 and c_2 are arbitrary constants.

Example 7.9. For the target solution $x(t) = \cos(t)$ and $y(t) = \sin(t)$ (unit circle), with f and g at most quadratic in x and y, we obtain

$$\dot{x}(t) = -y + c_1(x^2 + y^2 - 1), \ x(0) = 1,$$
$$\dot{y}(t) = x + c_2(x^2 + y^2 - 1), \ y(0) = 0,$$

where c_1 and c_2 are arbitrary constants. (These constants disappear when f and g are constrained to be at most linear in x and y.)

We now consider target solutions that are not known in closed form. Instead, they exist in the form of data series (x_i, y_i) from which are constructed approximate target solutions $(\tilde{x}(t), \tilde{y}(t))$. In the following two examples, the data series are obtained by numerical integration of a known two-dimensional system of ODEs.

Example 7.10. We first examine the Lotka-Volterra system

$$\dot{x}(t) = x - 2xy, \ x(0) = \frac{1}{3}, \tag{7.22}$$

$$\dot{y}(t) = 2xy - y, \ y(0) = \frac{1}{3}, \tag{7.23}$$

the solution of which is a periodic cycle. For $\tilde{x}(t)$ and $\tilde{y}(t)$ polynomials of degree 30, we obtain

$$\|x(t) - \tilde{x}(t)\|_2 \approx 0.00007 \text{ and } \|y(t) - \tilde{y}(t)\|_2 \approx 0.0001.$$

Given such a periodic cycle, there are a number of possibilities that can be explored, including:

1. Restricting the functional forms of the admissible vector fields $f(x, y)$ and $g(x, y)$ (e.g., Lotka-Volterra form) versus allowing the fields to range over a wider class of functions (e.g., quadratic or higher-degree polynomials). For example, an experimentalist may wish to find the "best" Lotka-Volterra system for a given data series.

2. Constraining the initial value (x_0, y_0) or allowing it to be a variable.

In Table 7.3 are presented numerical results for a few such possibilities applied to this example. In order to save space, we have defined $Ex = \|\tilde{x} - T\tilde{x}\|_2$, $Ey = \|\tilde{y} - T\tilde{y}\|_2$, $\Delta x = \|x - \tilde{x}\|_2$ and $\Delta y = \|y - \tilde{y}\|_2$, where $(\bar{x}, \bar{y})$ denotes the fixed point of the Picard operator T.

Table 7.3: Inverse problem results for the Lotka-Volterra system in Example 7.10.

Form of f and g	f and g	Initial conditions	Collage distances	Actual $\mathcal{L}^2$ error
$f = \lambda_1 x + \lambda_2 xy$ $g = \lambda_3 y + \lambda_4 xy$ x_0 constrained y_0 constrained	$f = 1.0000x - 2.0000xy$ $g = -1.0000y + 2.0000xy$	$x_0 = \frac{1}{2}$ $y_0 = \frac{1}{3}$	$Ex = 0.0002$ $Ey = 0.0002$	$\Delta x = 0.000002$ $\Delta y = 0.000002$
$f = \lambda_1 x + \lambda_2 xy$ $g = \lambda_3 y + \lambda_4 xy$ x_0 variable y_0 variable	$f = 1.0000x - 2.0000xy$ $g = -1.0000y + 2.0000xy$	$x_0 = 0.3333$ $y_0 = 0.3333$	$Ex = 0.0002$ $Ey = 0.0002$	$\Delta x = 0.000002$ $\Delta y = 0.000002$
f quadratic g quadratic x_0 constrained y_0 constrained	$f = -0.0008 + 1.0017x$ $+0.0016y - 0.0017x^2$ $-1.9999xy - 0.0015y^2$ $g = 0.0008 - 0.0018x$ $-1.0018y + 0.0017x^2$ $+2.0000xy + 0.0017y^2$	$x_0 = \frac{1}{2}$ $y_0 = \frac{1}{3}$	$Ex = 0.0002$ $Ey = 0.0002$	$\Delta x = 0.000004$ $\Delta y = 0.000004$
f quadratic g quadratic x_0 variable y_0 variable	$f = -0.0009 + 1.0020x$ $+0.0019y - 0.0019x^2$ $-2.0000xy - 0.0018y^2$ $g = 0.0008 - 0.0017x$ $-1.0016y + 0.0016x^2$ $+2.0000xy + 0.0015y^2$	$x_0 = 0.3333$ $y_0 = 0.3333$	$Ex = 0.0002$ $Ey = 0.0002$	$\Delta x = 0.000004$ $\Delta y = 0.000004$

From Table 7.3, it is clear that when the vector fields are constrained to be of Lotka-Volterra type (first two entries), the algorithm locates the original system in (7.22) and (7.23) to many digits of accuracy. Relaxing the constraint on the initial values (x_0, y_0) yields no significant change in the approximations. In the case where the vector fields are allowed to be quadratic, the "true" Lotka-Volterra system is located to two digits accuracy.

As well, as evidenced in the examples, it should come as no surprise that increasing the accuracy of the basis representation of the target function(s) leads to a resulting DE (or system) that is closer to the original. As an example, we present Table 7.4, which shows the effect of poorer polynomial approximations to $x(t)$ and $y(t)$ in the simplest case of the earlier Lotka-Volterra example, where we seek an f and g of the correct form.

Table 7.4: Effect of different quality basis representations for the Lotka-Volterra example.

Polynomial basis degree	$\|x(t) - \tilde{x}(t)\|_2$	$\|y(t) - \tilde{y}(t)\|_2$	f	g
10	0.0022853	0.0023162	$0.998918x - 1.997938xy$	$-0.999749y + 1.999531xy$
15	0.0001278	0.0001234	$1.000121x - 2.000197xy$	$-0.999974y + 1.999943xy$
20	0.0000162	0.0000164	$1.000022x - 2.000036xy$	$-0.999996y + 1.999987xy$
25	0.0000009	0.0000008	$0.999996x - 1.999994xy$	$-0.999998y + 1.999997xy$
30	0.0000002	0.0000002	$1.000003x - 2.000005xy$	$-0.999999y + 1.999999xy$

Note. The target functions $x(t)$ and $y(t)$ can also be approximated using trigonometric function series. Indeed, for the case of periodic orbits, one would expect better approximations. This is found numerically – the same accuracy of approximation is achieved with fewer trigonometric terms. However, there is a dramatic increase in computational time and memory requirements. This is due to the moment-type integrals, $g_k(t)$, that must be computed in (7.16) because of the polynomial nature of the vector fields. Our *Maple* algorithm computes these integrals using symbolic algebra. Symbolically, it is much easier to integrate products of polynomials in t than it is to integrate products of trigonometric series.

Example 7.11. The van der Pol equation

$$\dot{x}(t) = y, \qquad\qquad\qquad x(0) = 2, \qquad\qquad (7.24)$$
$$\dot{y}(t) = -x - 0.8(x^2 - 1)y, \; y(0) = 0, \qquad\qquad (7.25)$$

has a periodic cycle as its solution. Using 46th-degree polynomials to approximate the cycle, we find

$$\|x(t) - \tilde{x}(t)\|_2 \approx 0.0141 \text{ and } \|y(t) - \tilde{y}(t)\|_2 \approx 0.0894.$$

The results for this example are given in Table 7.5.

Table 7.5: Inverse problem results for the van der Pol system in Example 7.11.

Form of f and g	f and g	Initial conditions	Collage distances	Actual $\mathcal{L}^2$ error
$f = \lambda_1 y$ $g = \lambda_2 x + \lambda_3 y + \lambda_4 x^2 y$ x_0 constrained y_0 constrained	$f = 1.0001y$ $g = -1.0007x + 0.8003y$ $\quad -0.8012x^2 y$	$x_0 = 2$ $y_0 = 0$	$Ex = 0.0110$ $Ey = 0.0546$	$\Delta x = 0.0011$ $\Delta y = 0.0016$
$f = \lambda_1 y$ $g = \lambda_2 x + \lambda_3 y + \lambda_4 x^2 y$ x_0 variable y_0 variable	$f = 1.0000y$ $g = -1.0007x + 0.7993y$ $\quad -0.7999x^2 y$	$x_0 = 2.0000$ $y_0 = 0.0023$	$Ex = 0.0002$ $Ey = 0.0002$	$\Delta x = 0.0012$ $\Delta y = 0.0018$
f quadratic g quadratic$+\lambda_1 x^2 y$ x_0 constrained y_0 constrained	$f = 0.000176 - 0.000014x$ $\quad +1.000002y - 0.000052x^2$ $\quad +0.000038xy - 0.000041y^2$ $g = -0.000968 - 0.999941x$ $\quad +0.799944y + 0.000277x^2$ $\quad -0.000179xy + 0.000220y^2$ $\quad -0.799920x^2 y$	$x_0 = 2$ $y_0 = 0$	$Ex = 0.00043$ $Ey = 0.00043$	$\Delta x = 0.00026$ $\Delta y = 0.00239$
f quadratic g quadratic$+\lambda_1 x^2 y$ x_0 variable y_0 variable	$f = -0.000004 + 0.000004x$ $\quad +1.000002y + 0.000007x^2$ $\quad +0.000002xy - 0.000003y^2$ $g = -0.000591 - 0.999979x$ $\quad +0.799921y + 0.000151x^2$ $\quad -0.000105xy + 0.000142y^2$ $\quad -0.799898x^2 y$	$x_0 = 1.9999$ $y_0 = 0.0003$	$Ex = 0.00046$ $Ey = 0.00046$	$\Delta x = 0.00035$ $\Delta y = 0.00245$

Example 7.12. We examine the inverse problems studied in [125], wherein the parameter values of certain ecological models are estimated with a Nelder-Mead-type search method. In that paper, "synthetic data" are generated by numerically solving the differential equations of a proposed model with specified parameters. Gaussian noise is added to the numerical solution, which is then sampled at a number of uniformly distributed points. These sample points are fed into the parameter estimation process outlined in [125] in order to determine optimal parameter values for differential equations of the proposed form. We emphasize that, to the best of our understanding, the estimation method in [125] involves the minimization of the fixed-point approximation error $\|x - \bar{x}\|$ as opposed to minimization of the collage error $\|x - Tx\|$.

One particular case studied in [125] is the SML model

$$\frac{dS}{dt} = -K_s SX, \tag{7.26}$$

$$\frac{dX}{dt} = K_c SX - K_m \frac{X^2}{S}, \tag{7.27}$$

where $S(t)$ and $X(t)$ represent the substrate concentration and biomass at time t and the parameters K_s, K_c, and K_m are all positive. Of course, the Nelder-Mead-type search of [125] determines the near-optimal parameters for system (7.26)-(7.27) to have a solution as close as possible to the noised numerical solution. Increasing the standard deviation of the noise distribution decreases the quality of the fit to the parameters for the system with a zero-noise solution. The Gaussian distribution had zero mean and peak magnitude as large as the peak value of the variable to which it was added.

We employ collage coding on this problem using the same test parameters employed in [125]: $K_b = 0.0055$, $K_c = 0.0038$, and $K_m = 0.00055$. The system was solved numerically: 100 sampled data points (with Gaussian noise of low-amplitude ε added) were fitted to a 10-degree polynomial, and collage coding was then employed. The sampled data points for $S(t)$ were also used to fit the function $S(t)^{-1}$ that appears in (7.27).

We make two comments before presenting the results in Table 7.6: (i) Nelder-Mead-type searches are typically quite time- and resource-consuming, while the collage coding method is quite fast, and (ii) nowhere in [125] is the initial guess at the parameters (i.e. the seed for the algorithm) specified. This second point is quite important, as the results of the collage coding method provide an excellent initial guess for further optimization, which may not even be warranted because the results are so close.

Table 7.6: Collage coding results for the SML problem of [125].

ε	K_b	K_c	K_m
0.00	0.005500000	0.003800000	0.000550000
0.01	0.005520150	0.003739329	0.000531034
0.03	0.005560584	0.003617699	0.000492936
0.05	0.005601198	0.003495669	0.000454616
0.10	0.005703507	0.003188687	0.000357837

As shown in Table 7.4, the accuracy achieved by the method increases with the accuracy to which the target solutions $x(t)$ and $y(t)$ are approximated in terms of basis expansions. A significant additional increase in accuracy may be achieved if the inverse problem is partitioned in the time domain so that the target solutions are better approximated over smaller time intervals. We may also choose to partition the spatial domain and seek a piecewise-defined vector field, each piece of which minimizes the corresponding collage distance. Such an exploration appears in [101], and some related ideas will be discussed in Sect. 7.2.

7.1.3 Multiple, partial, and noisy data sets

In practical applications, it is most probable that not one but a number of (experimental) data curves or sets will be available, all of which are to be considered as solution curves of a single ODE or system of ODEs. It is relatively straightforward to accommodate multiple data sets in our inverse problem algorithm of Sect. 7.1.1.

For each target function (each possibly a basis representation of some numerical data), we can calculate the squared collage distance of (7.15). Each integral involves the polynomial coefficients λ_k of f since we assume that the data correspond to different solutions to the same DE. We then minimize the sum of the squared collage distances with respect to the variational parameters λ_k, $0 \leq k \leq N$. Notice that x_0 can no longer be treated as a parameter since the different solutions will correspond to different x_0 values. (It is also possible to minimize weighted sums of the squared collage distances if some of the solutions are presumed to be more or less accurate or perhaps more or less important.)

Minimizing the sum of the *squared* $\mathcal{L}^2$ collage distances is much easier to perform computationally than minimizing the sum of the collage distances. Certainly, from a theoretical perspective, the sum of the squared collage distances is *not* a metric. Nevertheless, squeezing this sum toward zero guarantees a squeezing of the individual collage distances, which is sufficient for applications. This situation is discussed in greater detail in [100].

Example 7.13. Consider the anharmonic oscillator system

$$\dot{x}(t) = y(t), \tag{7.28}$$
$$\dot{y}(t) = -x(t) - 0.1x^3(t). \tag{7.29}$$

The period of the resulting cycle depends on the initial conditions that are imposed. In what follows, we consider the solutions that correspond to the two sets of initial conditions $\{x(0) = 1, \ y(0) = 0\}$ and $\{x(0) = 2, \ y(0) = 0\}$. Solving each system numerically and fitting each component of the numerical solution with a polynomial gives us our input data. The results of minimizing the combined collage distances for various polynomial basis degrees are given in Table 7.7.

Table 7.7: Results for the anharmonic oscillator problem in Example 7.13 with two data sets.

Polynomial basis degree	f	g
5	$0.99894640y$	$-1.02120674x - 0.09696297x^3$
10	$0.99962727y$	$-1.00059610x - 0.09976199x^3$
15	$1.00001325y$	$-0.99999964x - 0.10000014x^3$
20	$0.99999417y$	$-0.99999985x - 0.10000008x^3$
25	$1.00000322y$	$-1.00000006x - 0.09999995x^3$
30	$1.00000322y$	$-1.00000006x - 0.09999997x^3$

It is also conceivable that only "partial data sets" are available; for example, only a portion of a solution curve $(x(t), y(t))$ that is suspected to be a periodic orbit of a two-dimensional system of ODEs. In this situation, we simply apply our algorithm to the available data.

Example 7.14. We again solve the Lotka-Volterra system of Example 7.10 numerically. This time, however, we only find a polynomial representation for a portion of the solution cycle and seek a system of ODEs that has this curve as a solution. Table 7.8 presents the results for both degree 10 and degree 30 polynomial fits and variously sized partial data sets when Lotka-Volterra type systems are sought. Table 7.9 presents similar results in the case of quadratic systems of ODEs. All coefficients and initial conditions are presented to six decimal places. At first glance, the tables paradoxically seem to suggest that using less data yields better results. This artifact is actually due to the improved basis representation of the numerical data as the partial data set is made smaller. In all cases, the Riemann sum that approximates the $\mathcal{L}^2$ distance between the numerical solution and the basis representation is computed using 80 partitions of the time interval under consideration. Therefore, as the time intervals are made smaller, the basis approximation is substantially improved. Comparing the results for degree 10 and degree 30 polynomial bases further confirms the observation that improved basis representations yield improved results. Of course, it bears mentioning that any of the earlier techniques for improving results can be used here. For example, partitioning the time domain at the turning points of $x(t)$ and $y(t)$ improves the basis representation and yields a better result, as expected.

The numerical examples in the earlier sections are highly idealized: the input to our algorithm is a polynomial representation of the exact numerical solution to a DE or system. In a practical application,

Table 7.8: Results for partial data sets and the Lotka-Volterra system.

Polynomial basis degree	Percentage of cycle used	Resulting Lotka-Volterra system	IC
10	80	$\dot{x}(t) = 1.000283x - 2.000482xy$ $\dot{y}(t) = -1.000078y + 2.000064xy$	$x_0 = 0.333311$ $y_0 = 0.333540$
10	50	$\dot{x}(t) = 0.999999x - 1.999998xy$ $\dot{y}(t) = -1.000002y + 2.000004xy$	$x_0 = 0.333333$ $y_0 = 0.333333$
10	20	$\dot{x}(t) = 1.000000x - 2.000000xy$ $\dot{y}(t) = -1.000000y + 2.000000xy$	$x_0 = 0.333333$ $y_0 = 0.333333$
30	80	$\dot{x}(t) = 0.999996x - 1.999993xy$ $\dot{y}(t) = -1.000001y + 2.000004xy$	$x_0 = 0.333334$ $y_0 = 0.333332$
30	50	$\dot{x}(t) = 0.999998x - 1.999995xy$ $\dot{y}(t) = -1.000001y + 2.000003xy$	$x_0 = 0.333333$ $y_0 = 0.333333$
30	20	$\dot{x}(t) = 1.000000x - 2.000000xy$ $\dot{y}(t) = -1.000000y + 2.000000xy$	$x_0 = 0.333333$ $y_0 = 0.333333$

it is far more likely that the interpolation of data points leads to a much rougher representation of the solution to the underlying DE or system. To simulate this imprecision, we can add Gaussian noise to our numerical solution before performing the basis fit. More precisely, the Riemann sum that approximates the $\mathcal{L}^2$ distance between our solution and the basis representation is constructed by adding a small amount of noise to each sample of the solution. We then look for a DE or system that generates this noisy solution. The earlier partitioning ideas could also be used here. The effect of noise is in itself a major subject for inquiry. We limit our discussion to one illustrative example, discussed in greater detail in [90].

Example 7.15. In [154], the authors consider how one might handle the additional challenge of chaotic solutions by focusing on a problem for the Lorenz system [118]

$$\frac{dx}{dt} = \sigma(y - x), \ x(0) = x_0,$$

$$\frac{dy}{dt} = (r - z)x - y, \ y(0) = y_0,$$

$$\frac{dz}{dt} = xy - \beta z, \ z(0) = z_0.$$

Parameters are chosen in the chaotic realm, and faux observational data are manufactured for the x-component with low-amplitude Gaussian noise added. To solve the inverse problem, a particular cost function is minimized. Such problems have been solved via Bock's multiple

Table 7.9: Results for partial data sets and the Lotka-Volterra system.

Polynomial basis degree	Percentage of cycle used	Resulting quadratic system
10	80	$\dot{x}(t) = -0.018442 + 1.037874x + 0.044724y$ $-0.040676x^2 - 1.991041xy - 0.046363y^2$ $\dot{y}(t) = 0.010396 - 0.020771x - 1.027131y$ $+0.023885x^2 + 1.992760xy + 0.029110y^2$ $x_0 = 0.333071$ $y_0 = 0.333846$
10	50	$\dot{x}(t) = -0.000675 + 1.001050x + 0.002024y$ $-0.000937x^2 - 2.000340xy - 0.001716y^2$ $\dot{y}(t) = -0.000070 - 0.000121x - 0.999811y$ $-0.000178x^2 + 2.000219xy - 0.000367y^2$ $x_0 = 0.333330$ $y_0 = 0.333334$
10	20	$\dot{x}(t) = -0.000001 + 1.000002x + 0.000003y$ $-0.000003x^2 - 1.999996xy - 0.000006y^2$ $\dot{y}(t) = 0.000001 - 0.000001x - 1.000002y$ $+0.000002x^2 + 1.999997xy + 0.000003y^2$ $x_0 = 0.333333$ $y_0 = 0.333333$
30	80	$\dot{x}(t) = 0.000019 + 0.999963x - 0.000040y$ $+0.000026x^2 - 1.999992xy + 0.000036y^2$ $\dot{y}(t) = -0.000020 + 0.000042x - 0.999966y$ $-0.000039x^2 + 2.000010xy - 0.000032y^2$ $x_0 = 0.333333$ $y_0 = 0.333334$
30	50	$\dot{x}(t) = -0.000001 + 1.000001x + 0.000002y$ $-0.000003x^2 - 1.999994xy - 0.000002y^2$ $\dot{y}(t) = -0.000001 - 0.000000x - 0.999999y$ $-0.000000x^2 + 2.000005xy - 0.000001y^2$ $x_0 = 0.333333$ $y_0 = 0.333333$
30	20	$\dot{x}(t) = 0.000000 + 1.000000x + 0.000001y$ $-0.000001x^2 - 2.000000xy - 0.000001y^2$ $\dot{y}(t) = 0.000000 - 0.000001x - 1.000001y$ $+0.000001x^2 + 1.999999xy + 0.000001y^2$ $x_0 = 0.333333$ $y_0 = 0.333333$

shooting approach [26]. This iterative method requires repeated numerical solving of the system on each subinterval of a chosen partition followed by the solution of a large-scale constrained optimization problem. The researchers comment on the difficulty in and benefit of finding a good initial guess of the parameters.

As a specific example of an inverse problem for the Lorenz system, we consider a problem posed by researchers at the Oxford Maths Institute, who provided the data set described below. Let

$$\frac{dx}{dt} = 10(y - x), \ x(0) = x_0,$$

$$\frac{dy}{dt} = (r - z)x - y, \ y(0) = y_0,$$

$$\frac{dz}{dt} = xy - \frac{8}{3}z, \ z(0) = z_0,$$

with four parameters: r, x_0, y_0, and z_0. The system is solved numerically for fixed choices of the parameters, and the data for solution component $y(t)$ are sampled on intervals of length 0.1, with low-amplitude Gaussian noise added in. We are given the earlier system and 8192 noised data values for $y(t)$, and our job is to determine the value of the parameter r.

Using multiple shooting, one obtains $r = 26.6179 \pm 0.0114$. The behaviour-based method yields $r = 26.6$ along with the 90% confidence interval $[26, 27.5]$. The agreement of the two results to the first decimal place is coincidental.

We approach the problem with collage coding in mind. Since the differential equation for $y(t)$ involves both $x(t)$ and $z(t)$, in order to construct the collage distance for y we will need values for $x(t)$ and $z(t)$. We discretize the known DEs for $x(t)$ and $z(t)$, with a time step of size 0.1, and use the data values for $y(t)$ to manufacture this new data,

$$x(t_{i+1}) = x(t_i) + 10\,(y(t_i) - x(t_i))\,0.1 = y(t_i),$$

$$z(t_{i+1}) = z(t_i) + \left(x(t_i)y(t_i) - \frac{8}{3}z(t_i)\right)0.1,$$

subject to $x(t_0) = x(0) = x_0$ and $y(t_0) = y(0) = y_0$. In general, the values of $x(t_i)$ and $z(t_i)$, $i \geq 1$, will depend on the unknowns x_0 and z_0, but a simplification occurs here, as the x-values are merely a shift of the y-values. Following the treatment in [101], we would next like to fit polynomials (of some chosen degree) to the data, construct the collage distance of interest, and minimize it. Unfortunately, the dependence of the z-data values on the unknown z_0 causes some trouble. We proceed instead by discretizing the squared $\mathcal{L}^2$ collage distances, written here for N data points:

$$\Delta_x^2 = \sum_{k=0}^{N-1} \left[x_k - x_0 - \sum_{i=0}^{k} 10(y_i - x_i)0.1 \right]^2 0.1, \qquad (7.30)$$

$$\Delta_y^2 = \sum_{k=0}^{N-1} \left[y_k - y_0 - \sum_{i=0}^{k} (rx_i - z_i x_i - y_i)0.1 \right]^2 0.1, \qquad (7.31)$$

$$\Delta_z^2 = \sum_{k=0}^{N-1} \left[z_k - z_0 - \sum_{i=0}^{k} \left(x_i y_i - \frac{8}{3} z_i \right) 0.1 \right]^2 0.1, \qquad (7.32)$$

where $x_k = x(t_k)$, $y_k = y(t_k)$, and $z_k = z(t_k)$. Now, Δ_x^2 depends quadratically on x_0, and imposing $\frac{\partial \Delta_x^2}{\partial x_0} = 0$ determines the minimizing value, namely

$$\sum_{k=0}^{N-1} \left[x_k - x_0 - \sum_{i=0}^{k} (y_i - x_i) \right] = 0$$

$$\Rightarrow x_0 = \frac{1}{N} \left(\sum_{k=0}^{N-1} \left[x_k - \sum_{i=0}^{k} (y_i - x_i) \right] \right).$$

Next, we minimize Δ_z^2, which now depends only on z_0, imposing $\frac{\partial \Delta_z^2}{\partial z_0} = 0$ and recalling that each z_k depends on z_0. Finally, we minimize Δ_y^2 by imposing $\frac{\partial \Delta_y^2}{\partial y_0} = 0$ and $\frac{\partial \Delta_y^2}{\partial r} = 0$ to find the minimizing y_0- and r-values.

Table 7.10 presents the value of r we obtain when different amounts of input data are fed into the discrete collage coding machinery described above. Notice that our collage coding interval can consist of any collection of consecutive data points; we can even combine the collage distances for disconnected sets of data points or multiple runs of data gathering.

It is perhaps interesting to note that on occasion we obtain a value of r that lies inside the 90% confidence interval obtained via the behaviour method. Otherwise, it lies just to the left of that confidence interval. Since collage-coding is cheap and fast, we suggest that a collage-coding pretreatment provides an excellent initial guess for further optimization with other algorithms. Using the trapezoid rule to manufacture the data for x and z leads to similar results.

We can generalize our approach for partial data sets. Our framework consists of an N-dimensional system, $\frac{dx_i}{dt} = f_i(t, x_1, \ldots, x_N)$, $x_i(0) = x_{i0}$, $i = 1, \ldots, N$, that admits the following partitioning:

Table 7.10: r-values obtained via collage coding, to three decimal places.

Data points used	Minimal-collage r	Data points used	Minimal-collage r
1–512	23.625	1–512	23.625
513–1024	24.971	1–1024	24.543
1025–1536	24.968	1–1536	25.000
1537–2048	25.345	1–2048	25.178
2049–2560	26.066	1–2560	25.841
2561–3072	25.343	1–3072	26.393
3073–3584	23.596	1–3584	26.640
3585–4096	24.205	1–4096	24.990
4097–4608	25.182	1–4608	24.183
4609–5120	25.608	1–5120	24.022
5121–5632	25.265	1–5632	23.982
5633–6144	23.488	1–6144	23.981
6145–6656	24.538	1–6656	24.017
6657–7168	24.470	1–7168	24.047
7169–7680	23.609	1–7680	24.060
7681–8192	24.829	1–8192	24.083

- data sets are obtainable for the components x_i, $i = 1, \ldots, M$, $M < N$, and
- all unknown coefficients or parameters occur in f_i, $i = 1, \ldots, M$.

In this setting, data can be manufactured for the components x_i, $i = M+1, \ldots, N$; in general, such data will depend on the initial values x_{i0}, $i = M + 1, \ldots, N$. The resulting system of discretized squared collage distances $\Delta_{x_i}^2$, $i = M + 1, \ldots, N$, can be solved for the minimizing values of these latter initial conditions. The remaining system $\Delta_{x_i}^2$, $i = 1, \ldots, M$, will then depend only on the first M initial conditions and the unknown parameters, and the minimizing values can be calculated.

7.2 Two-point boundary value problems

We consider two-point boundary value problems of the form

$$\frac{d^2u}{dx^2}(x) = f\left(x, u(x), \frac{du}{dx}(x)\right), \quad A < x < B, \quad (7.33)$$

$$a_1 u(A) + a_2 \frac{du}{dx}(A) = \alpha, \tag{7.34}$$

$$b_1 u(B) + b_2 \frac{du}{dx}(B) = \beta, \tag{7.35}$$

where $u(x) : [A, B] \mapsto \mathbb{R}$, a_1, a_2, b_1, b_2, α, and β are real, and $f : \mathbb{R}^3 \mapsto \mathbb{R}$ is continuous. Boundary value problems of this type arise in numerous applications; methods of direct solution and requirements for (7.33)–(7.35) to have a unique solution are well-known. Here we consider an inverse problem:

> *Given a function $u(x)$ defined on [A,B], find a two-point boundary value problem (7.33)–(7.35), possibly with the form of f prescribed, that admits $u(x)$ as an approximate solution as closely as desired.*

We will typically suppose that the coefficients in the boundary conditions (namely, a_1, a_2, b_1, b_2) are known. The value of α or β may be known; while the form of $f(x, u(x), u'(x))$ might be known, it will contain unknown parameters. The reader will observe that a direct method of solution to our question is quite unlikely to be successful since the given $u(x)$, generated from noisy observational data, is in general not a true solution of any such model. Based on the discussion in Sect. 7.1, we propose a collage coding process that incorporates a novel approach to coping with the two boundary conditions in place of the single initial condition.

In order to treat the boundary value problem, replace the boundary conditions in (7.33)–(7.35) with two possibly undetermined initial conditions in order to take advantage of the theory of initial value problems (see [84] for other uses of this idea). Thus, we consider the initial value problem

$$\frac{d^2u}{dx^2} = f\left(x, u(x), \frac{du}{dx}(x)\right), \quad u(0) = u_0, \quad \frac{du}{dx}(0) = u'_0, \tag{7.36}$$

where $u : \mathbb{R} \mapsto \mathbb{R}$, $f : \mathbb{R}^3 \mapsto \mathbb{R}$ is continuous, and without loss of generality the initial conditions are stated at $x = 0$. Under suitable conditions on f, existence and uniqueness of solutions can of course be established by converting this second-order equation into a first-order system. Integrating (7.36) with respect to x from 0 to x twice and interchanging the order of the integrals yields

$$u(x) = u_0 + u_0'x + \int_0^x (x - s)f(s, u(s), u'(s))ds.$$

The one-dimensional Picard operator associated with (7.36), acting on functions with continuous derivatives, is then defined by

$$(Tu)(x) = u_0 + u_0'x + \int_0^x (x - s)f\left(s, u(s), \frac{du}{dx}(s)\right) ds. \qquad (7.37)$$

In what follows we set $u_0 = u_0' = 0$ without loss of generality since nonzero values of these parameters can always be accommodated by the transformation $v(x) = (u(x) - u_0) - xu_0'$. Following the setup in Sect. 7.1, let $I = [0, \delta]$, $\delta > 0$, and $C(I)$ $(C^1[I])$ be the space of continuous (continuously differentiable) functions on I. Define

$$d_\infty(u, v) = \sup_{x \in I}\{|u - v| + |u' - v'|\}, \ \forall \ u, v \in C^1(I), \qquad (7.38)$$

$$\bar{C}^1(I) = \{u \in C^1(I) \mid \|u\|_\infty \leq M\},$$
$$D = \{(x, u, u') \mid x \in I, \|u\|_\infty \leq M\}, \text{ and}$$

$$d_2(u, v) = \left(\int_0^\delta (|u - v| + |u' - v'|)^2 dx\right)^{\frac{1}{2}}, \ \forall \ u, v \in \bar{C}^1(I). \ (7.39)$$

We require that

(i) f satisfy

$$\max_{(x,u,u') \in D} |f(x, u, u')| \leq \frac{M}{(\delta + 1)\delta} \text{ and}$$

(ii) f also satisfy the following Lipschitz condition on D: for (x, u, u') and (x, v, v') in D, real numbers $K_1, K_2 \geq 0$ exist such that

$$|f(x, u, u') - f(x, v, v')| \leq K_1|u - v| + K_2|u' - v'|. \qquad (7.40)$$

With this setup, $T : \bar{C}^1(I) \mapsto \bar{C}^1(I)$. Let $\Pi(I)$ denote the set of all Picard operators of the form (7.37) with f satisfying (i) and (ii) above. A proof similar to that of Theorem 7.2 (see [92]) establishes the following.

Theorem 7.16. *For sufficiently small δ, every Picard operator $T \in \Pi(I)$ is contractive with respect to the* sup *metric (7.38) and the* $\mathcal{L}^2$ *metric (7.39) on* $\bar{C}^1(I)$.

The final result recasts the continuity theorem in this setting. We define the following norm on D

$$d_D(f_1(x, u, u'), f_2(x, u, u')) = \sup_{(x,u,u')\in D} |f_1(x, u, u') - f_2(x, u, u')|.$$

Theorem 7.17. *Let* $T_1, T_2 \in \Pi(I)$ *with*

$$(T_i u)(x) = (u_0)_i + (u_0')_i x + \int_0^x (x - s) f_i \left(s, u(s), \frac{du}{dx}(s) \right) ds$$

$t \in I$, $i = 1, 2$, *with contraction factors* c_1 *and* c_2, *respectively. Let* $\bar{u}_i(x) \in \bar{C}(I)$ *be the fixed point of* T_i. *Then with* $c = \min(c_1, c_2)$

$$d_\infty(\bar{u}_1, \bar{u}_2) \le \frac{1}{1 - c} \left[|(u_0)_1 - (u_0)_2| + \delta|(u_0')_1 - (u_0')_2| + \frac{\delta^2}{2} d_D(f_1, f_2) \right].$$

In summary, in terms of the Picard map our inverse problem is: *Given a function* $u(x)$ *defined on* $[A, B]$, *find a* $T \in \Pi(I)$ *such that* $\bar{u}(x)$, *the fixed point of* T, *satisfies* $d_2(\bar{u}, u) < \epsilon$, *with* ϵ *chosen as small as desired.* Using the collage theorem, we realize that $d_2(\bar{u}, u) < \epsilon$ can be satisfied if the collage distance $d_2(Tu, u)$ can be made arbitrarily small. Thus, we seek to minimize this collage distance in some systematic way.

7.2.1 Inverse problem for two-point BVPs

Given a target function $u(x)$, which may be the interpolation of experimental data points, it is not obvious how to pick the functional form of $f(t, u, u')$ in (7.33) (and hence the Picard operator). We present a result (similar to Theorem 7.4) that shows that restricting ourselves to polynomial f still allows us to make the collage distance arbitrarily small on some subinterval of I. The proof follows a similar path. This result will be the basis for a partitioning scheme described later.

Theorem 7.18. *Pick* $\delta > 0$, *defining* $I = [0, \delta]$. *Let* $u(x)$ *be a solution to (7.36) with* $u_0 = u_0' = 0$. *Then, given any* $\epsilon > 0$, *there exists an* $I_1 \subseteq I$ *and an operator* $T_1 \in \Pi(I)$ *such that* $d_\infty(u, T_1 u) < \epsilon$ *on* I_1.

We work on $[A, B]$, focusing on the initial value problem (7.36) to start. For ease of computation in this discussion, we look for a second-order linear differential equation with polynomial coefficients, but the algorithm may be applied to other choices. This means that our inverse boundary value problem is: Given $u(x)$ defined on $[A, B]$, find a boundary value problem of the form

$$\frac{d^2u}{dx^2}(x) = f(x, u, u')$$

$$= -p(x)u - q(x)u', \quad A < x < B, \quad (7.41)$$

$$a_1u(A) + a_2u'(A) = \alpha, \text{ and} \quad (7.42)$$

$$b_1u(B) + b_2u'(B) = \beta, \quad (7.43)$$

where $p(x) = \sum_{i=0}^{N} p_i x^i$ and $q(x) = \sum_{i=0}^{N} q_i x^i$. The associated Picard operator is

$$Tu(x) = u_A + u'_A - \int_A^x (x - s) \left(\sum_{i=0}^{N} p_i s^i u'(s) - \sum_{i=0}^{N} q_i s^i u(s) \right) ds,$$

and the squared $\mathcal{L}^2$ collage distance is given by

$$\Delta^2 = \int_A^B (u(x) - (Tu)(x))^2 dx.$$

Notice that Δ^2 is quadratic in the parameters u_A, u'_A, p_i, q_i, $0 \leq i \leq N$. Minimizing Δ^2 with respect to these variational parameters by imposing that the partial derivative with respect to each of them is zero yields a linear system.

7.2.2 Practical considerations and examples

The minimization of the squared collage distance Δ^2 determines values for the parameters in f as well as u_A and u'_A. Let $\mathcal{A} = \{a_1, a_2, \alpha\}$ and $\mathcal{B} = \{b_1, b_2, \beta\}$. There are cases dependent on the form of the boundary conditions.

Type I. $\mathcal{A}$ and $\mathcal{B}$ each have at least one unspecified member. Then, having found $u(A) = u_A$ and $u'(A) = u'_A$, (7.42) can be satisfied with appropriate choices of the unspecified elements in $\mathcal{A}$. The solution to the initial value problem can then be used to determine $u(B)$ and $u'(B)$, with the unspecified members of $\mathcal{B}$ then being chosen to satisfy (7.43).

Type II.   Only one of $\mathcal{A}$ or $\mathcal{B}$ has all of its members specified.
If all of the members of $\mathcal{B}$ are specified, depending on whether
$a_1 = 0$, when we minimize Δ^2 we do not treat u_A or u'_A as a
variational parameter. Either the first or second row and column in the matrix $\mathcal{M}$ above is deleted, with the corresponding
entries in y and $\mathcal{C}$ deleted. If a_1 is nonzero, we solve for u'_A and
the parameters in f in terms of u_A. The resulting initial value
problem (with $u(A) = u_A$ and $u'(A) = u'_A$) is solved in terms of
u_A, and then the fully specified boundary condition at $x = B$ is
satisfied as well as possible by a choice of u_A. A similar calculation can be performed when $a_1 = 0$. If all of the members of $\mathcal{A}$
are specified, we perform a similar calculation with our Picard
operator constructed at the point $x = B$.

Type III.   All of the members of $\mathcal{A}$ and $\mathcal{B}$ are specified. If $a_1 a_2$ is
nonzero, using (7.42), we eliminate $u'(A) = u'_A$ from our Picard
operator and leave u_A free during the minimization process, as
in the previous case. We solve the resulting initial value problem
and use (7.43) to determine the best choice for u_A. If a_1 or a_2
is zero, a similar calculation can be performed.

Forcing the solution to the initial value problem to satisfy the omitted
boundary condition may be computationally difficult, particularly in
the case when the solution depends on an unspecified parameter and
numerical methods are inapplicable. On the other hand, it is quite
simple to force the image of the target at the endpoint involved in the
boundary condition, say $x = B$, to satisfy the boundary condition:
$b_1(Tu)(B) + b_2(Tu)'(B) = \beta$. In the three cases above, working with
the solution of the initial value problem may be replaced by imposing
such a condition on the map.

If the three equations (7.41), (7.42), and (7.43) are suitably incom-
patible, or the target function cannot be fit well by a solution to a
linear differential equation, the method outlined above will generate a
poor result. The collage theorem assures that finding a Picard operator with a small contraction factor that induces an extremely small
collage distance is sufficient to guarantee that the actual error in our
approximation (the distance between the target and the solution to
the resulting boundary value problem) is also very small. Two possible ways of reducing the minimal collage distance achieved by the
algorithm above are (i) increasing the complexity of f and (ii) partitioning the interval $[A, B]$ and performing the algorithm above on each
subinterval.

Increasing the degree of the polynomial coefficients in a linear f adds parameters to the problem, which may result in a smaller collage distance. If the target function is expected to be the solution of a boundary value problem with f having some particular nonlinear form, this technique can be used to approximate the coefficients in that form.

If the interval $I = [A, B]$ is partitioned into n subintervals, $I_i = [x_i, x_{i+1}]$, $i = 0, \ldots, n-1$, with $x_0 = A$ and $x_n = B$, the algorithm outlined earlier may be applied to each subinterval, with minor changes, to obtain a piecewise-defined operator $T = T_i$ on I_i. A modified algorithm might begin in any one of the n subintervals and then proceed to neighbouring subintervals until all of them have been treated. We impose that T is continuous at the partition points: $T_i(x_{i+1}) = T_{i+1}(x_{i+1})$. Note that this means that the n inverse problems are not independent. We might choose to impose conditions on the continuity of f as well. Such continuity conditions determine the values of u_0 and u_0' in the Picard operator on a neighbouring interval. Cases similar to those listed above have to be considered for the first and final subintervals (where the boundary conditions apply).

Example 7.19. For the target function $u(x) = e^{-x} + e^{4x}$ on $[0, 1]$, we wish to find a boundary value problem of the form

$$w''(x) = aw(x) + bw'(x), \ w(0) + w'(0) = 5, \ w(1) + w'(1) = 5e^4.$$

The fully specified boundary conditions are satisfied by the target function, so our question amounts to determining optimal values of the parameters a and b in the linear differential equation. One could, of course, plug the target function into the differential equation, equate coefficients of the different exponential terms, and solve the resulting system. In this case, we obtain the values $a = 4$ and $b = 3$. On the other hand, minimizing the collage distance gives to five decimal places the linear system

$$2.46776u_0 + 2.01901(5 - u_0) + 15.76212b + 4.75925a = 77.31593,$$
$$0.78209u_0 + 0.62672(5 - u_0) + 4.75925b + 1.44122a = 23.48700.$$

Solving for a and b in terms of u_0 gives

$$a = -4.7303u_0 + 13.46622 \text{ and } b = 1.39979u_0 + 0.19869.$$

Forcing the Picard operator T to satisfy $(Tu)(1) + (Tu)'(1) = 5e^4$ lets us determine that $u_0 = 2.00167$ and hence $a = 3.99780$ and $b = 3.00062$.

Example 7.20. Consider the target functions $u_n(x) = \sin(n\pi x)$ on $[0,1]$, where $n = 1,2,3,\dots$. For each n, we seek a boundary value problem of the form

$$w_n''(x) = a_n w_n(x) + b_n w_n'(x), \quad w_n(0) = 0, \quad w_n(1) = 0,$$

that admits $u_n(x)$ as an approximate solution. The associated collage distance Δ_n^2 for the Picard operator T_n depends on a_n, b_n, and $(w_n)_0'$, a parameter we introduce and use to satisfy the boundary condition at $x = 1$. Imposing the usual stationarity conditions with respect to a_n and b_n lets us obtain a complicated expression for each coefficient in terms of $(w_n)_0'$:

$$a_n = \frac{(-n\pi + (w_n)_0')(n^2\pi^2(2(-1)^n + 7) + 18((-1)^n - 1))}{3(n\pi(n^2\pi^2 + 3 + 12(-1)^n))}, \quad (7.44)$$

$$b_n = n\pi \frac{(w_n)_0' n^2\pi^2 + (5n\pi - 2(w_n)_0') + 4(n\pi + 2(w_n)_0')(-1)^n}{n^2\pi^2 + 3 + 12(-1)^n}, (7.45)$$

Additionally imposing that $(T_n w_n)(1) = 0$ allows us to determine that $(w_n)_0' = n\pi$. Plugging this value into (7.44) and (7.45) gives $a_n = 0$ and $b_n = (n\pi)^2$, as we expect. This example highlights a potential use of this method in the investigation of inverse eigenvalue problems. The traditional inverse eigenvalue problem for Sturm-Liouville equations supposes that the coefficients of the differential equation involve an eigenvalue and some unknown function of x. The inverse problem seeks to determine the unknown function when a complete set of eigenvalues is known. For example, see [5].

Example 7.21. The function $u(x) = 2(2-x)^{-1}$ on $[0,1]$ does not satisfy a linear differential equation. We look for the minimal-collage linear BVP
$$w''(x) = aw(x) + bw'(x), \quad w(0) = 1, \quad w(1) = 2,$$

where the boundary conditions have been chosen to agree with the target function and a and b are the unknown parameters for which we seek to find the best values. If we build our operator at $x = 0$, we might attempt to satisfy the boundary condition by either (i) making the Picard operator T satisfy $(Tw)(1) = 2$ or (ii) making the solution to the associated initial value problem satisfy the boundary condition at $x = 1$. In case (i), to five decimal places, the coefficients obtained are $a = 2.97395$ and $b = -1.08057$, with collage distance 0.44330. The actual error, the $\mathcal{L}^2$ distance between the target and the solution to the induced boundary value problem, is 0.00377. In case (ii), we get to five decimal places $a = 2.76652$ and $b = -0.93830$, with a collage distance of 0.00308. The actual error is 0.00061. Of course, if we look for differential equations of the form

$$w''(x) = aw(x) + bw'(x) + cw(x)w'(x), \; w(0) = 1, \; w(1) = 2,$$

we get $a = 0$, $b = 0$, and $c = 1$, the exact solution.

Example 7.22. The collage method of solving such inverse problems is reasonably robust. To simulate the interpolation of a set of data points, we take the curve $y(x) = xe^x$ and generate 20 uniformly spaced data points on $[0, 1]$. Gaussian noise with small amplitude ε is added to the data point values, and the noised points are fit to a polynomial, $u(x)$. First, we look for a boundary value problem of the form

$$w''(x) = aw(x) + bw'(x), \; 4w(0) - w'(0) = \alpha, \; 2w(1) - w'(1) = \beta,$$

which exhibits $u(x)$ as an approximate solution. We proceed as in the earlier examples, this time treating a, b, w_0, and w'_0 as our collage coding variables in the Picard operator. For each run, let T_1 be the minimal-collage Picard operator we obtain, with fixed points $\bar{u}_1$.

Next, we instead seek to satisfy with $u(x)$ the BVP

$$w''(x) = aw(x) + bw'(x), \; 4w(0) - w'(0) = -0.5, \; 2w(1) - w'(1) = 0.5.$$

Notice that the boundary conditions are not satisfied by our target function, so we expect this case to struggle. In order to satisfy the boundary condition at $x = 1$, we treat only a and b as optimization parameters, using the boundary condition at $x = 0$ to express w'_0 in terms of w_0. Minimization of the collage distance in this case yields expressions for a and b in terms of w_0. Demanding that the Picard operator T satisfy $2(Tw)(1) - (Tw)'(1) = 0.5$ determines w_0 and hence a and b. Let the minimal-collage operator thus obtained be T_2, with fixed point $\bar{u}_2$.

The results of various cases are presented in Table 7.11. In Fig. 7.2, the graphs of $u(x)$ and the solution to the generated boundary value problem for the cubic fit with $\varepsilon = 0.03$ are presented for each of the two collage coding schemes above.

Let us repeat the process above with the modified target function $y(x) = x^2 e^x$. Unlike our xe^x, observe that this target function is not in the range of the class of Picard operators we allow. Repeating the procedure, we arrive at Table 7.12.

Because each row of each table deals with a different approximation of one of our two target functions, it is difficult to draw sweeping conclusions. As we might have expected, the true errors for our second

Table 7.11: Collage coding results for $y(x) = xe^x$ in Example 7.22.

Degree	ε	$d_2(y,u)$	$d_2(u,T_1(u))$	$d_2(u,\bar{u}_1)$	$d_2(u,T_2(u))$	$d_2(u,\bar{u}_2)$
2	0	0.01962	0.00428	0.08150	0.02927	0.12363
2	0.03	0.02103	0.00465	0.09744	0.02808	0.12968
2	0.07	0.03005	0.00516	0.12134	0.02674	0.13946
2	0.10	0.03921	0.00556	0.14067	0.02589	0.14798
3	0	0.00167	0.00131	0.01279	0.01556	0.21763
3	0.03	0.01771	0.00327	0.06185	0.02009	0.14938
3	0.07	0.04078	0.01132	0.35998	0.06389	0.08046
3	0.10	0.05811	0.01818	0.73042	0.11076	0.11931
4	0	0.00011	0.00010	0.00031	0.01777	0.20602
4	0.03	0.01806	0.01090	0.16009	0.01921	0.07696
4	0.07	0.04217	0.02610	0.51586	0.05195	0.10126
4	0.10	0.06025	0.03771	0.83912	0.17073	0.24650

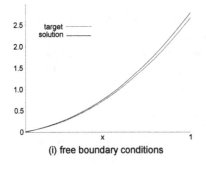

(i) free boundary conditions

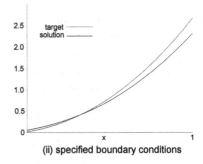

(ii) specified boundary conditions

Fig. 7.2: Graph of $u(x)$ and the solution of the BVP for Example 7.22 with degree $= 3$ and $\varepsilon = 0.03$.

target function were greater than those for the first target because the first target is in the range of our Picard operators and, for the second target function, some collage coding runs with constrained boundary values did quite poorly. This can be seen as good news! In an experimental setting, one might propose a model for the phenomenon under study and then attempt to determine parameters in the model from observational data, perhaps huge amounts of data from many experimental runs. Indeed, the collage coding approach can be applied to multiple data sets by constructing and minimizing the sum of the squared collage distances for the individual data sets. One can use a weighted sum to give more credence to more reliable data. Such an approach can render negligible the impact of a few bad data sets. If

Table 7.12: Collage coding results for $y(x) = x^2 e^x$ in Example 7.22.

Degree	ε	$d_2(y, u)$	$d_2(u, T_1(u))$	$d_2(u, \bar{u}_1)$	$d_2(u, T_2(u))$	$d_2(u, \bar{u}_2)$
2	0	0.05286	0.02977	3.48485	0.04538	1.96423
2	0.03	0.05227	0.03003	3.16067	0.04538	1.60518
2	0.07	0.05508	0.03039	2.79473	0.04537	1.26306
2	0.10	0.05957	0.03065	2.56191	0.04560	1.11703
3	0	0.00611	0.00349	0.11527	0.10519	58.86074
3	0.03	0.01937	0.01136	1.24661	0.08689	21.29959
3	0.07	0.04196	0.02235	2.20382	0.09649	9.97591
3	0.10	0.05917	0.03040	2.54766	0.04477	1.05438
4	0	0.00052	0.00484	0.51538	0.12643	148.19929
4	0.03	0.01801	0.01714	1.65440	0.24462	1652.98125
4	0.07	0.04211	0.03359	2.57270	0.12742	1.89151
4	0.10	0.06019	0.04580	3.02624	0.10414	2.90807

the collage coding results are poor, the suggestion is that the model is incorrect.

Example 7.23. Consider the target function $u(x) = \sin\left(2x + \frac{\pi}{4}\right)$ on $[0, 1]$. We seek a boundary value problem of the form

$$w''(x) = aw(x) + bw'(x), \quad 2w(0) - w'(0) = 0, \quad 4w(1) + w'(1) = -1,$$

which admits $u(x)$ as an approximate solution. Notice that the boundary condition at $x = 1$ is not satisfied by the target function. Operating on $[0, 1]$, our method returns the boundary value problem (coefficients to five decimal places)

$$w''(x) = -3.79565w(x) + 0.79760w'(x), \quad 0 < x < 1,$$
$$2w(0) - w'(0) = 0.3,$$
$$4w(1) + w'(1) = -1,$$

with collage distance 0.01102 and actual error 0.01381. Partitioning at $x = 0.5$ gives

$$w''(x) = \begin{cases} -4.00000w(x), & 0 < x < \frac{1}{2}, \\ 0.50131w(x) + 2.81711w'(x), & \frac{1}{2} \leq x < 1, \end{cases}$$
$$2w(0) - w'(0) = 0,$$
$$4w(1) + w'(1) = -1,$$

with collage distance 0.00445 and actual error 0.00910. Partitioning uniformly into four subintervals gives

$$w''(x) = \begin{cases} -4.00000w(x), \ 0 < x < \frac{1}{4}, \\ -4.00000w(x), \ \frac{1}{4} \le x < \frac{1}{2}, \\ -4.00000w(x), \ \frac{1}{2} \le x < \frac{3}{4}, \\ 21.48987w(x) + 9.20272w'(x), \ \frac{3}{4} \le x < 1, \end{cases}$$

$$2w(0) - w'(0) = 0,$$
$$4w(1) + w'(1) = -1,$$

with collage distance 0.00171 and actual error 0.00463. Figure 7.3 gives graphs of the target function, the image under the Picard operator, and the solution to the resulting boundary value problem in each case.

7.3 Quasilinear partial differential equations

Consider the general first-order quasilinear partial differential equation,

$$u_t(x,t) + B(x,t,u(x,t))u_x(x,t) = F(x,t,u(x,t)) \qquad (7.46)$$
$$u(x,0) = g(x), \qquad (7.47)$$

where we assume that B, F and g are C^1 functions; in our applications, we will in fact have polynomial B and F. The system (7.46)-(7.47) can be written as

$$(1, B, F)) \cdot (u_t(x,t), u_x(x,t), -1) = 0 \qquad (7.48)$$
$$u(x,0) = g(x), \qquad (7.49)$$

whence we employ the method of characteristics. For this problem, the characteristics satisfy

$$\frac{dt}{dr} = 1, \ t(0) = 0 \Rightarrow t = r \qquad (7.50)$$

$$\frac{dx}{dr} = B(x,t,u(x,t)), \ x(0) = s \qquad (7.51)$$

$$\frac{du}{dr} = F(x,t,u(x,t)), \ u(0) = g(s), \qquad (7.52)$$

where s parametrizes the initial data curve and r parametrizes the solution along each resulting characteristic curve.

Having solved (7.50), we replace t by r in the remaining two equations. The solution to the resulting system in x and u (for each particular fixed choice of s) exists locally and depends continuously on the initial data.

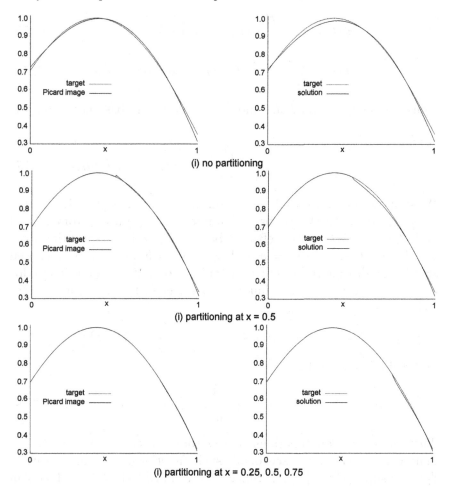

Fig. 7.3: Graph of $u(x)$ and $(Tu)(x)$, as well as $u(x)$ and the solution of the BVP for Example 7.23.

Now, let $x(t) = \psi(t)$ be the base characteristic satisfying $x(0) = \psi(0) = s$; then $h(t) = u(\psi(t), t)$ is the corresponding characteristic on the solution surface. Thus,

$$\frac{d\psi(r)}{dr} = B(\psi(r), r, u(\psi(r), r)), \tag{7.53}$$

$$\frac{dh(r)}{dr} = F(\psi(r), r, u(\psi(r), r)). \tag{7.54}$$

We integrate (7.53) and (7.54) with respect to r from $r = 0$ to $r = t$ to obtain

$$\psi(t) = s + \int_0^t B(\psi(r), r, u(\psi(r), r))dr, \tag{7.55}$$

$$h(t) = h(0) + \int_0^t F(\psi(r), r, u(\psi(0), r))dr. \tag{7.56}$$

Thus the solution to (7.46) and (7.47) along the base characteristic $x = \psi(t)$ is

$$u(\psi(t), t) = g(\psi(0)) + \int_0^t F(\psi(r), r, u(\psi(r), r))dr. \tag{7.57}$$

We suppose that our initial data lie on the interval $x \in [a, b]$. Let $x(t) = \psi_{right}(t)$ be the rightmost base characteristic (sprouting from $x = a$) and $x(t) = \psi_{left}(t)$ be the leftmost base characteristic (sprouting from $x = b$). That is, $\psi_{left}(0) = a$ and $\psi_{right}(0) = b$. Let $\tau > 0$ be chosen small enough that the base characteristics do not cross on $[0, \tau]$ (this can be done thanks to continuous dependence). Define a compact set G by

$$G = \{(x, t) | \psi_{left}(t) \leq x \leq \psi_{right}(t), 0 \leq t \leq \tau\},$$

and define a map T by

$$(Tu)(x, t) = g(x) + \int_0^t F(x, r, u(x, r))dr. \tag{7.58}$$

That is, T acts on functions $u(x, t)$ that are continuous on G; we denote by $C(G)$ the space of all such functions. For continuous g, $\|g\|_\infty = \sup_{x \in G} |g(x)| \leq K_g$. Since F is assumed to be C^1, its Jacobian matrix has a bounded sup norm on G: $\|DF\|_\infty \leq K_{DF}$. We can pick τ smaller if necessary to satisfy $\tau K_{DF} = c < 1$. Furthermore, we can find $K_F > 0$ such that

$$\|F\|_\infty = \sup_{(x,t) \in G} |F(x, t, u(x, t))| \leq \frac{K_F}{\tau}.$$

Now, define $\bar{C}(G) = \{u | u \in C(G), \|u\|_\infty \leq K_g + K_F\}$. Then

$$T : C(G) \mapsto C(G)$$

and

$$\|Tu\|_\infty = \sup_{(x,t)\in G} \left| g(x) + \int_0^t F(x,r,u(x,r))dr \right|$$

$$\leq \sup_{(x,t)\in G} |g(x)| + \sup_{(x,t)\in G} \left| \int_0^t F(x,r,u(x,r))dr \right|$$

$$\leq K_g + \frac{K_F}{\tau} \left| \sup_{(x,t)\in G} \int_0^t dr \right|$$

$$\leq K_g + \frac{K_F}{\tau}\tau$$

$$= K_g + K_F.$$

We conclude that $T : \bar{C}(G) \mapsto \bar{C}(G)$. The following theorem states that T is contractive with respect to the sup norm.

Theorem 7.24. *For u and v in $C(G)$, T defined in (7.58) is contractive in the* sup *metric; namely,*

$$\|Tu - Tv\|_\infty \leq c\|u - v\|_\infty,$$

where $c = \tau K_{DF} < 1$.

Proof. We have

$$\|Tu - Tv\|_\infty = \sup_{(x,t)\in G} \left| \int_0^t \Big(F(x,r,u(x,r)) - F(x,r,v(x,r)) \Big) dr \right|$$

Using the mean value theorem, we have that

$$F(x,r,u(x,r)) - F(x,r,v(x,r)) = DF(x,r,\bar{u}(x,r))\, (u(x,r) - v(x,r)),$$

where $\bar{u}$ lies on the line segment connecting u and v. Hence,

$$\|Tu - Tv\|_\infty = \sup_{(x,t)\in G} \left| \int_0^t D_F(x,r,\bar{u}(x,r))\, (u(x,r) - v(x,r))\, dr \right|$$

$$\leq K_{DF}\|u - v\|_\infty \sup_{(x,t)\in G} \left| \int_0^t dr \right|$$

$$\leq c\|u - v\|_\infty.$$

$\square$

We can prove that T is in fact contractive with respect to the $\mathcal{L}^2$ norm.

Theorem 7.25. *For u and v in $C(G)$, T defined in (7.58) is contractive in the $\mathcal{L}^2$ metric, with*

$$\|Tu - Tv\|_2 \leq \frac{c}{\sqrt{2}}\|u - v\|_2$$

and $c = \tau K_{DF} < 1$.

Proof. Using the Cauchy-Schwarz inequality and the mean value theorem, we have

$$\|Tu - Tv\|_2^2$$
$$= \iint_G (Tu - Tv)^2 dA$$
$$= \iint_G \left(\int_0^t F(x, r, u(x, r)) - F(x, r, v(x, r)) dr \right)^2 dA$$
$$= \iint_G \left(\int_0^t DF(x, r, \bar{u}(x, r))(u(x, r) - v(x, r)) dr \right)^2 dA$$
$$\leq \iint_G \left(\int_0^t DF(x, r, \bar{u}(x, r))^2 dr \right) \left(\int_0^t (u(x, r) - v(x, r))^2 dr \right) dA$$
$$\leq (K_{DF})^2 \iint_G \left(\int_0^t dr \right) \left(\int_0^t (u - v)^2 dr \right) dA$$
$$\leq (K_{DF})^2 \iint_G \left(t \int_0^t (u - v)^2 dr \right) dA.$$

Define a rectangle R lying in G by

$$R = \{(x, t) | x_{left} \leq x \leq x_{right}, t_{bottom} \leq t \leq t_{top}\}.$$

On R, we have

$$\iint_R (Tu - Tv)^2 dA$$

$$\leq (K_{DF})^2 \int_{x_{left}}^{x_{right}} \int_{t_{bottom}}^{t_{top}} \int_0^t t(u-v)^2 dr dt dx$$

$$\leq (K_{DF})^2 \int_{x_{left}}^{x_{right}} \left[\int_0^{t_{bottom}} \int_{t_{bottom}}^{t_{top}} t(u-v)^2 dt dr \right.$$

$$\left. + \int_{t_{bottom}}^{t_{top}} \int_{t_r}^{t_{top}} t(u-v)^2 dt dr \right] dx$$

$$\leq (K_{DF})^2 \int_{x_{left}}^{x_{right}} \left[\int_0^{t_{bottom}} \frac{(t_{top} - t_{bottom})^2}{2} (u-v)^2 dr \right.$$

$$\left. + \int_{t_{bottom}}^{t_{top}} \frac{(t_{top} - r)^2}{2} (u-v)^2 dr \right] dx$$

$$\leq (K_{DF})^2 \left(\frac{\tau^2}{2} \right) \int_{x_{left}}^{x_{right}} \left[\int_0^{t_{bottom}} (u-v)^2 dr + \int_{t_{bottom}}^{t_{top}} (u-v)^2 dr \right] dx$$

$$\leq \frac{c^2}{2} \iint_R (u-v)^2 dA.$$

Notice that the inequality holds for a subset of G consisting of the union of two or more such rectangles. We can approximate G increasingly better by the union of ever smaller rectangles, with the inequality holding for each such approximation. As the number of rectangles used to approximate G approaches infinity, the error in the approximation approaches 0, and in this limit we prove that the inequality holds on G. We conclude that

$$\|Tu - Tv\|_2 \leq \frac{c}{\sqrt{2}} \|u - v\|_2.$$

$\square$

In practice, we work with the $\mathcal{L}^2$ collage distance even though $\bar{C}(G)$ is not complete with respect to it. Since the fixed point of T lies in $C(G)$, this is not a problem. In addition, the collage coding framework that we have proved for a single quasilinear PDE readily extends to systems of first-order quasilinear PDEs.

7.3.1 Inverse problems for traffic and fluid flow

We model one-dimensional flow in the x direction with the first-order quasilinear PDE

$$\frac{\partial}{\partial t}\rho(x,t) + \frac{\partial}{\partial x}\left(\rho(x,t)v(x,t,\rho(x,t))\right) = 0, \quad \rho(x,0) = g(x), \quad (7.59)$$

where $\rho(x,t)$ is the density at position x and time t of the substance under consideration, and $v(x,t,\rho)$ is the velocity at that place and time, possibly also dependent upon the density there. $g(x)$ is the initial density function. This equation is used to model the density of a pollutant or oxygen suspended in a fluid flowing in one dimension. By taking a continuum viewpoint, the same equation is used to model the density of automobile traffic on a straight stretch of road. We focus here on such flow applications.

Notice that if $v = c$, a positive constant, then the solution to (7.59) is $\rho(x,t) = g(x - ct)$, a forward wave. One makes the reasonable suggestion that the velocity function v should pass through two points: (i) when $\rho = 0$, we assume that $v = c$, where c is the maximum velocity attainable in the medium under study, and (ii) there is a maximum value of ρ that we call ρ_{max}, where $v = 0$. We suppose further that $v = v(\rho)$, and seek a functional form such that the points $(v, \rho) = (c, 0)$ and $(v, \rho) = (0, \rho_{max})$ lie on the curve. We consider the family of power functions

$$v = c\left(1 - \frac{\rho}{\rho_{max}}\right)^n, \quad \text{where } n > 0,$$

which is nonnegative for $\rho \in [0, \rho_{max}]$. Plugging into (7.59), our PDE becomes

$$\frac{\partial}{\partial t}\rho(x,t) + c\left(1 - \frac{\rho}{\rho_{max}}\right)^{n-1}\left(1 - n\frac{\rho}{\rho_{max}}\right)\frac{\partial}{\partial x}\rho(x,t) = 0, \quad (7.60)$$

where $\rho(x,0) = g(x)$, $c > 0$ is the maximum velocity, and $\rho_{max} > 0$ is the maximum density.

Equation (7.60) is a first-order quasilinear PDE. We can solve it by using the method of characteristics. In the case $n = 1$, the solution can be found in closed form. For other values of n, we might well have to solve numerically; in fact, finding the base characteristics in such cases will quite possibly require solving numerically for each individual base characteristic. In all cases, the base characteristics can cross, leading to the formation of a shock. We call the first such crossing time the blowup time. In our formulation, the set G is chosen so that it does not include the blowup time.

The inverse problem we consider is as follows. Suppose we set up cameras at various positions x_i, $i = 1, \ldots, N$, on a straight stretch of road. The cameras take pictures of the road traffic at times t_j, $j = 1, \ldots, M$. Somebody examines the pictures to count the number of cars at each position at each time, thus producing the observational data values $\rho(x_i, t_j)$. From these data, we wish to estimate the physically meaningful parameters c and ρ_{max}.

We simulate this experiment by first solving (7.60) with chosen values of c, ρ_{max}, n, and $g(x)$. Next, we gather faux observational data by sampling the solution at a grid of values, possibly adding low-amplitude Gaussian noise. We use our collage coding machinery to find the minimal collage parameter values, which hopefully lie close to our chosen values. Of course, one would never try to perform such a process by hand. It turns out that we encountered one particular difficulty along the way, to which Maple delivered a one-line solution. Partway through the collage coding process, one needs to solve the base characteristic ODE

$$\frac{d\psi(r)}{dr} = B(\psi(r), r, \rho(\psi(r), r)),$$

where B is a polynomial with unknown coefficients and $\rho(\psi(r), r)$ is a polynomial that has been fitted to the observational data. Even in the case where B is linear, the result is a complicated nest of integrals. In order to calculate our collage distance, we will need an expression for $\psi(r)$ from which we can easily obtain values in terms of the parameters. Since we make the first time at which data is gathered $t = 0$, we expand $\psi(r)$ as a series at $r = t = 0$.

Example 7.26. We set $n = 1$ and $g(x) = 1 + x$. In Table 7.13, we present some results for our experiment. It is strangely striking that the case $c = 1$ and $\rho_{max} = 1$ struggles so much more to do well compared with the other cases in the table, but all of the results are quite good and robust to the presence of noise.

The method works as expected. For example, taking data from a solution for the case $n = 2$ and seeking the minimal collage distance with $n = 1$ leads to a striking mismatch: the minimal collage parameters do not produce a small collage distance, suggesting that the data do not fit this model.

7.4 Urison integral equations

Let $I = [a, a + \delta]$, $\delta > 0$, and define $(C(I), \|\cdot\|_\infty)$ as the complete metric space of continuous functions on I with sup norm:

$$\text{for } x_1, x_2 \in C(I), \quad \|x_1 - x_2\|_\infty = \sup_{t \in I} |x_1(t) - x_2(t)|.$$

We consider the Urison integral map [89]

$$(Tx)(t) = \varphi(t) + \int_a^t K(t, s, x(s)) ds, \tag{7.61}$$

where (i) $\varphi(t)$ is continuous for $t \in I$, with $\|\varphi(t)\|_\infty \le N$ on I, (ii) we define $D = \{(t, s, x) \mid a \le s \le t \le a + \delta, |x| \le 2N\}$, and (iii) on D, $K(t, s, x)$ is continuous and satisfies the Lipschitz condition

$$|K(t, s, x_1) - K(t, s, x_2)| \le M|x_1 - x_2|. \tag{7.62}$$

We define a norm for continuous functions r on D:

$$\|r_1 - r_2\|_D = \sup_{(t,s,x) \in D} |r_1(t, s, x) - r_2(t, s, x)|.$$

Let

$$\bar{C}(I) = \{x \in C(I) \mid \|x\|_\infty \le 2N\},$$

and choose δ sufficiently small that $\delta M < 1$ and $\|K\|_D \le \frac{N}{\delta}$. Then

$$\|Tx\|_\infty = \sup_{t \in I} \left| \varphi(t) + \int_a^t K(t, s, x(s)) ds \right|$$

$$\le \|\varphi(t)\|_\infty + \sup_{t \in I} \left(\int_a^t |K(t, s, x(s))| ds \right)$$

$$\le N + \frac{N}{\delta} \delta = 2N,$$

so $T : \bar{C}(I) \mapsto \bar{C}(I)$. Let $\mathcal{U}(I)$ be the set of all Urison integral operators (7.61) satisfying the conditions above.

Thinking ahead, the sup norm is not a computationally desirable metric; we would like to enjoy the pleasures of the $\mathcal{L}^2$ norm:

$$\text{for } x_1, x_2 \in \mathcal{L}^2(I), \quad \|x_1 - x_2\|_2 = \left(\int_a^{a+\delta} (x_1(t) - x_2(t))^2 dt \right)^{\frac{1}{2}}.$$

The metric space $(\bar{C}(I), \|\cdot\|_2)$ is not complete but, as we'll establish by the contractivity with respect to the sup norm, the fixed point of T lies in $\bar{C}(I)$. Our first result establishes that a Urison integral operator is contractive in both the sup and $\mathcal{L}^2$ norms.

Theorem 7.27. $T \in \mathcal{U}(I)$ *is contractive in the* sup *and* $\mathcal{L}^2$ *norms,*

$$\|Tx_1 - Tx_2\| \le c\|x_1 - x_2\|, \text{ for all } x_1, x_2 \in \bar{C}(I),$$

where $c = \delta M < 1$ *and* $\| \cdot \|$ *is either the* sup *norm or the* $\mathcal{L}^2$ *norm.*

Proof. Observe that

$$|(Tx_1)(t) - (Tx_2)(t)| = \left| \int_a^t (K(t, s, x_1(s)) - K(t, s, x_2(s))) \, ds \right|$$

$$\le M \int_a^t |x_1(s) - x_2(s)| \, ds, \qquad (7.63)$$

from which the result for the *sup* norm follows quickly. To obtain the result for the $\mathcal{L}^2$ norm, use the Cauchy-Schwarz inequality in (7.63) to get

$$|(Tx_1)(t) - (Tx_2)(t)| \le M \left| \left(\int_a^t (x_1(s) - x_2(s))^2 ds \right)^{\frac{1}{2}} \right| \left| \int_a^t ds \right|^{\frac{1}{2}}$$

$$\le \delta^{-1}(t - a)d_2(x_1, x_2)$$

$$= M\delta^{\frac{1}{2}} d_2(x_1, x_2). \qquad (7.64)$$

This gives

$$\|Tx_1 - Tx_2\|_2 = \left(\int_a^t |(Tx_1)(t) - (Tx_2)(t)|^2 dt \right)^{\frac{1}{2}}$$

$$\le M\delta^{\frac{1}{2}} \|x_1 - x_2\|_2 \delta^{\frac{1}{2}}$$

$$= M\delta \|x_1 - x_2\|_2. \qquad (7.65)$$

$\square$

In a framework designed to discuss eigenvalue problems for such integral operators, [43] in fact showed that Volterra integral operators have a unique fixed point for any positive choice of δ.

Our next result establishes that the fixed points of two operators in $\mathcal{U}(I)$ depend continuously on the parameters in the maps.

Theorem 7.28. *Let* $T_1, T_2 \in \mathcal{U}(I)$, *with*

$$(T_i x)(t) = \varphi_i(t) + \int_a^t K_i(t, s, x(s))ds, \ t \in I, \ i = 1, 2,$$

with contraction factors c_1 *and* c_2, *respectively. Let* $\bar{x}_i(t) \in \bar{C}(I)$ *be the fixed point of* T_i. *Then*

$$\|\bar{x}_1 - \bar{x}_2\|_\infty \le \frac{1}{1 - \min(c_1, c_2)} [\|\varphi_1 - \varphi_2\|_\infty + \delta\|K_1 - K_2\|_D].$$

7.4.1 Inverse problem for Urison integral equations

The inverse problem of interest is:

> *Given a continuous target solution $x(t)$ on a closed interval I, determine functions $K(t, s, x)$ and $\varphi(t)$, restricted to a particular class of functional forms, such that the induced Urison integral operator has a solution $\bar{x}(t)$ with $\|\bar{x} - x\|_2 < \epsilon$ for ϵ as small as desired.*

We solve the problem above by instead finding a map T that yields a minimal collage distance $\|Tx - x\|_2$. We would like to be able to calculate the fixed point $\bar{x}$ of T in order to calculate $\|\bar{x} - x\|_2$, the actual error in the solution to our inverse problem. Unfortunately, a general solution to the direct problem (given T, find $\bar{x}$) is unavailable. The common iterative scheme of a direct solution for Volterra integral operators ([174]) sets

$$x(t) = \sum_{n=0}^{\infty} x_n(t), \tag{7.66}$$

where

$$x_0(t) = \varphi(t) \text{ and } x_n(t) = \int_a^t K_1(t, s) x_{n-1}(s) ds, \ n \geq 1.$$

An upper bound for the nth term in the series solution (7.66) is

$$\|x_n\|_\infty \leq \frac{N^{n+1}}{n!}, \tag{7.67}$$

giving us a rough way to check the magnitude of terms in the tail of the series.

Of course, if our target solution $x(t)$ is the interpolation of some experimental data points, there will be some error appearing in x. Denote by x^* the actual function that models the data, and suppose that $\|x - x^*\|_2 < \varepsilon_1$. Then

$$\|x^* - \bar{x}\|_2 \leq \|x^* - x\|_2 + \|x - \bar{x}\|_2 < \varepsilon_1 + \varepsilon.$$

If ε_1 and ε are small, then the fixed point of the operator we find is close to the actual solution.

For the sake of discussion and ease of presentation, we suppose that φ and K are restricted to be polynomials of some specified degree. Clearly, if x is a polynomial, then there are simple choices of polynomials φ and K for which

$$x(t) = \varphi(t) + \int_a^t K(t, s, x(s))ds, \ t \in I = [a, a + \delta], \qquad (7.68)$$

admits x as an exact solution; such an exact solution to our inverse problem may be possible provided φ is allowed to be of the required degree. This case is mentioned again later. In general, we may not be able to find operators in this class that make the collage distance arbitrarily small. The following result establishes that, by partitioning I, we are able to find a piecewise-defined integral operator with polynomials $\varphi(t)$ and $K(t, s, x)$ and corresponding collage distance as small as we like. The collage theorem then implies that the fixed point of the resulting operator is close to the target solution. Let $\mathcal{U}_p(I)$ be the set of all Urison integral operators (7.61) with polynomials φ and $K(t, s, x)$.

Theorem 7.29. *Pick a and $\delta > 0$, defining $I = [a, a + \delta]$. Let $x(t)$ be a solution to (7.68). Then, given any $\epsilon > 0$, there exists an $I_1 \subseteq I$ and an operator $T_1 \in \mathcal{U}_p(I_1)$ such that $\|x - T_1 x\|_\infty < \epsilon$ on I_1.*

Proof. δ is chosen and $x(t)$ is given. ϵ is also given. Let $T \in \mathcal{U}(I)$ be the operator with fixed point $x(t)$, defined by functions $\varphi(t)$ and $K(t, s, x)$, with contraction factor c. Choose a_1 and $\delta_1 > 0$ such that $I_1 = [a_1, a_1 + \delta_1] \subseteq I$ (observe that all smaller choices of $\delta_1 > 0$ preserve this relationship), and let $D_1 = \{(t, s, x) \mid a_1 \le s \le t \le a_1 + \delta_1, |x| \le 2N\}$. By the Weierstrass approximation theorem, there exist polynomials $\varphi_1(t)$ and $K_1(t, s, x)$ of some degree such that on I_1

$$\|\varphi - \varphi_1\|_\infty < \frac{\epsilon}{8} \text{ and } \|K - K_1\|_{D_1} < \frac{\epsilon}{8\delta_1 N}.$$

If necessary, pick $\delta_1 > 0$ smaller to ensure that $c_1 = \delta_1 \|K_1\|_{D_1} \le \frac{1}{2}$. By the proof of Theorem 7.27, the associated operator T_1 is contractive in sup norm on $\bar{C}(I_1)$ with contraction factor c_1. Let $x_1(t)$ be its fixed point. Using Theorem 7.28, on I_1,

$$\begin{aligned}
\|x - T_1 x\|_\infty &\le \|x - x_1\|_\infty + \|x_1 - T_1 x\|_\infty \\
&= \|x - x_1\|_\infty + \|T_1 x_1 - T_1 x\|_\infty \\
&\le (1 + c_1)\|x - x_1\|_\infty \\
&< 2\|x - x_1\|_\infty \\
&\le \frac{2}{1 - \min(c, c_1)} [\|\varphi_1 - \varphi\|_\infty + \delta_1 N \|K_1 - K\|_{D_1}] \\
&\le 4 \left[\frac{\epsilon}{8} + \delta_1 N \frac{\epsilon}{8\delta_1 N} \right] = \epsilon.
\end{aligned}$$

$\square$

We let

$$\varphi(t) = \sum_{i=0}^{N_\varphi} a_i t^i, \tag{7.69}$$

$$K(t, s, x) = \sum_{i=0}^{N_K} \sum_{j=0}^{N_K} \sum_{l=0}^{N_K} c_{i,j,l} s^i t^j x^l, \tag{7.70}$$

for some $N_\varphi, N_K > 0$. Then

$$(Tx)(t) = \sum_{i=0}^{N_\varphi} a_i t^i + \sum_{i=0}^{N_K} \sum_{j=0}^{N_K} \sum_{l=0}^{N_K} c_{i,j,l} t^j g_{i,l}(t) ds, \tag{7.71}$$

where

$$g_{i,l}(t) = \int_a^t (x(s))^l s^i, \quad i, l = 0, \ldots, N_K. \tag{7.72}$$

The squared $\mathcal{L}^2$ collage distance,

$$\Delta^2 = \int_I [x(t) - (Tx)(t)]^2 dt, \tag{7.73}$$

is quadratic in the parameters a_i and $c_{i,j,l}$. One can minimize the squared collage distance, Δ^2, by setting the partial derivatives with respect to each of these variational parameters to zero to get a linear system for the unknown parameters.

Increasing the degree of K and/or φ increases the number of parameters in the problem. Although this can decrease the collage distance and allow a closer approximation of the target, it can also increase the computation time significantly. Consider again the trivial case mentioned earlier: if the target function $x(t)$ is a polynomial (or data points fitted by a polynomial) of degree N_T with coefficients α_i, $i = 1, \ldots, N_T$, then the collage distance can be made to be zero by either (i) setting $N_\varphi = N_T$, $a_i = \alpha_i$, and $c_{i,j,l} = 0$ (that is, $\varphi(t) = x(t)$ and $K(t, s, x) = 0$) or (ii) setting $N_K = N_T$ and matching coefficients of the same power of t (note that $g_{i,l}$ is a polynomial) in the double sum and the target. Other functional forms of K and φ can also be used.

If we partition the interval I into n subintervals $I_i = [t_i, t_{i+1}]$, $i = 0, \ldots, n-1$, where $t_0 = a$ and $t_n = a + \delta$, then the earlier algorithm can be applied to each subinterval. Minimization of the collage distance on subinterval I_i yields an operator T_i with the overall operator $T = T_i$ on I_i.

7.5 Hammerstein integral equations

The Hammerstein integral operator is

$$(Tu)(x) = \int_0^1 K(x, y, u(y))dy + l(x).$$

In imaging applications, u represents an input signal (or image) and the kernel K describes how this signal has been blurred. Of course, the kernel can take on many forms, as an image might be blurred for a wide variety of reasons ranging from effects of motion capture to focus problems to diffraction effects of the medium in which a photograph was taken. The term $l(x)$ represents the contribution of additive noise to the recorded image. The result of applying the Hammerstein integral operator, Tu, is a blurred and noised image.

We discuss the inverse problem: given a recorded signal as well as l, denoise and deblur the image by recovering the parameters in K (and possibly its functional form). The reader is referred to [94, 95] for more discussion.

We assume that $l : [0, 1] \mapsto \mathbb{R}$ is a given function and $K : \mathbb{R} \times \mathbb{R} \times \mathbb{R} \mapsto \mathbb{R}$ satisfies

$$|K(\alpha_1, \beta, \gamma) - K(\alpha_2, \beta, \gamma)| \leq C_1(\beta, \gamma)|\alpha_1 - \alpha_2| \quad (7.74)$$

$$\text{and } |K(\alpha, \beta, \gamma_1) - K(\alpha, \beta, \gamma_2)| \leq C_2(\alpha, \beta)|\gamma_1 - \gamma_2|. \quad (7.75)$$

Furthermore, suppose that for $u \in L^2([0, 1])$ the function $\xi(y) := C_1(y, u(y))$ belongs to $L^2([0, 1])$. These next results are quite simple, and we mention them for the benefit of the reader. The first result states the existence and uniqueness of the solution in the space $C([0, 1])$.

Theorem 7.30. *Let $l : [0, 1] \mapsto \mathbb{R}$ be a continuous function. Then*

(i) $(Tu)(x)$ is a continuous function.
(ii) If $\|C_2\|_\infty := \sup_{\alpha, \beta \in [0,1]} |C_2(\alpha, \beta)| < 1$, then $T : C([0, 1]) \mapsto C([0, 1])$ is a contraction.

Proof. Given a continuous function $u \in C([0, 1])$, we have

$$|(Tu)(x) - (Tu)(z)|$$

$$\leq \int_0^1 |K(x, y, u(y)) - K(z, y, u(y))|dy + |l(x) - l(z)|$$

$$\leq |x - z| \int_0^1 |C_1(y, u(y))|dy + |l(x) - l(z)|$$

$$\leq |x - z| \left(\int_0^1 |C_1(y, u(y))|^2 dy \right)^{\frac{1}{2}} + |l(x) - l(z)|, \qquad (7.76)$$

and this shows the continuity of the function $(Tu)(x)$. To prove contractivity, we have

$$
\begin{aligned}
d_\infty(Tu, Tv) &= \sup_{x \in [0,1]} |(Tu)(x) - (Tv)(x)| \\
&\leq \sup_{x \in [0,1]} \int_0^1 |K(x, y, u(y)) - K(x, y, v(y))| dy \\
&\leq \sup_{x \in [0,1]} \int_0^1 |C_2(x, y)||u(y) - v(y)| dy \\
&\leq \|C_2\|_\infty d_\infty(u, v).
\end{aligned}
$$

$\square$

Consider now the operator T on the space $L^2([0, 1])$.

Theorem 7.31. *Suppose* $l \in C([0, 1])$ *and* $C_2 \in L^2([0, 1] \times [0, 1])$ *with* $\|C_2\|_2 < 1$. *Then* $T : L^2([0, 1]) \mapsto L^2([0, 1])$ *is a contraction.*

Proof. We first prove that $Tu \in L^2([0, 1])$. For each fixed $x_1, x_2 \in [0, 1]$, we compute

$$
\begin{aligned}
&|Tu(x_1) - Tu(x_2)| \\
&\leq \int_0^1 |K(x_1, y, u(y)) - K(x_2, y, u(y))| dy + |l(x_1) - l(x_2)| \\
&\leq |x_1 - x_2| \int_0^1 |C_1(y, u(y))| dy + |l(x_1) - l(x_2)| \\
&\leq |x_1 - x_2| \left(\int_0^1 |C_1(y, u(y))|^2 dy \right)^{\frac{1}{2}} + |l(x_1) - l(x_2)|,
\end{aligned}
$$

so the function $(Tu)(x)$ is bounded on $[0, 1]$ and then belongs to $L^2([0, 1])$. To prove contractivity, we have

$$
\begin{aligned}
&\|Tu - Tv\|_2^2 \\
&= \int_0^1 |(Tu)(x) - (Tv)(x)|^2 dx \\
&\leq \int_0^1 \left| \int_0^1 K(x, y, u(y)) dy + l(x) - \int_0^1 K(x, y, v(y)) dy - l(x) \right|^2 dx \\
&\leq \int_0^1 \left(\int_0^1 |K(x, y, u(y)) - K(x, y, v(y))| dy \right)^2 dx
\end{aligned}
$$

$$\leq \int_0^1 \left(\int_0^1 |C_2(x,y)||u(y) - v(y)|dy \right)^2 dx$$

$$\leq \int_0^1 \|C_2\|_2^2 \|u - v\|_2^2 dx = \|C_2\|_2^2 \|u - v\|_2^2.$$

$\square$

Corollary 7.32. *If $C_2 \in L^2([0,1] \times [0,1])$ and $\|C_2\|_2 < 1$, then the fixed-point equation $Tu = u$ has a unique solution $\bar{u}$ belonging to $L^2([0,1])$ and for all $u_0 \in L^2([0,1])$ the sequence $u_n = Tu_{n-1}$ converges to $\bar{u}$. If $\|C_2\|_\infty < 1$, then $\bar{u}$ is continuous.*

7.5.1 Inverse problem for Hammerstein integral equations

For each fixed $u \in \mathbb{R}$, we have $K(x,y,\cdot) \in L^2([0,1] \times [0,1])$, and then

$$K(x,y,u) = \sum_{i,j=1}^{\infty} a_{i,j}(u)\phi_j(x)\phi_i(y), \qquad (7.77)$$

where $\phi_i(y)$ is an orthonormal basis in $L^2([0,1])$. For $u \in C([0,1])$, we suppose that the functions $a_{i,j}(u)$ satisfy

(i) $a_{i,j}(u) \in L^2([u_{min}, u_{max}])$, where $u_{min} = \min_{0 \leq x \leq 1} u(x)$ and $u_{max} = \max_{0 \leq x \leq 1} u(x)$, so that

$$a_{i,j}(u) = \sum_{k=1}^{\infty} a_{i,j,k}\psi_k(u),$$

where $\psi_k(u)$ is a basis for $L^2([u_{min}, u_{max}])$.

(ii) Let $|a_{i,j}(u) - a_{i,j}(v)| = \left| \sum_{k=1}^{\infty} a_{i,j,k}(\psi_k(u) - \psi_k(v)) \right| \leq b_{i,j}|u-v|$. Then

$$|K(x,y,u) - K(x,y,v)| = \left| \sum_{i,j}(a_{i,j}(u) - a_{i,j}(v))\phi_j(x)\phi_i(y) \right|$$

$$\leq \left| \sum_{i,j} b_{i,j}\phi_j(x)\phi_i(y) \right| |u-v|,$$

so $C_2(x, y) = \left| \sum_{i,j} b_{i,j} \phi_j(x) \phi_i(y) \right|$ in (7.75).

We truncate each sum at upper limit n. Given a target solution $u \in C([0, 1])$ and the function $l(x)$, the inverse problem consists of finding coefficients $a_{i,j,k}$ in the expansion of K such that u is approximately admitted as the fixed point of the corresponding Hammerstein integral operator. Our collage distance is

$$\|u - Tu\|_2^2$$

$$= \int_0^1 \left| u(x) - \int_0^1 K(x, y, u(y)) dy - l(x) \right|^2 dx$$

$$= \int_0^1 \left| -\sum_{j=1}^n u_j \phi_j(x) + \sum_{j=1}^n l_j \phi_j(x) \right.$$

$$\left. + \int_0^1 \sum_{i,j,k=1}^n a_{i,j,k} \psi_k(u(y)) \phi_i(y) \phi_j(x) dy \right|^2 dx$$

$$= \int_0^1 \left| \sum_{j=1}^n \phi_j(x) \left(-u_j + l_j + \sum_{i,k=1}^n \int_0^1 a_{i,j,k} \sum_{l=1}^n c_{k,l} \phi_l(y) \phi_i(y) dy \right) \right|^2 dx$$

$$= \int_0^1 \left| \sum_{j=1}^n \phi_j(x) \left(-u_j + l_j + \sum_{i,k,l=1}^n a_{i,j,k} c_{k,l} \right) \right|^2 dx$$

$$\leq \int_0^1 \left[\sum_{j=1}^n \phi_j^2(x) \right] \left[\sum_{j=1}^n \left(-u_j + l_j + \sum_{i,k,l=1}^n a_{i,j,k} c_{k,l} \right) \right]^2 dx$$

$$= \left[\int_0^1 \sum_{j=1}^n \phi_j^2(x) dx \right] \left[\sum_{j=1}^n \left(-u_j + l_j + \sum_{i,k,l=1}^n a_{i,j,k} c_{k,l} \right) \right]^2$$

$$= n \sum_{j=1}^n \left(-u_j + l_j + \sum_{i,k,l=1}^n a_{i,j,k} c_{k,l} \right)^2,$$

where $c_{k,l} = c_{k,l}(u_1, \ldots, u_n)$ have known values since ψ_k are chosen and $u = u(y)$ is our known target function. Regarding the constraints, we have that

$$\int_0^1 \int_0^1 (C_2(x,y))^2 dxdy$$

$$= \int_0^1 \int_0^1 \left(\sum_{i,j} b_{i,j} \phi_j(x)\phi_i(y) \right)^2 dxdy$$

$$= \int_0^1 \int_0^1 \sum_{i,j} (b_{i,j}\phi_j(x)\phi_i(y))^2 \, dxdy$$

$$+ 2\int_0^1 \int_0^1 \sum_{i,j,h,k} \phi_j(x)\phi_i(y)\phi_h(x)\phi_k(y) dxdy$$

$$= \sum_{i,j} b_{i,j}^2 + 2 \sum_{i,j,h,k} \int_0^1 \phi_j(x)\phi_h(x)dx \int_0^1 \phi_i(y)\phi_k(y)dy$$

$$= \sum_{i,j} b_{i,j}^2,$$

so now the inverse problem becomes

$$\min_A \|u - Tu\|^2, \tag{7.78}$$

where $A = \left\{ \sum_{i,j} b_{i,j}^2 \le C_A < 1 \right\}$.

Example 7.33. We set $l(x) = x$ and set $K_{true}(x,y,u) = k_{true}(x,y)u$. In this case, $a_{i,j}(u) = a_{i,j}u$, so that

$$K_{true}(x,y,u) = \sum_{i,j=1}^\infty a_{i,j}\phi_j(x)\phi_i(y),$$

$$\|u - Tu\|_2^2 \le \sum_{i=1}^n \left(-u_j + l_j + \sum_{i=0}^n a_{i,j}u_j \right)^2,$$

and $b_{i,j} = a_{i,j}$. We define

$$k_{true}(x,y) = \begin{cases} \pi(-1-x+y), & x-y < -\frac{1}{2}, \\ \pi(x-y), & -\frac{1}{2} \le x-y \le \frac{1}{2}, \\ \pi(1-x+y), & \frac{1}{2} < x-y. \end{cases}$$

Let

$$\phi(\cdot) \in \{\phi_s(\cdot), \phi_c(\cdot)\} = \left\{ \sqrt{2}\sin(\pi\cdot), \sqrt{2}\cos(\pi\cdot) \right\}. \tag{7.79}$$

In this rather limited finite basis, the representation of $k_{true}(x,y)$ is

$$k_{basis}(x, y) = \frac{2}{\pi}\phi_s(x)\phi_c(y) - \frac{2}{\pi}\phi_c(x)\phi_s(y) = \frac{4}{\pi}\sin(\pi(x - y)).$$

We build the Hammerstein integral operator

$$(T_{true}u)(x) = \int_0^1 k_{true}(x, y)u(y)\, dy + l(x)$$

and construct an approximation of its fixed point $\bar{u}_{true}(x)$ by iterating the map 20 times on the initial function $u_0(x) = 0$. We set $u_{target}(x) = T^{20}u_0(x)$; see Fig. 7.4. Thus, our example inverse problem

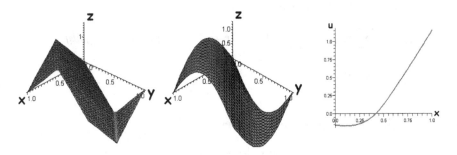

Fig. 7.4: (left to right) The graphs of $z = k_{true}(x, y)$, $z = k_{basis}(x, y)$, and $y = u_{target}(x)$.

is: Given $u_{target}(x)$ and $l(x)$, find a representation of the kernel $k(x, y)$ in our finite basis such that the corresponding Hammerstein integral operator admits $u_{target}(x)$ as an approximate solution. Following our development, we minimize the constrained squared L^2 collage distance

$$\Delta^2 = \|u - Tu\|_2^2 + \lambda\left(\sum_{i,j} a_{i,j}^2 - C_A\right),$$

where $C_A \in (0, 1)$ is a chosen fixed value. We denote the kernel that minimizes Δ^2 by $k_{collage}(x, y)$, and we observe that the desired result is $k_{collage}(x, y) = k_{basis}(x, y)$. To achieve this result, we would have to set

$$C_A = 2\left(\frac{2}{\pi}\right)^2 \approx 0.81057.$$

Instead, we systematically set $C_A = \frac{i}{10}$, $i = 1, 2, \ldots, 9$, and minimize each corresponding Δ^2. Perhaps one might anticipate that the minimum of the nine minimal collage distances we find will occur when $C_A = 0.8$.

In Table 7.14, we present the results obtained when we restrict our basis further so that it consists of precisely the two functions $\psi_1(x, y) = 2\sin(\pi x)\cos(\pi y)$ and $\psi_2(x, y) = 2\cos(\pi x)\sin(\pi y)$, which appear in $k_{true}(x, y)$. The pleasing result is that the minimum of the minimal collage distances occurs when $C_A = 0.8$. Indeed,

$$a_1 \text{ and } -a_2 \approx \frac{2}{\pi} \approx 0.63661,$$

the true best result.

Note that when we minimize Δ^2, treating C_A as an additional variable, we obtain $\lambda = 0$, a linear system in the coefficients a_1 and a_2, and a constraint equation, as opposed to an inequality. Still, on solving the linear system for a_1 and a_2, we can check whether these minimizing values lead to a value of $C_A < 1$. For this example, we obtain $a_1 = 0.63644$ and $a_2 = 0.63675$, giving $C_A = 0.81051$.

When we add the basis function $\phi_s(2x)\phi_s(y)$ with coefficient a_3, we obtain $a_1 = 0.63675$, $a_2 = -0.55321$, and $a_3 = 0.04580$, with $C_A = 0.71359$. The quality of the approximation diminishes in part because in this example $\|C_2\|_2 = 0.90690$, meaning the operator is not very contractive.

If we modify κ_{true}, replacing each π by $\frac{\pi}{8}$, then the new value of $\|C_2\|_2$ is the smaller 0.113362. We obtain $a_1 = 0.07957$, $a_2 = -0.07495$, and $a_3 = 0.00722$, compared with the desired values $a_1 = -a_2 = \frac{1}{4\pi} \approx 0.07958$ and $a_3 = 0$.

7.6 Random fixed-point equations

For notational convenience, we consider the case of scalar random integral equations, but analogous results can be proved in similar ways for the vector-valued case. The reader is referred to [97, 98]. Let $(\Omega, \mathcal{F}, P)$ be a given probability space. Let $\phi : \Omega \times \mathbb{Z} \times \mathbb{Z} \mapsto \mathbb{Z}$ satisfy the conditions

1. $|\phi(\omega, t, z_1) - \phi(\omega, t, z_2)| \leq K_\phi(\omega)|z_1 - z_2|,$
2. $\sup_{t \in [t_0, t_0 + \delta]} |\phi(\omega, t, 0)| = \tilde{K}_\phi(\omega),$

a.e. $\omega \in \Omega$, where K_ϕ and $\tilde{K}_\phi$ are random variables with known distributions. Given $\alpha \in [0, 1]$, let K_α be defined such that

$$\Omega_\alpha = \left\{ w \in \Omega : K_\phi(w), \tilde{K}_\phi(w), x_0(w) \in [-K_\alpha, K_\alpha] \right\},$$

so that $P(\Omega_\alpha) \geq \alpha$. Let $M > K_\alpha$, and consider now the space $X = \{x \in C([t_0, t_0 + \delta]) : \|x\|_\infty \leq M\}$ endowed with the usual d_∞ metric.

Proposition 7.34. *For a.e. $\omega \in \Omega_\alpha$, let $(T_\omega x)(t) = \int_{t_0}^{t} \phi(\omega, s, x(s))ds + x_0(\omega)$, $t \in [t_0, t_0 + \delta]$. If δ is small enough, then $T_\omega : X \mapsto X$.*

Consider now the operator $\widetilde{T} : Y \mapsto Y$, where $Y = \{x : \Omega_\alpha \mapsto X, x \text{ is measurable}\}$ and

$$\left[(\widetilde{T}x)(\omega)\right](t) = \int_{t_0}^{t} \phi(\omega, s, [x(\omega)](s))ds + x_0(\omega) \qquad (7.80)$$

for a.e. $\omega \in \Omega_\alpha$. We use the notation $[x(\omega)]$ to emphasize that for a.e. $\omega \in \Omega_\alpha$ $[x(\omega)]$ is an element of X. We have the following result.

Proposition 7.35. *Let $E_\alpha(K_\phi) = \int_{\Omega_\alpha} K_\phi(\omega)dP(\omega)$. If $\delta E_\alpha(K_\phi) < 1$, then $\widetilde{T}$ is a contraction on Y. In particular, if $\delta \mathbb{E}(K_\phi) < 1$, then $\widetilde{T}$ is a contraction on Y.*

By Banach's theorem we have the existence and uniqueness of the solution of the equation $\widetilde{T}x = x$. For a.e. $(\omega, t) \in \Omega_\alpha \times [t_0, t_0 + \delta]$, we have

$$x(\omega, t) = \int_{t_0}^{t} \phi(\omega, s, x(\omega, s))ds + x_0(\omega). \qquad (7.81)$$

7.6.1 Inverse problem for random DEs

We can solve inverse problems for specific classes of random differential equations by reducing the problem to inverse problems for ordinary differential equations. Let us consider the system of random equations

$$\begin{cases} \dfrac{d}{dt}X_t = AX_t + B_t, \\ x(0) = x_0, \end{cases} \qquad (7.82)$$

where $X : \mathbb{R} \times \Omega \mapsto \mathbb{R}^n$, A is a (deterministic) matrix of coefficients, and B_t is a classical vector Brownian motion. As above, an inverse problem for this kind of equation can be formulated as: Given an i.d. sample of observations of $X(t, \omega)$, say $X(t, \omega_1), \ldots, X(t, \omega_n)$, get an estimation of the matrix A. For this purpose, let us take the integral over Ω of both sides of the previous equation and suppose that $X(t, \omega)$ is sufficiently regular; recalling that $B_t \sim \mathcal{N}(0, t)$, we have

$$\int_\Omega \frac{dx}{dt} dP(\omega) = \frac{d}{dt} \mathbb{E}(X(t, \cdot)) = A\mathbb{E}(X(t, \cdot)). \qquad (7.83)$$

This is a deterministic differential equation in $\mathbb{E}(X(t, \cdot))$. From the sample of observations of $X(t, \omega)$, we can then get an estimation of $\mathbb{E}(X(t, \cdot))$ and then use of approach developed for deterministic differential equations to solve the inverse problem for A.

Example 7.36. Consider the first-order system

$$\frac{d}{dt} x_t = a_1 x_t + a_2 y_t + b_t,$$

$$\frac{d}{dt} y_t = b_1 x_t + b_2 y_t + c_t.$$

Setting $a_1 = 0.8$, $a_2 = -0.7$, $b_1 = 0.9$, $b_2 = 0.6$, $x_0 = 1.2$, and $y_0 = 1$, we construct observational data values for x_t and y_t for $t_i = \frac{i}{N}$, $1 \leq i \leq N$, for various values of N. For each of M data sets, different pairs of Brownian motion are simulated for b_t and c_t. Figure 7.5 presents several plots of b_t and c_t for $N = 100$. In Fig. 7.6, we present some

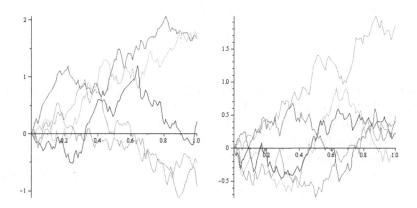

Fig. 7.5: Example plots of b_t and c_t for $N = 100$.

plots of our generated x_t and y_t, as well as phase portraits for x_t versus y_t. For each sample time, we construct the mean of the observed data values, $x_{t_i}^*$ and $x_{t_i}^*$, $1 \leq i \leq N$. We minimize the squared collage distances

$$\Delta_x^2 = \frac{1}{N} \sum_{i=1}^{N} \left(x_{t_i}^* - x_0 - \frac{1}{N} \sum_{j=1}^{i} \left(a_1 x_{t_j}^* + a_2 y_{t_j}^* \right) \right)^2$$

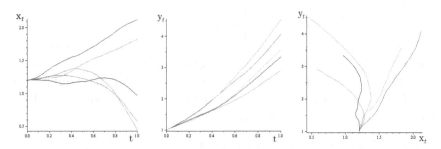

Fig. 7.6: Example plots of x_t, y_t, and x_t versus y_t for $N = 100$.

and

$$\Delta_y^2 = \frac{1}{N} \sum_{i=1}^{N} \left(y_{t_i}^* - y_0 - \frac{1}{N} \sum_{j=1}^{i} \left(b_1 x_{t_j}^* + b_2 y_{t_j}^* \right) \right)^2$$

to determine the minimal collage parameters a_1, a_2, b_1, and b_2. The results of the process are summarized in Table 7.15.

In what follows, we suppose that $\phi(\omega, t, z)$ has the following polynomial form in t and z:

$$\phi(\omega, t, z) = a_0(\omega) + a_1(\omega)t + a_2(\omega)z + a_3(\omega)t^2 + a_4(\omega)tz$$
$$+a_5(\omega)z^2 + \cdots . \tag{7.84}$$

Suppose that x_0 and each a_i are random variables defined on the same probability space $(\Omega, \mathcal{F}, P)$. Let μ and σ^2 be the unknown mean and variance of x_0 and ν_i and σ_i^2 be the unknown mean and variance of a_i. Given data from independent realizations $x(\omega_j, t)$, $j = 1, \ldots, N$, of the random variable x, the fixed point of T, we wish to recover the means. Each realization $x(\omega_j, t)$, $j = 1, \ldots, N$, of the random variable is the solution of a fixed-point equation

$$x(\omega_j, t) = x_0(\omega_j) + \int_0^t \phi(\omega_j, s, x(\omega_j, s)) ds$$

$$= x_0(\omega_j) + \int_0^t \Big[a_0(\omega_j) + a_1(\omega_j)s + a_2(\omega_j)x(\omega_j, s) + a_3(\omega_j)s^2$$

$$+a_4(\omega_j)sx(\omega_j, s) + a_5(\omega_j)(x(\omega_j, s))^2 + \cdots \Big] ds. \tag{7.85}$$

Thus, for each target function $x(\omega_j, t)$, we can find samples of realizations for $x_0(\omega_j)$ and $a_i(\omega_j)$ via the collage coding method for polynomial deterministic integral equations. Upon treating each realization, we will have determined $x_0(\omega_j)$ and $a_i(\omega_j)$, $i = 1, \ldots, M$, $j = 1, \ldots, N$. We then construct the approximations

$$\mu \approx \mu_N = \frac{1}{N} \sum_{j=1}^{N} x_0(\omega_j) \text{ and } \nu_i \approx (\nu_i)_N = \frac{1}{N} \sum_{j=1}^{N} a_i(\omega_j), \quad (7.86)$$

where we note that results obtained from collage coding each realization are independent. Using our approximations of the means, we can also calculate that

$$\sigma^2 \approx \sigma_N^2 = \frac{1}{N-1} \sum_{j=1}^{N} (x_0(\omega_j) - \mu_N)^2,$$

$$\sigma_i^2 \approx (\sigma_i)_N^2 = \frac{1}{N-1} \sum_{j=1}^{N} (a_i(\omega_j) - (\nu_i)_N)^2.$$

Example 7.37. We consider the linear case $\phi(\omega, t, z) = a_0(\omega) + a_1(\omega)t + a_2(\omega)z$. The realizations are calculated by numerically solving the related differential equation, sampling the solution at ten uniformly distributed points, and fitting a sixth-degree polynomial $x(\omega_j, t)$ to the data. Figure 7.7 illustrates some of the realizations. Table 7.16 lists the distributions from which the parameters that generate each realization are selected. The results of the preceding approach to the inverse problem are presented in Table 7.17.

We include an example where the parameters are selected from χ^2 distributions. This example shows that we can avoid the technical details that define the maximal allowed value of δ by instead just choosing δ very small. In this example, we pick $\delta = 0.1$.

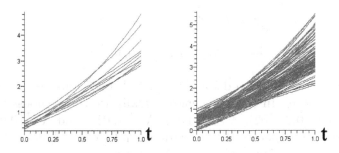

Fig. 7.7: Graphs for Example 7.37, with linear ϕ. left to right: ten realizations and 100 realizations.

Example 7.38. We suppose that $\phi(\omega, t, z)$ is quadratic z, namely

$$\phi(\omega, s, z) = a_0(\omega) + a_1(\omega)z + a_5(\omega)z^2.$$

The realizations are calculated by numerically solving the related differential equation, sampling the solution at ten uniformly distributed points, and fitting a polynomial $x(\omega_j, t)$ to the data. Figure 7.8 presents some of the realizations. Table 7.18 lists the distributions used for the parameters, and the collage coding results are presented in Table 7.19. Although the theoretical presentation of the sections above deals with a single equation, the results extend naturally to systems. In the following final example, we return to the damped oscillator problem from Example 7.37.

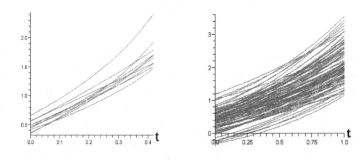

Fig. 7.8: Graphs for Example Example 7.38, with quadratic ϕ. left to right: ten realizations and 100 realizations.

Example 7.39. We consider the damped harmonic oscillator system

$$\frac{dx_1}{dt} = x_2, \ x_1(0) = x_{10}, \tag{7.87}$$

$$\frac{dx_2}{dt} = -bx_2 - kx_1, \ x_2(0) = x_{20}, \tag{7.88}$$

where the coefficients b and k and the initial conditions x_{10} and x_{20} are random variables. In order to generate realizations, we choose the distributions $\mathcal{N}(0.2, 0.02)$, $\mathcal{N}(0.5, 0.02)$, $\mathcal{N}(\frac{1}{3}, 0.01)$, and $\mathcal{N}(\frac{1}{3}, 0.01)$, respectively. We generate N realizations and fit a polynomial target to uniformly sampled data points. Next, we collage code each target and calculate the mean parameter values and corresponding mean operator's fixed point. Results are presented in Table 7.20. Figure 7.9 presents some visual results. In Fig. 7.9, the left pictures show graphs in phase space of both the realizations and the fitted polynomials; these orbits are coincident at the resolution of the picture, but there are in fact slight errors. The graphs on the right show the target polynomials along with the fixed points calculated via collage coding for deterministic integral equations; once again, at the resolution of the picture, the orbits appear to be coincident. The thicker orbit in each graph on the right is the fixed point of the operator defined by mean parameter values.

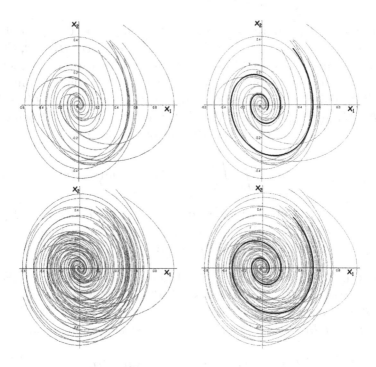

Fig. 7.9: Graphs for Example 7.39. Top left: ten realizations and targets. Top right: targets, collage coding fixed points, and fixed point of the operator defined by the mean parameter values (thicker curve). Bottom: similar results for 30 realizations.

Note that we can formulate collage coding frameworks for random analogues of the other deterministic inverse problems considered in this chapter.

7.7 Stochastic differential equations

Let $(\Omega, \mathcal{F}, P)$ be a probability space, $\{\mathcal{F}_t\}_{t \geq 0}$ be a filtration, $\{B_t\}_{t \geq 0}$ be a classical $\mathbb{R}^d$ Brownian motion, X_0 be a $\mathcal{F}_0 - \mathbb{R}^d$-measurable random vector, and $g : \mathbb{R}^d \times \mathbb{R}^d \mapsto \mathbb{R}^d$. We look at the solution of the SDE:

$$\begin{cases} dX_t = \displaystyle\int_{\mathbb{R}^d} g(X_t, y) d\mu_t(y) dt + dB_t, \\ X_{t=0} = X_0, \end{cases} \qquad (7.89)$$

where $\mu_t = P_{X_t}$ is the law of X_t. Given $T > 0$, consider the complete metric space $(C([0,T]), d_\infty)$ and the space $M(C([0,T]))$ of probability measures on $C([0,T])$. It is well-known that associated with each process X_t one can define a random variable from Ω to $C([0,T])$. This then induces a probability measure on $M(C([0,T]))$. Let $\Phi : M(C([0,T])) \mapsto M(C([0,T]))$, the function that associates to each element $m \in M(C([0,T]))$ the law of the process

$$X_0 + B_t + \int_0^t \int_{C([0,T])} g(X_s, w_s) dm(w_s) ds. \qquad (7.90)$$

If X_t is a solution of (7.91) then its law on $C([0,T])$ is a fixed point of Φ, and vice versa. We have the following theorem that states an existence and uniqueness result for (7.91).

Theorem 7.40. *[161] Let $(\Omega, \mathcal{F}, P)$ be a probability space, $\{\mathcal{F}_t\}_{t \geq 0}$ be a filtration, $\{B_t\}_{t \geq 0}$ be a classical $\mathbb{R}^d$ Brownian motion, X_0 be a $\mathcal{F}_0 - \mathbb{R}^d$-measurable random vector, and $g : \mathbb{R}^d \times \mathbb{R}^d \mapsto \mathbb{R}^d$ be a bounded Lipschitz function. Consider the following stochastic differential equation:*

$$\begin{cases} dX_t = \displaystyle\int_{\mathbb{R}^d} g(X_t, y) d\mu_t(y) dt + dB_t, \\ X_{t=0} = X_0. \end{cases} \qquad (7.91)$$

We have that

(i) for $t \leq T$, $m_1, m_2 \in M(C([0,T]))$,

$$D_t(\Phi(m_1), \Phi_2(m_2)) \leq c_T \int_0^t D_s(m_1, m_2) \, ds$$

where c_T is a constant and D_s is the distance between the images of m_1, m_2 on $C([0,s])$;

(ii) Φ is eventually contractive since there is a $k > 0$ so that

$$D_T(\Phi^k(m_1), \Phi^k(m_2)) \leq \frac{c_T^k T^k}{k!} D_T(m_1, m_2) = c_{T,k} D_T(m_1, m_2)$$

with $c_{T,k} < 1$.

7.7.1 Inverse problem for SDEs

The aim of the inverse problem consists of finding an estimation of g starting from a sample of observations of X_t. Let $(X_t^1, X_t^2, \ldots X_t^n)$, $t \in [0, T]$, be an independent sample (i.d.) and μ_n the estimated law of the process. We have the following trivial corollary of the collage theorem.

Corollary 7.41. *Let $\mu_n \in M(C[0, T])$ be the estimated law of the process. If μ is the law of the process X_t of (7.91), then there exists a constant C such that the following estimate holds:*

$$D_T(\mu, \mu_n) \leq C D_T(\Phi(\mu_n), \mu_n). \tag{7.92}$$

The inverse problem is then reduced to the minimization of the collage distance $D_T(\Phi(\mu_n), \mu_n)$, which is a function of the unknown coefficients of g.

7.8 Applications

Many models in biology and economics can be formulated in terms of deterministic and random differential equations. We now present how to solve inverse problems for some dynamical models by using the collage method described above.

7.8.1 Population dynamics

The importance of population dynamics in mathematical biology is well known and has a long history that goes back to the exponential law of Malthus and the Malthusian growth model. Many authors in the literature (see, for instance, [32, 104] and the references therein) have highlighted the impact of population size on economic growth; it can affect the production function GDP as well as modify aggregate labor supply and savings, and so on. Links between demographic indicators and economic growth have been studied by many authors in the literature; according to up-to-date demographic forecasts (United Nations webpage, http://www.un.org), the world population annual growth rate is expected to fall gradually from 1.8% (1950–2000) to 0.9% (2000–2050) before reaching a value of 0.2% between the years 2050 and 2100. Following Lotka [119], population dynamics can be described through a nonautonomous differential equation as $L(t) = L(t)g(L(t))$ (if $L(t) = 0$, then there is no population growth). We are interested in the estimation of the function g through the solution of a parameter identification problem. In many papers in the literature, g is supposed to be constant (which leads to exponential population growth) or be equal to $n - dL(t)$ (which implies logistic behavior), with $n, d > 0$. Here we wish to solve an inverse problem for this kind of differential equation using the collage method above and real data available at the Angus Maddison webpage (http://www.ggdc.net/MADDISON/oriindex.htm). We suppose now that $L(t)$ is driven by the differential equation

$$\frac{dL(t)}{dt} = L(t)g(L(t)). \tag{7.93}$$

We use data from six continents (Africa, Asia, Australia, Europe, South America, and North America) over the period 1870–2008 and look for a polynomial solution of the form

$$g(u) = \sum_{i=0}^{m} g_i u^i, \tag{7.94}$$

where the g_i are constant coefficients. This leads to

$$\frac{dL(t)}{dt} = \sum_{i=1}^{m+1} g_i L^i(t). \tag{7.95}$$

The results to eight decimal digits (using third-order polynomials) are provided in Table 7.21. We now provide empirical evidence to model population size by using higher-order polynomials (which include the logistic process) and real data. To do this, we estimate the unknown parameters through the solution of an inverse problem. The solution of the inverse problem suggests that a good fitting curve for Australia, Europe, and North America for these data is the logistic one (see

Fig. 7.10), while South America shows exponential behavior ($g_0, g_1 >$ 0). Africa and Asia show a negative coefficient g_0, which can be justified by migration effects.

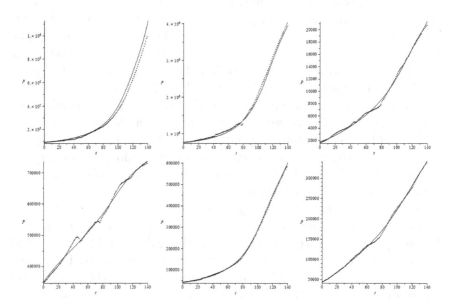

Fig. 7.10: (left to right, top to bottom) Population dynamics in Africa, Asia, Australia, Europe, South America, and North America. The origin corresponds to the year 1870.

7.8.2 mRNA and protein concentration

We consider a two-stage model of mRNA $x(t)$ and protein concentration $y(t)$ [156]. The system of ODEs modeling the concentrations features an environmental input that is assumed to be a Markov process switching between two states,

$$\frac{dx}{dt} + \mu_r x = R(t), \tag{7.96}$$

$$\frac{dy}{dt} + \mu_p y = r_p x, \tag{7.97}$$

where μ_r and μ_p are the decay rates of mRNA and protein, respectively, r_p is the environment independent rate at which protein is translated from active mRNA, and $R(t)$ is the continuous-time two-state Markov process modeling the switching environmental conditions. That is, $R(t)$ is assumed to switch between two states, r_0 and r_1. The two-state model corresponds to a system where the gene's enhancer sites can be bound to by different transcription factors or cofactors. Once bound, transcription occurs at a constant rate, and the mRNA product breaks down at a constant rate μ_r. On the other hand, the protein production depends on the concentration of mRNA. The model gives rise to a joint stochastic process for the levels of mRNA and protein.

Given observations of numerous realizations of the stochastic variables, the goal of the inverse problem is to recover the two decay rates and the mRNA-to-protein translation rate. As the system (7.98) below shows – thanks to linearity – this can be done by taking the expectations of both sides and then estimating the expected values by using a finite set of observations.

As an experiment, we set $\alpha_2 = 0.125$, $\mu_r = 0.2$, $\mu_p = 0.1$, and $r_p = 0.3$. Furthermore, we set $r_0 = 0.2$ and $r_1 = 0.1$ and define the Markov process to select state r_0 with probability 0.25. For each of 100 realizations, we generate 50 data values one time unit apart. The values of five realizations as well as the mean values of the 100 realizations are plotted in Fig. 7.11. The expected values of x and y satisfy the

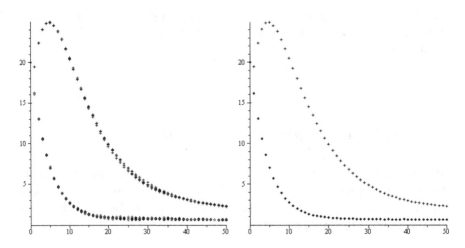

Fig. 7.11: Realizations (left) and mean values (right) for the mRNA (cross) and protein (diamond) systems.

system

$$\frac{d\mathbb{E}(x)}{dt} + \mu_r \mathbb{E}(x) = p_0 r_0 + (1 - p_0) r_1, \tag{7.98}$$

$$\frac{d\mathbb{E}(y)}{dt} + \mu_p \mathbb{E}(y) = r_p \mathbb{E}(x). \tag{7.99}$$

Hence we seek the parameters α_1, α_2, β_1, and β_2 of the system

$$\frac{dx}{dt} + \alpha_1 x = \alpha_2, \tag{7.100}$$

$$\frac{dy}{dt} + \beta_1 y = \beta_2 x, \tag{7.101}$$

that minimize the corresponding squared L_2 collage distances

$$\Delta_x^2 = \int_0^{50} \left(x(t) - \int_0^t (\alpha_2 - \alpha_1 x(s)) \, ds \right)^2 dt, \tag{7.102}$$

$$\Delta_y^2 = \int_0^{50} \left(y(t) - \int_0^t (\beta_2 x(s) - \beta_1 y(s)) \, ds \right)^2 dt. \tag{7.103}$$

We solve a penalized version of this model that provides the following results

$$\alpha_1 = 0.200, \quad \alpha_2 = 0.127, \quad \beta_1 = 0.100, \quad \alpha_2 = 0.300. \tag{7.104}$$

Increasing the number of realizations moves us even closer to this value.

7.8.3 Bacteria and amoeba interaction

In [70], the authors develop a model for the pathogenesis mechanism of the opportunistic human pathogen *Pseudomonas aeruginosa* in co-culture with *Dictyostelium* amoebae. With $u(t)$ and $v(t)$ being the number of bacteria and amoeba cells, respectively, at time t, the model takes the form

$$\frac{du}{dt} = r \left(1 - \frac{u}{K} \right) u - auv, \tag{7.105}$$

$$\frac{dv}{dt} = -mv + duv - \frac{buv}{1 + bTv}. \tag{7.106}$$

In the absence of the amoeba, the bacteria undergo logistic growth with intrinsic growth rate r and carrying capacity K. In the absence of bacteria, the amoeboid population undergoes an exponential growth decay with death rate m, as the bacteria are assumed to be the unique food source for the amoeba. The amoeba cells feed on bacteria through a mass-action mechanism with attack rate a. The amoeba growth rate is proportional to the uptake of bacteria, with proportionality constant d. Finally, bacteria attack and kill amoeba cells according to a Holling-type function with handling time T and attack rate b.

We consider the inverse problem of recovering all of the constant rates in the model equations from observations of the bacteria and amoeba populations.

For a simulation, we set $r = 0.6$, $K = 10$, $a = 0.6$, $m = 0.52$, $d = 0.6$, $b = 0.44$, and $T = 3.25$. The system admits the stable equilibrium point $(1.286, 0.871)$ to three decimal places. The solution trajectories with $u(0) = 8$ and $v(0) = 6$ are illustrated in Fig. 7.12. We simulate the

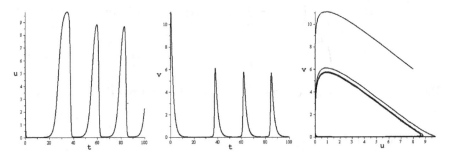

Fig. 7.12: Plots of the number of bacteria u and the number of amoeba cells v for our simulation.

gathering of observational data $u_i = u(t_i)$, $v_i = v(t_i)$ times $t_i = 0.1i$, $0 \le 1 \le 400$, in order to capture the behaviour of the two "spikes." Given these two sets of data, we seek to recover the parameters α_j, $j = 1, 2, 3$, and β_j, $j = 1, 2, 3, 4$, so that the system

$$\frac{du}{dt} = \alpha_1 (1 + \alpha_2 u) u + \alpha_3 uv, \qquad (7.107)$$

$$\frac{dv}{dt} = \beta_1 v + \beta_2 uv + \frac{\beta_3 uv}{1 + \beta_3 \beta_4 v}. \qquad (7.108)$$

admits the data sets as an approximate solution. The associated squared L^2 collage distances are

$$\Delta_u^2 = \int_0^{40} \left(u(t) - u(0) - \int_0^t (\alpha_1 (1 + \alpha_2 u(s)) u(s) + \alpha_3 u(s) v(s)) \, ds \right)^2 dt$$

$$\Delta_v^2 = \int_0^{40} \left(v(t) - v(0) - \int_0^t \left(\beta_1 v(s) + \beta_2 u(s) v(s) + \frac{\beta_3 u(s) v(s)}{1 + \beta_4 v(s)} \right) ds \right)^2 dt.$$

A penalization method allows inclusion of the constraints in the objective function; we discretize the problem in accordance with the observational data interval and seek to find the parameters that minimize the objective function. For Δ_u^2, the method yields, to four decimal places, $\alpha_1 = 0.6012$, $\alpha_2 = -0.0603$, and $\alpha_3 = -0.5940$. The squared collage distance Δ_v^2 is a more complicated function of its parameters, so we use particle swarm ant colony optimization to minimize it. We obtain $\beta_1 = -0.5182$, $\beta_2 = 0.5795$, $\beta_3 = -0.5131$, and $\beta_4 = 2.5274$. The values for the α_i, β_1, and β_2 agree well with the known values. The values for β_3 and β_4 are further away from the expected values of -0.44 and 1.43.

7.8.4 Tumor growth

At each time t, we suppose that a tumor occupies the region $u(t)$. The tumor grows or shrinks due to cell proliferation or death according to the level of a diffusing nutrient concentration. In [68], the problem is cast as a free-boundary problem in three spatial dimensions, with the assumption that the tumor is always spherical. Here, we assume that the tumor always occupies a convex region that encloses the origin, and we work in two spatial dimensions for convenience: the ideas extend to three dimensions. Given observational data, we wish to recover a model in the differential form

$$u(t + dt) = u(t) + [(Au)(t) + b]dt, \tag{7.109}$$

where, for each fixed t, we have that $u(t) \in \mathbb{H}_c$ is a compact and convex subset of $\mathbb{R}^2$ and the sum is understood in the Minkowski sense. The convexity assumption means that we can codify this model in terms of an infinite-dimensional operator A and an infinite-dimensional vector b that capture the growth rate information in every direction. The real-world problem is: Given observational data in $N < \infty$ directions, recover an N-dimensional model

$$\frac{du_N}{dt} = A_N u_N + b_N \tag{7.110}$$

that approximates well the evolution of the tumor. As an example, we construct an always-convex region by considering the intersection of several rotated, time-varying ellipses. Three of the frames in the region's evolution are given in Fig. 7.13. To simulate the data gathering process, we calculate the position of the frontier of the tumor along

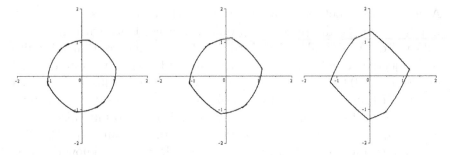

Fig. 7.13: Snapshots in the evolution of our simulated tumor.

N uniformly distributed rays through the origin. This corresponds to an N-dimensional model of (7.110), with dimension i corresponding to the growth along ray i. In Fig. 7.14, we present visual results for the case $N = 40$; for the same three times illustrated in Fig. 7.13, we plot the data points produced by the model we have recovered, connecting them with straight line segments. Denoting by $A(t)$ the area of the

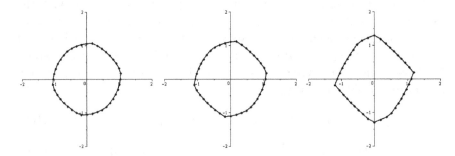

Fig. 7.14: Snapshots produced by our recovered model with $N = 40$.

tumor at time t, we can use the model to calculate and forecast the relative growth rate of the area of the tumor, $\frac{A'(t)}{A(t)}$. In Table 7.22, we present some results for different values of N. In this case, the models show that the relative growth rate of the tumor is decreasing.

Table 7.13: Collage coding results. For each row, we use the given values of c and ρ_{max} and set the initial condition to $g(x) = 1 + x$. In all cases, we seek the minimal collage parameters for linear $v(\rho)$ and g.

Target solution			Noise	Minimal collage parameters		
c	ρ_{max}	$g(x)$ from data?	ε	c	ρ_{max}	$g(x)$
1	1	yes	0	1.18075	1.11353	$1.00022+1.00043x$
1	1	no	0	1.31521	1.18766	$1.02101+0.97752x$
1	1	yes	0.05	1.24822	1.16554	$1.00970+0.99809x$
1	1	no	0.05	1.34192	1.20133	$1.02243+0.97576x$
1	1	yes	0.1	1.31449	1.21543	$1.01919+0.99574x$
1	1	no	0.1	1.36120	1.20900	$1.02286+0.97432x$
1	4	yes	0	1.00001	4.00000	$1.00000+1.00000x$
1	4	no	0	1.00001	4.00000	$1.00000+1.00000x$
1	4	yes	0.05	1.07105	4.17463	$1.00948+0.99765x$
1	4	no	0.05	1.02710	4.07272	$1.00311+0.99925x$
1	4	yes	0.1	1.14214	4.34042	$1.01897+0.99531x$
1	4	no	0.1	1.05385	4.14351	$1.00617+0.99851x$
3	4	yes	0	3.01990	4.00247	$1.00000+1.00000x$
3	4	no	0	3.03402	4.00475	$1.00190+0.99865x$
3	4	yes	0.05	3.08765	4.04955	$1.00943+0.99771x$
3	4	no	0.05	3.06499	4.01672	$1.00490+0.99735x$
3	4	yes	0.1	3.15563	4.09561	$1.01891+0.99537x$
3	4	no	0.1	3.09551	4.02805	$1.00782+0.99607x$

Table 7.14: Collage coding results with $k_{collage}(x,y) = a_1\phi_1(x,y) + a_2\phi_2(x,y)$ and $a_1^2 + a_2^2 = C_A$.

C_A	Minimal Δ^2	a_1	a_2
0.1	0.17153	0.29898	−0.10301
0.2	0.12759	0.40955	−0.17964
0.3	0.09614	0.48283	−0.25860
0.4	0.07136	0.53421	−0.33856
0.5	0.05088	0.57113	−0.41690
0.6	0.03356	0.59838	−0.49187
0.7	0.01929	0.61908	−0.56280
0.8	0.01120	0.63526	−0.62964
0.9	0.01613	0.64823	−0.69267

Table 7.15: Minimal collage distance parameters for different N and M.

N	M	a_1	a_2	b_1	b_2
10	10	0.6893	−0.4456	0.8753	0.3813
10	100	0.8617	−0.6617	0.7834	0.6269
100	10	0.8499	−0.5981	0.9055	0.5757
100	100	0.6842	−0.6163	0.9319	0.5823

Table 7.16: Distributions used in the inverse problem of Example 7.37.

Label	True values			
	a_0	a_1	a_2	x_0
1	$\mathcal{N}(1,0.04)$	$\mathcal{N}(0.7,0.04)$	$\mathcal{N}(0.3,0.04)$	$\mathcal{N}(0.4,0.09)$
2	$\mathcal{N}(2,0.09)$	$\mathcal{N}(0.5,0.09)$	$\mathcal{N}(0.4,0.09)$	$\mathcal{N}(0.5,0.04)$
3	$\chi^2(6)$	$\chi^2(5)$	$\chi^2(6)$	$\chi^2(4)$

Table 7.17: Results for the inverse problem of Example 7.37. The first column indicates the distributions used from Table 7.16. N is the number of realizations, and the final four columns give in parentheses the mean and variance obtained via collage coding for each parameter.

		Minimal collage values			
Label	N	a_0	a_1	a_2	x_0
1	10	$(0.983, 0.066)$	$(0.680, 0.028)$	$(0.307, 0.029)$	$(0.379, 0.020)$
1	100	$(0.974, 0.045)$	$(0.691, 0.032)$	$(0.345, 0.034)$	$(0.411, 0.084)$
1	1000	$(0.998, 0.037)$	$(0.698, 0.038)$	$(0.296, 0.041)$	$(0.389, 0.090)$
2	10	$(1.974, 0.147)$	$(0.470, 0.062)$	$(0.411, 0.065)$	$(0.486, 0.009)$
2	100	$(1.962, 0.101)$	$(0.487, 0.073)$	$(0.468, 0.076)$	$(0.507, 0.037)$
2	1000	$(1.997, 0.083)$	$(0.497, 0.086)$	$(0.394, 0.092)$	$(0.493, 0.040)$
3	10	$(6.133, 9.801)$	$(5.895, 10.902)$	$(3.488, 3.446)$	$(3.415, 6.002)$
3	100	$(5.987, 11.467)$	$(5.025, 9.232)$	$(5.615, 10.340)$	$(4.551, 9.910)$
3	1000	$(6.153, 12.397)$	$(4.711, 8.290)$	$(5.966, 11.051)$	$(3.934, 7.830)$

Table 7.18: Distributions used in the inverse problem of Example 7.38.

	True values			
Label	a_0	a_1	a_5	x_0
1	$\mathcal{N}(1, 0.04)$	$\mathcal{N}(0.7, 0.04)$	$\mathcal{N}(0.3, 0.04)$	$\mathcal{N}(0.4, 0.09)$
2	$\mathcal{N}(2, 0.09)$	$\mathcal{N}(0.5, 0.09)$	$\mathcal{N}(0.4, 0.09)$	$\mathcal{N}(0.5, 0.04)$

Table 7.19: Results for the inverse problem of Example 7.38. The first column indicates the distribution from Table 7.18 from which realizations are generated. N is the number of realizations, and the final four columns give in parentheses the mean and variance obtained via collage coding for each parameter.

		Minimal collage values			
Label	N	a_0	a_1	a_5	x_0
1	10	$(0.992, 0.067)$	$(0.671, 0.027)$	$(0.310, 0.030)$	$(0.379, 0.020)$
1	100	$(0.978, 0.045)$	$(0.688, 0.032)$	$(0.346, 0.034)$	$(0.411, 0.084)$
1	1000	$(1.000, 0.037)$	$(0.696, 0.038)$	$(0.296, 0.041)$	$(0.389, 0.089)$
2	10	$(1.975, 0.147)$	$(0.470, 0.062)$	$(0.411, 0.064)$	$(0.486, 0.009)$
2	100	$(1.962, 0.101)$	$(0.487, 0.073)$	$(0.468, 0.076)$	$(0.507, 0.037)$
2	1000	$(1.997, 0.084)$	$(0.498, 0.088)$	$(0.393, 0.093)$	$(0.493, 0.040)$

Table 7.20: Results for the random damped oscillator inverse problem of Example 7.39. The results for the random variables are given in parentheses as mean and variance.

	Minimal collage values			
N	b	k	x_{10}	x_{20}
10	$(0.281, 0.021)$	$(0.592, 0.023)$	$(0.333, 0.005)$	$(0.414, 0.004)$
30	$(0.224, 0.021)$	$(0.518, 0.023)$	$(0.300, 0.013)$	$(0.372, 0.007)$
100	$(0.215, 0.023)$	$(0.506, 0.021)$	$(0.344, 0.009)$	$(0.328, 0.012)$

Table 7.21: Minimal collage distance parameters

	g_0	g_1	g_2	g_3
Africa	-0.00763537	0.00000018	-0.00000000	0.00000000
Asia	-0.02926752	0.00000005	-0.00000000	0.00000000
Australia	0.03003342	-0.00000383	0.00000000	-0.00000000
Europe	0.12968633	-0.00000070	0.00000000	-0.00000000
South America	0.00203530	0.00000025	-0.00000000	0.00000000
North America	0.03432962	-0.00000030	0.00000000	-0.00000000

Table 7.22: Model calculations for the relative growth rate $\frac{A'(t)}{A(t)}$ of the tumor.

Time t	$N = 20$	$N = 40$	$N = 60$
0.2	0.0434	0.0453	0.0458
0.4	0.0411	0.0430	0.0434
0.6	0.0390	0.0408	0.0411
0.8	0.0370	0.0387	0.0390
1.0	0.0352	0.0368	0.0352
2.0	0.0291	0.0304	0.0304
3.0	0.0236	0.0244	0.0243
4.0	0.0196	0.0201	0.0198
5.0	0.0168	0.0170	0.0166

Chapter 8
Further Developments and Extensions

8.1 Generalized collage theorems for PDEs

In this section, we formulate collage theorems to treat inverse problems for *elliptic*, *parabolic*, and *hyperbolic* partial differential equations. Such problems have understandably received a great deal of attention over the years in the study of distributed systems (see, for example, [142]).

The structure of the framework can be outlined by considering the guiding example of linear steady-state heat or fluid flow,

$$\nabla \cdot (\kappa(\mathbf{x})\nabla u(\mathbf{x})) = f(\mathbf{x}), \quad \mathbf{x} \in \Omega \subset \mathbb{R}^n, \tag{8.1}$$

with appropriate boundary conditions, where $\kappa(\mathbf{x})$ represents the diffusivity and $f(\mathbf{x})$ the source term. When $f = 0$, (8.1) is the well-known conductivity equation. The inverse problem is to determine $\kappa(\mathbf{x})$ from a sufficient set of observations of $u(\mathbf{x})$ [69, 173, 85, 77, 78, 79]. (There is, of course, the important question of sufficient conditions for the existence and uniqueness of solutions to (8.1) —see, for example, [144, 59]. We assume that these conditions are satisfied.)

We consider a variational form of (8.1) that is obtained by integrating both sides with respect to elements of a suitable set of basis functions (for example, the finite element "hat" functions). With (8.1) in mind, this produces a linear system of equations involving the known values u_k associated with unknown values κ_k over a set of grid points. The "solution" of this system is often accomplished by minimizing an appropriate least-squares functional, possibly with the inclusion of additional *penalty functions* for the purpose of *regularization*. This procedure is described nicely in the recent monograph by Vogel [165].

Suppose that we step back a bit from these traditional practical approaches and view the inverse problem for this linear boundary value problem in terms of the *Lax-Milgram representation theorem*. Very briefly, we recall that in the forward problem it is a standard procedure to multiply both sides of a linear PDE such as (8.1) with an element $v \in H$, where H is a suitable Hilbert space of functions (i.e., $H_0^1(\Omega)$), and integrate over Ω to obtain the equation

$$a(u, v) = \phi(v), \qquad v \in H. \tag{8.2}$$

Here $\phi(v) = \int_\Omega fv\,d\mathbf{x}$ and $a(u, v)$ is a bilinear form on $H \times H$.

The inverse problem may now be viewed as follows. Suppose that we have an observed solution u and a given (restricted) family of bilinear functionals $a_\lambda(u, v)$, $\lambda \in \mathbb{R}^n$. (In (8.1), for example, the λ_k could correspond to the unknown diffusivity values on grid points or the unknown coefficients of a multinomial expansion assumed for $\kappa(\mathbf{x})$.) We now seek to find "optimal" values of λ, for example, those that minimize the function

$$F(\lambda) = \sup_{v \in H, \|v\| = 1} |a_\lambda(u, v) - \phi(v)|. \tag{8.3}$$

Now suppose that

$$\lambda^* = \operatorname{argmin} F(\lambda). \tag{8.4}$$

Then, by the Lax-Milgram theorem, there exists a unique function u_{λ^*} such that

$$a_{\lambda^*}(u_{\lambda^*}, v) = \phi(v) \quad \text{for all} \ \ v \in H. \tag{8.5}$$

As we shall show in this chapter, the error in the approximation of our given solution u with u_{λ^*} is given by

$$\| u - u_{\lambda^*} \| \ \leq \ \frac{1}{m_{\lambda^*}} F(\lambda^*), \tag{8.6}$$

where the constant m_λ characterizes the expansivity of the bilinear form a_λ.

Following our earlier studies (given in Chapter 7) on the inverse problems using fixed points of contraction mappings, we shall refer to the minimization of the functional $F(\lambda)$ in (8.3) as a *generalized collage method*. We view (8.6) and the collage theorem as variations on a theme. We emphasize that this framework provides a rigorous basis for most practical methods of solving inverse problems in boundary value problems.

Definition 8.1. A linear functional on a real Hilbert space H is a linear map from H to $\mathbb{R}$. A linear functional ϕ is bounded, or continuous, if there exists a constant M such that

$$|\phi(x)| \leq M\|x\| \tag{8.7}$$

for all $x \in H$.

By the linearity of ϕ, it is trivial to prove that we may choose

$$M = \max_{x \in H, \|x\|=1} \phi(x). \tag{8.8}$$

Theorem 8.2. *(Riesz representation) Let H be a Hilbert space and ϕ be a bounded linear functional. Then there is a unique vector $x \in H$ such that $\phi(y) =< x, y >$ for all $y \in H$, where $< \cdot, \cdot >$ denotes an inner product in H.*

Theorem 8.3. *(Lax-Milgram representation) Let H be a Hilbert space and ϕ be a bounded linear nonzero functional. Suppose that $a(u, v)$ is a bilinear form on $H \times H$ that satisfies the following:*

- *there exists a constant $M > 0$ such that $|a(u, v)| \leq M\|u\|\|v\|$ for all $u, v \in H$,*
- *there exists a constant $m > 0$ such that $a(u, u) \geq m\|u\|^2$ for all $u \in H$.*

Then there is a unique vector $u^* \in H$ such that $\phi(v) = a(u^*, v)$ for all $v \in H$.

Example 8.4. Consider the following second-order ordinary differential equation in $u(x)$:

$$\begin{cases} -\frac{d}{dx}\left(k(x)\frac{du}{dx}\right) = f(x), \ x \in (0, 1), \\ u(0) = u(1) = 0. \end{cases} \tag{8.9}$$

It is well-known that on the Hilbert space $H_0^1([0, 1])$ built with all $\mathcal{L}^2$ functions that have a weak derivative in $\mathcal{L}^2$, this problem can be reformulated in a variational form as

$$\int_0^1 k(x)u'(x)v'(x)dx = \int_0^1 f(x)v(x)dx \tag{8.10}$$

for all $v \in H_0^1([0, 1])$. In terms of the notation used above, this equation can be rewritten in the form

$$a_\lambda(u, v) = \phi(v). \tag{8.11}$$

By the Lax-Milgram theorem, we have the well-known result that this problem has a unique solution in $H_0^1[0, 1]$.

8.1.1 Elliptic PDEs

Suppose that we have a given Hilbert space H, a "target" element $u \in H$ and a family of bilinear functionals a_λ. Then, by the Lax-Milgram theorem, there exists a unique vector u_λ such that $\phi(v) = a_\lambda(u_\lambda, v)$ for all $v \in H$. We would like to determine if there exists a value of the parameter λ such that $u_\lambda = u$ or, more realistically, such that $\|u_\lambda - u\|$ is small enough. The following theorem will be useful for the solution of this problem.

Theorem 8.5. *(Generalized collage theorem) Suppose that $a_\lambda(u, v)$: $\Lambda \times H \times H \to \mathbb{R}$ is a family of bilinear forms for all $\lambda \in \Lambda$ and $\phi : H \to \mathbb{R}$ is a given linear functional. Let u_λ denote the solution of the equation $a_\lambda(u, v) = \phi(v)$ for all $v \in H$, as guaranteed by the Lax-Milgram theorem. Given a target element $u \in H$, then*

$$\|u - u_\lambda\| \le \frac{1}{m_\lambda} F(\lambda), \tag{8.12}$$

where

$$F(\lambda) = \sup_{v \in H,\, \|v\|=1} |a_\lambda(u, v) - \phi(v)|. \tag{8.13}$$

Proof. For each λ, we know by the Lax-Milgram theorem that there exists a unique $u_\lambda \in H$ such that $a_\lambda(v, u_\lambda) = \phi(v)$ for all $v \in H$. From Theorem 8.3, we then have

$$\begin{aligned}
m_\lambda \|u - u_\lambda\|^2 &\le |a_\lambda(u - u_\lambda, u - u_\lambda)| \\
&= |a_\lambda(u - u_\lambda, u) - a_\lambda(u - u_\lambda, u_\lambda)| \\
&= |a_\lambda(u - u_\lambda, u) - \phi(u - u_\lambda)|
\end{aligned}$$

such that

$$\|u - u_\lambda\| \le \left(\frac{1}{m_\lambda}\right) \sup_{v \in H,\, \|v\|=1} |a_\lambda(u, v) - \phi(v)|. \tag{8.14}$$

$\square$

Example 8.6. Consider (8.10). Let

$$\phi(v) = \int_0^1 f(x) v(x)\, dx \tag{8.15}$$

and

$$a_\lambda(u, v) = \int_0^1 k(x)u'(x)v'(x)dx, \tag{8.16}$$

where $k : \mathbb{R} \mapsto \mathbb{R}$ is a parameter function. Given a function $u \in H_0^1([0, 1])$ and using the generalized collage theorem, Theorem 8.5, the inverse problem can then be formulated as

$$\inf_{\lambda \in \Lambda} \frac{1}{m_\lambda} F(\lambda), \tag{8.17}$$

where $F(\lambda)$ is defined in (8.13).

In order to ensure that the approximation u_λ is close to a target element $u \in H$, we can, by the generalized collage theorem, try to make the term $F(\lambda)/m_\lambda$ in (8.17) as close to zero as possible. The appearance of the m_λ factor complicates the procedure, as does the factor $1/(1-c)$ in the standard collage theorem. As such, we shall follow the usual practice and ignore the m_λ factor, assuming, of course, that all allowable values are bounded away from zero. So, if $\inf_{\lambda \in \Lambda} m_\lambda \geq m > 0$, then the inverse problem can be reduced to the minimization of the function $F(\lambda)$ on the space Λ; that is,

$$\inf_{\lambda \in \Lambda} F(\lambda). \tag{8.18}$$

Now let $\langle e_i \rangle \subset H$ be a basis of the Hilbert space H, not necessarily orthogonal, such that each element $v \in H$ can be written as $v = \sum_i \alpha_i e_i$. Computing, we have, for all $v \in H$,

$$|a_\lambda(u, v) - \phi(v)|^2 = \left| a_\lambda \left(u, \sum_i \alpha_i e_i \right) - \phi \left(\sum_i \alpha_i e_i \right) \right|^2$$

$$\leq \left(\sum_i \alpha_i |a_\lambda(u, e_i) - \phi(e_i)| \right)^2$$

$$\leq \left[\sum_i \alpha_i^2 \right] \left[\sum_i |a_\lambda(u, e_i) - \phi(e_i)|^2 \right],$$

Inequality (8.14) can be rewritten as

$$\inf_{\lambda \in \Lambda} \| u - u_\lambda \|$$

$$\leq \inf_{\lambda \in \Lambda} \left(\frac{1}{m_\lambda} \right) \left(\sup_{v \in H, \|v\|=1} \left[\sum_i \alpha_i^2 \right] \right) \left[\sum_i |a_\lambda(u, e_i) - \phi(e_i)|^2 \right]$$

$$= \frac{1}{m} \sup_{v \in H, \|v\|=1} \left[\sum_i \alpha_i^2 \right] \inf_{\lambda \in \Lambda} \left[\sum_i |a_\lambda(u, e_i) - \phi(e_i)|^2 \right].$$

8.1.1.1 Approximating the problem

Let $V_n =< e_1, e_2, \ldots, e_n >$ be the finite-dimensional vector space generated by e_i, $V_n \subset H$. Given a target $u \in H$, let $\Pi_{V_n} u$ be the projection of u on the space V_n and consider the following problem: Find $u_\lambda \in V_n$ such that $\|\Pi_{V_n} u - u_\lambda\|$ is as small as possible. Following the same analysis that led to (8.1.1), we have

$$
\|\Pi_{V_n} u - u_\lambda\| \leq \left(\frac{1}{m_\lambda}\right) \max_{v \in V_n,\ \|v\|=1} |a_\lambda(\Pi_{V_n} u, v) - \phi(v)|
$$

$$
\leq \frac{1}{m_\lambda} \max_{v=\sum_{i=1}^n \alpha_i e_i \in V_h, \|v\|=1} \left[\sum_{i=1}^n \alpha_i^2\right] \left[\sum_i |a_\lambda(u, e_i) - \phi(e_i)|^2\right]
$$

$$
= \frac{M}{m} \left[\sum_i |a_\lambda(u, e_i) - \phi(e_i)|^2\right],
$$

where $M = \max_{v=\sum_{i=1}^n \alpha_i e_i \in V_h, \|v\|=1} \sum_{i=1}^n \alpha_i^2$, so we have reduced the problem to the minimization of the function

$$
\inf_{\lambda \in \Lambda} \|\Pi_{V_n} u - u_\lambda\| \leq \frac{M}{m} \inf_{\lambda \in \Lambda} \sum_{i=1}^n |a_\lambda(u, e_i) - \phi(e_i)|^2
$$

$$
= \frac{M}{m} (F_n(\lambda))^2. \tag{8.19}
$$

8.1.1.2 One-dimensional steady-state diffusion

As an application of the preceding method, we consider the *one-dimensional steady-state diffusion equation*

$$
-\frac{d}{dx}\left(\kappa(x)\frac{du}{dx}\right) = f(x),\ 0 < x < 1, \tag{8.20}
$$

$$
u(0) = u_{left}, \tag{8.21}
$$

$$
u(1) = u_{right}, \tag{8.22}
$$

where the *diffusivity* $\kappa(x)$ varies in x. The inverse problem of interest is: Given $u(x)$, possibly in the form of an interpolation of data points, and $f(x)$ on $[0,1]$, determine an approximation of $\kappa(x)$. In [165], this problem is studied and solved via a regularized least-squares minimization problem. It is important to stress that the approach in [165] seeks to directly minimize the error between the given $u(x)$ and the solutions $v(x)$ Eq. (8.20). The collage coding approach allows us to perform a different minimization to solve the inverse problem. A natural goal is to recover $\kappa(x)$ from observations of the response $u(x)$ to a point source $f(x) = \delta(x - x_s)$, a Dirac delta function at $x_s \in (0, 1)$.

In what follows, we consider $u_{left} = u_{right} = 0$, although the approach can be modified to treat nonzero values. We multiply (8.20) by a test function $\xi_i(x) \in H_0^1([0, 1])$ and integrate by parts to obtain $a(u, \xi_i) = \phi(\xi_i)$, where

$$a(u, \xi_i) = \int_0^1 \kappa(x)u'(x)\xi_i'(x)\, dx - \xi_i(x)\kappa(x)u'(x)\Big|_0^1 \tag{8.23}$$

$$= \int_0^1 \kappa(x)u'(x)\xi_i'(x)\, dx, \text{ and} \tag{8.24}$$

$$\phi(\xi_i) = \int_0^1 f(x)\xi_i(x)\, dx. \tag{8.25}$$

For a fixed choice of n, introduce the partition of $[0, 1]$

$$x_i = \frac{i}{n+1}, \quad i = 0, \ldots, n+1,$$

with n interior points, and define for $j = 0, 1, 2, \ldots$

$$V_n^r = \{v \in C[0, 1] : v(0) = v(1) = 0 \text{ and } v \text{ is a polynomial}$$
$$\text{of degree } r \text{ on } [x_{i-1}, x_i], \ i = 1, \ldots, n+1\}.$$

Denote a basis for V_n^r by $\{\xi_1, \ldots, \xi_n\}$. When $r = 1$, our basis consists of the hat functions

$$\xi_i(x) = \begin{cases} (n+1)(x - x_{i-1}), \ x_{i-1} \leq x \leq x_i, \\ -(n+1)(x - x_{i+1}), \ x_i \leq x \leq x_{i+1}, \ i = 1, \ldots, n, \\ 0, \text{ otherwise,} \end{cases}$$

and when $r = 2$, our hats are replaced by parabolas with

$$\xi_i(x) = \begin{cases} (n+1)^2(x - x_{i-1})(x - x_{i+1}), \ x_{i-1} \leq x \leq x_{i+1}, i = 1, \ldots, n, \\ 0, \text{ otherwise.} \end{cases}$$

Suppose that $\kappa(x) > 0$ for all $x \in [0, 1]$. Then the m_λ in our formulation, which we denote by m_κ, can be chosen equal to $\inf_{x \in [0,1]} \kappa(x)$. In fact, we have

$$a(u, u) = \int_0^1 \kappa(x)u'(x)u'(x)\, dx$$

$$\geq \inf_{x \in [0,1]} \kappa(x) \int_0^1 (u'(x))^2 \, dx$$
$$= m_\kappa \|u\|_{H_0^1}^2,$$

where the norm on H_0^1 is defined by the final equality. As a result, because we divide by m_κ, we expect our results will be good when $\kappa(x)$ is bounded away from 0 on $[0,1]$.

We shall consider two different scenarios: (i) a continuous framework and (ii) a discretized framework. In a final discussion (iii), we consider the case of $f(x)$ being a point source in each of the two frameworks. Finally, we discuss the incorporation of multiple data sets by our method.

8.1.1.3 Continuous framework

Assume that we are given data points u_i measured at various x-values having no relation to our partition points x_i. These data points are interpolated to produce a continuous target function $u(x)$, a polynomial, say. Let us now assume a polynomial representation of the diffusivity,

$$\kappa(x) = \sum_{j=0}^N \lambda_j x^j. \tag{8.26}$$

In essence, this introduces a regularization into our method of solving the inverse problem. Working on V_n^r, we have

$$a_\lambda(u, \xi_i) = \sum_{j=0}^N \lambda_j A_{ij}, \text{ with } A_{ij} = \int_{x_{i-1}}^{x_{i+1}} x^j u'(x) \xi_i'(x) \, dx. \tag{8.27}$$

Letting

$$b_i = \int_0^1 f(x)\xi_i(x) \, dx = \int_{x_{i-1}}^{x_{i+1}} f(x)\xi_i(x) \, dx, \quad i = 1, \ldots, n, \tag{8.28}$$

we now minimize

$$(F_n(\lambda))^2 = \sum_{i=1}^n \left[\sum_{j=0}^N \lambda_j A_{ij} - b_i \right]^2. \tag{8.29}$$

Various minimization techniques can be used; in this work we used the quadratic-program solving package in Maplesoft's Maple.

As a specific experiment, consider $f(x) = 8x$ and $\kappa_{true}(x) = 2x + 1$, in which case the solution to the steady-state diffusion equation is $u_{true}(x) = x - x^2$. We shall sample this solution at ten data points, add Gaussian noise of small amplitude ε to these values and then fit a polynomial of degree 2 to the data points, to be denoted as $u_{target}(x)$. Given $u_{target}(x)$ and $f(x)$, we seek a degree 10 polynomial $\kappa(x)$ with coefficients λ_i such that the steady-state diffusion equation admits $u_{target}(x)$ as an approximate solution. We now construct $F_{30}(\lambda)$ and minimize it with respect to the λ_i. Table 8.1 presents the results. In all cases, the recovered coefficients for all terms of degree 2 and higher are zero to five decimal places, so we do not report them in the table. Furthermore, d_2 denotes the standard $\mathcal{L}^2$ metric on $[0, 1]$.

Table 8.1: Collage coding results when $f(x) = 8x$, $\kappa_{true}(x) = 1 + 2x$, data points = 10, number of basis functions = 30, and degree of $\kappa_{collage} = 10$. In the first four rows, we work on V_{30}^1; in the last four rows, we work on V_{30}^2. In each case, $F_{30}(\lambda)$ is equal to zero to ten decimal places.

Noise ε	$d_2(u_{true}, u_{target})$	$\kappa_{collage}$	$d_2(\kappa_{collage}, \kappa_{true})$
0.00	0.00000	$1.00000 + 2.00000x$	0.00000
0.01	0.00353	$1.03050 + 2.05978x$	0.06281
0.05	0.01770	$1.17365 + 2.33952x$	0.35712
0.10	0.03539	$1.42023 + 2.81788x$	0.86213
0.00	0.00000	$1.00000 + 2.00000x$	0.00000
0.01	0.00353	$1.00832 + 2.03967x$	0.03040
0.05	0.01770	$1.03981 + 2.21545x$	0.16011
0.10	0.03539	$1.07090 + 2.48292x$	0.34301

8.1.1.4 Discretized framework

In a practical example, we generally obtain discrete data values for u. If we are given values u_i at the partition points x_i, $i = 1, \ldots, n$, and set $u_0 = u_{left} = 0$ and $u_{n+1} = u_{right} = 0$, then, working on V_n^r, we write

$$u(x) = \sum_{l=1}^{n} u_l \xi_l(x).$$

Then (8.27) becomes

$$A_{ij} = \sum_{l=1}^{n} u_l \int_{x_{i-1}}^{x_{i+1}} x^j \xi_l'(x) \xi_i'(x)\, dx.$$

Notice that we face a problem regardless of our approach. In the earlier approach, we interpolate the points to obtain a target $u(x)$ to use in the formulas above; it is quite possible that small errors in that interpolation can lead to large errors in the derivative $u'(x)$ that we need to calculate the A_{ij}. Here as well, small errors in our data values u_i can be amplified. If, in addition, we are given values of $f(x_i) = f_i$, $i = 0, \ldots, n+1$, then we extend our basis of V_n^r by adding the two "half-hat" functions at the end-points. Namely, when $r = 1$, we add the functions

$$\xi_0(x) = \begin{cases} -(n+1)\,(x - x_1),\ x_0 \leq x \leq x_1, \\ \\ 0,\ \text{otherwise}, \end{cases}$$

$$\xi_{n+1}(x) = \begin{cases} (n+1)\,(x - x_n),\ x_n \leq x \leq x_{n+1}, \\ \\ 0,\ \text{otherwise}, \end{cases}$$

and when $r = 2$, we add

$$\xi_0(x) = \begin{cases} (n+1)^2 \left(x + \frac{1}{n+1}\right)(x - x_1),\ x_0 \leq x \leq x_1, \\ \\ 0,\ \text{otherwise}, \end{cases}$$

$$\xi_{n+1}(x) = \begin{cases} (n+1)^2\,(x - x_n)\left(x - \frac{n+2}{n+1}\right),\ x_n \leq x \leq x_{n+1}, \\ \\ 0,\ \text{otherwise}. \end{cases}$$

We represent $f(x)$ in this extended basis, writing

$$f(x) = \sum_{l=0}^{n+1} f_l \xi_l(x)$$

to approximate b_i and, thereafter, c_i.

In Table 8.2, we repeat the same experiment as in framework (i), this time without interpolating the data points and instead approximating the A_{ij} as discussed above.

Table 8.2: Collage coding results when $f(x) = 8x$, $\kappa_{true}(x) = 1 + 2x$, data points = 100, number of basis functions = 40, and degree of $\kappa_{collage} = 4$, working on V_{40}^1.

Noise ε	$\kappa_{collage}$
0.00000	$1.00000 + 2.00000x + 0.00000x^2 + 0.00000x^3 + 0.00000x^4$
0.00001	$1.00001 + 2.00058x - 0.05191x^2 + 0.10443x^3 - 0.05928x^4$
0.0001	$1.00004 + 2.05768x - 0.52113x^2 + 1.04778x^3 - 0.59470x^4$

8.1.1.5 Point sources

Finally, we consider the case where $f(x)$ is a point source at one of our partition points,

$$f(x) = \delta(x - x_s), \quad \text{where } s \in \{1, 2, \ldots, n\}.$$

Working on V_n^r, (8.28) becomes

$$b_i = \phi(\xi_i) = \int_0^1 f(x)\xi_i(x)\,dx = \begin{cases} 1, \text{ if } i = s, \\ 0, \text{ otherwise.} \end{cases}$$

In framework (i), we can use (8.29) to solve our inverse problem, where the right-hand side of the equation now simplifies to A_{sk}. A similar change occurs in framework (ii).

But suppose that we seek an expansion of $\kappa(x)$ in the extended ξ_i basis,

$$\kappa(x) = \sum_{j=0}^{n+1} \lambda_j \xi_j(x). \tag{8.30}$$

Plugging in (8.30) and the basis expansion for $u(x)$, we get

$$A_{ij} = \sum_{k=1}^n u_k \int_0^1 \xi_j(x)\xi_k'(x)\xi_i'(x)\,dx.$$

Clearly, $A_{ij} = 0$ if $|i-j| > 1$; the corresponding matrix A is tridiagonal. The problem is once again to minimize $(F_n(\lambda))^2$, as given in (8.19). And again, a number of techniques can be used. (Note, however, that if we impose the stationarity condition $\frac{d(F_n(\lambda))^2}{d\lambda_k} = 0$, we obtain the linear system presented by Vogel in [165].)

As an example, we set $f(x) = \delta\left(x - \frac{3}{10}\right)$ and pick n such that $\frac{3}{10}$ is a partition point. We choose $\kappa_{true}(x)$, solve the boundary value problem, sample the data at the x_i, and possibly add noise to produce our observed data. The goal is now to recover $\kappa_{true}(x)$.

Table 8.3 presents some results in the case where we seek a polynomial representation of $\kappa_{collage}(x)$. Figure 8.1 shows the exact solution and the data points in the cases we consider.

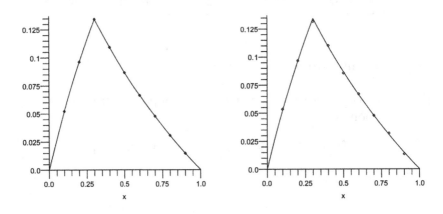

Fig. 8.1: The solution to our test problem and the nine data points used to generate the results of Table 8.3. Right to left): no noise added and Gaussian noise with amplitude 0.001 added.

On the other hand, we might seek a representation of $\kappa_{collage}(x)$ in terms of the extended ξ_i basis. When we choose $\kappa_{true}(x) = 1$ and $n = 9$, we determine that $\kappa_{collage}(x) = 1$. Figure 8.2 shows the results when we choose $\kappa_{true}(x) = 1 + 2x$ and work with $r = 1$.

In this case, the number of data points must be increased to achieve very good results. The results are similarly good when very low-amplitude noise is added to the observed data values. However, even with low amplitude noise, when the visual agreement between the data and true solution seems strong and the $\mathcal{L}^2$ error between the two seems small, the algorithm may encounter difficulty for low values of n.

As a final experiment, we consider the case where

$$\kappa_{true}(x) = \begin{cases} 1, & x \le \frac{1}{3}, \\ 2, & x > \frac{1}{3}, \end{cases}$$

Table 8.3: Collage coding results when $f(x) = \delta\left(x - \frac{3}{10}\right)$, $\kappa_{true}(x) = 1 + 2x$, data points $= 10$, and number of basis functions $= 10$. The form of $\kappa_{collage}$ indicates which method of discussion (iii) has been used.

noise ε	space	$\kappa_{collage}$
0.00	V_9^1	$0.996829 + 2.00363x$
0.00	V_9^1	$0.99669 + 2.00572x - 0.006617x^2 + 0.00266x^3$
0.001	V_9^1	$1.20541 + 1.61207x$
0.001	V_9^1	$0.959467 + 5.70939x - 12.9070x^2 + 8.08981x^3$

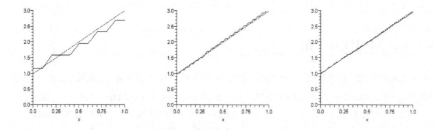

Fig. 8.2: $\kappa_{true}(x) = 1 + 2x$ (dashed) and $\kappa_{collage}(x)$ (solid) for different cases: (left to right) $n = 9$, $n = 39$, and $n = 69$. A point source is placed at $x_s = \frac{3}{10}$, and no noise is added to the observed data.

simulating a scenario where two rods with different constant diffusivities are melded at $x = \frac{1}{3}$. We solve several instances of our boundary value problem, each with a single point source at a distinct location, and then produce observational data by sampling the solution and possibly adding low-amplitude Gaussian noise. As one can see in Fig. 8.3, collage coding with 40 uniformly placed unnoised data values taken from the solution with a point source at $x_s = \frac{3}{10}$ produces a $\kappa_{collage}(x)$ in fine agreement with our $\kappa_{true}(x)$. However, when low-amplitude noise is added to data points, the result worsens quite dramatically, as we see in the rightmost plot of the figure.

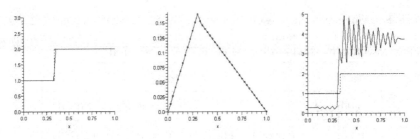

Fig. 8.3: $\kappa_{true}(x)$ (dashed) and $\kappa_{collage}(x)$ (solid) for 40 data points with no noise added (left) and low amplitude Gaussian noise added (right). In each case, the data were generated from the solution with a single point source at $x_s = \frac{3}{10}$. The center plot shows the solution and the 40 noisy data points. The $\mathcal{L}^2$ error is 0.0006.

8.1.1.6 Multiple data sets

It is most likely that several experiments would be performed in order to determine the conductivity $\kappa(x)$ – for example, measuring the steady-state responses to point sources located at several positions x_i, $i = 1, 2, \cdots, M$, on the rod. These results could be combined into one determination by considering the minimization of a linear combination of squared collage errors of the form (8.29),

$$(G_n(\lambda))^2 = \sum_{k=1}^{M} \mu_k (F_n^{(k)}(\lambda))^2$$

$$= \sum_{k=1}^{M} \mu_k \left(\sum_{i=1}^{n} \left[\sum_{j=0}^{N} \lambda_j A_{ij}^{(k)} - b_i^{(k)} \right]^2 \right), \qquad (8.31)$$

where the kth set of elements $A_{ij}^{(k)}$ and $b_i^{(k)}$, $k = 1, 2, \cdots, M$, is obtained from the response to the source at x_k. The nonnegative μ_k are weighting factors.

In the almost certain case of noisy data sets, it is possible that such a combination of data may improve the estimates of $\kappa(x)$ obtained by minimizing (8.31). This is indeed demonstrated by the plots in Fig. 8.4, where a number of equally noisy sets of observed data have been combined. (Here $\mu_k = 1$.)

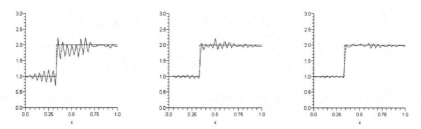

Fig. 8.4: $\kappa_{true}(x)$ (dashed) and $\kappa_{collage}(x)$ (solid) using multiple noised data sets, each produced from the solution with one point source. Point sources are placed at: $\frac{3}{10}$ and $\frac{7}{10}$ (left); the tenths, excluding 0 and 1 (center); and the twentieths, excluding 0 and 1 (right).

8.1.1.7 Two-dimensional steady-state diffusion

We extend our work to an inverse problem for the two-dimensional steady-state diffusion equation. With $D = \{0 < x, y < 1\}$,

$$-\nabla \cdot (\kappa(x,y)\nabla u(x,y)) + q(x,y)u(x,y) = f(x,y), \ (x,y) \in D, \quad (8.32)$$

$$u(x,y) = 0, \ (x,y) \in \partial D, \quad (8.33)$$

where the diffusivity $\kappa(x,y)$ and radiativity $q(x,y)$ vary in both x and y. Given $u(x,y)$, $q(x,y)$, and $f(x,y)$ on $[0,1]^2$, we wish to find an approximation of $\kappa(x,y)$. We multiply (8.32) by a test function $\xi_{ij}(x,y) \in H_0^1([0,1]^2)$ and then integrate over D to get, suppressing the dependence on x and y,

$$\iint_D f\xi_{ij}\, dA = -\iint_D \nabla \cdot (\kappa\nabla u)\, \xi_{ij}\, dA + \iint_D qu\xi_{ij}\, dA$$

$$= -\iint_D (\nabla\kappa \cdot \nabla u)\xi_{ij},\, dA - \iint_D \kappa\xi_{ij}\nabla^2 u,\, dA$$

$$+ \iint_D qu\xi_{ij}\, dA. \quad (8.34)$$

Upon application of Green's first identity, with $\hat{n}$ denoting the outward unit normal to ∂D, (8.34) becomes

$$\iint_D f\xi_{ij}\,dA = -\iint_D (\nabla\kappa\cdot\nabla u)\xi_{ij}\,dA - \left(\int_{\partial D}\kappa\xi_{ij}(\nabla u\cdot\hat{n})\,ds\right.$$

$$\left. -\iint_D \nabla(\kappa\xi_{ij})\cdot\nabla u\,dA\right) + \iint_D qu\xi_{ij}\,dA$$

$$= \iint_D \kappa\nabla\xi_{ij}\cdot\nabla u\,dA - \int_{\partial D}\kappa\xi_{ij}(\nabla u\cdot\hat{n})\,ds$$

$$+ \iint_D qu\xi_{ij}\,dA. \tag{8.35}$$

Equation (8.35) can be written as $a(u,\xi_{ij}) = \phi(\xi_{ij})$, with

$$a(u,\xi_{ij}) = \iint_D \kappa\nabla\xi_{ij}\cdot\nabla u\,dA - \int_{\partial D}\kappa\xi_{ij}(\nabla u\cdot\hat{n})\,ds$$

$$+ \iint_D qu\xi_{ij}\,dA, \tag{8.36}$$

$$\phi(\xi_{ij}) = \iint_D f\xi_{ij}\,dA. \tag{8.37}$$

For N and M fixed natural numbers, we define $h_x = \frac{1}{N}$ and $h_y = \frac{1}{M}$, as well as the $(N+1)(M+1)$ nodes in $[0,1]^2$:

$$(x_i, y_j) = (ih_x, jh_y), \ 0 \le i \le N, \ 0 \le j \le M.$$

The corresponding finite-element basis functions $\xi_{ij}(x,y)$ are pyramids with hexagonal bases, such that $\xi_{ij}(x_i, y_j) = 1$ and $\xi_{ij}(x_k, y_l) = 0$ for $k \ne i$, $l \ne j$. If i or j is 0, the basis function restricted to D is only a portion of such a pyramid.

Now, if we expand $\kappa(x,y)$ in this basis, writing

$$\kappa(x,y) = \sum_{k=0}^{N}\sum_{l=0}^{M}\lambda_{kl}\xi_{kl}(x,y),$$

then

$$a(u,\xi_{ij}) = \sum_{k=0}^{N}\sum_{l=0}^{M}\left[\lambda_{kl}\left(\iint_D \xi_{kl}\nabla\xi_{ij}\cdot\nabla u\,dA\right.\right.$$

$$\left.\left. -\int_{\partial D}\xi_{kl}\xi_{ij}(\nabla u\cdot\hat{n})\,ds\right) + \iint_D qu\xi_{ij}\,dA\right].$$

Defining

$$A_{klij} = \iint_D \xi_{kl}\nabla\xi_{ij}\cdot\nabla u\,dA - \int_{\partial D}\xi_{kl}\xi_{ij}(\nabla u\cdot\hat{n})\,ds$$

and

$$b_{ij} = \iint_D f\xi_{ij}\, dA - \iint_D qu\xi_{ij}\, dA$$

means that we must minimize

$$(F_{NM}(\lambda))^2 = \sum_{i=0}^{N}\sum_{j=0}^{M}\left[\sum_{k=0}^{N}\sum_{l=0}^{M}\lambda_{kl}A_{klij} - b_{ij}\right]^2. \tag{8.38}$$

Example 8.7. We set $u(x,y) = \sin(\pi x)\sin(\pi y)$, $q(x,y) = 0$ and use the function $\kappa(x,y) = 1 + 6x^2y(1-y)$ to determine $f(x,y)$ via (8.32). Now, given the functions $u(x,y)$, $f(x,y)$, and $q(x,y)$, we seek to approximate $\kappa(x,y)$. This inverse problem is treated as Example 3 in [85] using a modified Uzawa algorithm. We plug our known functions into (8.38) and find the minimizing values of λ_{kl} using Maple's quadratic program solver. In Fig. 8.5, we present graphs of our actual $\kappa(x,y)$ as well as the results obtained by minimizing (8.38) with $N = M = 3, 4, 5$.

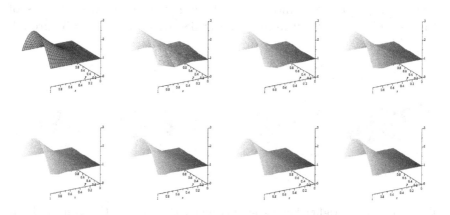

Fig. 8.5: (left to right, top to bottom) For two-dimensional Example 1, the graphs of our actual $\kappa(x,y)$ and the collage-coded approximations of κ with $N = M = 3$ through $N = M = 9$.

Next, we perturb the target function $u(x,y)$, leaving $f(x,y)$ and $q(x,y)$ exact. Table 8.4 presents the $\mathcal{L}^2$ error $\|u - u_{noisy}\|$ between the true solution u and the noised target u_{noisy} and the resulting error $\|\kappa - \kappa_{collage}\|$ between the true κ and the collage-coded approximation $\kappa_{collage}$ for numerous cases of N and M. Note that $\|\kappa\|_2 = 1.38082$ and $\|u\|_2 = 0.5$.

Table 8.4: Numerical results for the inverse problem with different levels of noise.

| | | $\|\kappa - \kappa_{collage}\|$ | |
| | | | |
$N = M$	$u_{noisy} = u$	$\|u - u_{noisy}\| = 0.025$	$\|u - u_{noisy}\| = 0.05$
3	0.06306	0.09993	0.17050
4	0.03480	0.07924	0.15561
5	0.02246	0.07275	0.15128
6	0.01564	0.07118	0.15065
7	0.01160	0.07051	0.15039
8	0.00902	0.07008	0.15014
9	0.00733	0.06981	0.14996

Example 2. We set $u(x, y) = \sin(\pi x)\sin(\pi y)$, $q(x, y) = 4 + \cos(\pi xy)$, and $\kappa(x, y) = (1 + x^2 + xy)/1000$. With these choices, we determine the function $f(x, y)$. The inverse problem is to estimate $\kappa(x, y)$ when given $u(x, y)$, $f(x, y)$, and $q(x, y)$. In Fig. 8.6, we present graphs of the results obtained by minimizing (8.38) with $N = M = 3$ through $N = M = 5$.

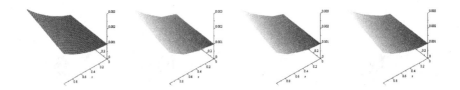

Fig. 8.6: (left to right) For two-dimensional Example 2, the graphs of our actual $\kappa(x, y)$ and the collage-coded approximations of κ with $N = M = 3, 4, 5$.

8.1.2 Parabolic PDEs

Suppose that we have a given Hilbert space H, and let us consider the abstract formulation of a parabolic equation,

$$\begin{cases} \langle \frac{d}{dt}u, v \rangle = \psi(v) + a(u, v), \\ u(0) = f, \end{cases} \tag{8.39}$$

where $\psi : H \to \mathbb{R}$ is a linear functional, $a : H \times H \to \mathbb{R}$ is a bilinear form, and $f \in H$ is an initial condition. The aim of the inverse problem for the system of equations above consists of getting an approximation of the coefficients and parameters starting from a sample of observations of a target $u \in H$. To do this, let us consider a family of bilinear functionals a_λ and let u_λ be the solution to

$$\begin{cases} \langle \frac{d}{dt}u_\lambda, v \rangle = \psi(v) + a_\lambda(u_\lambda, v), \\ u_0 = f. \end{cases} \tag{8.40}$$

We would like to determine if there exists a value of the parameter λ such that $u_\lambda = u$ or, more realistically, such that $\|u_\lambda - u\|$ is small enough. To this end, Theorem 8.8 states that the distance between the target solution u and the solution u_λ of (8.40) can be reduced by minimizing a functional that depends on parameters.

Theorem 8.8. *Let $u : [0, T] \to L^2(D)$ be the target solution that satisfies the initial condition in (8.39), and suppose that $\frac{d}{dt}u$ exists and belongs to H. Suppose that $a_\lambda(u, v) : \Lambda \times H \times H \to \mathbb{R}$ is a family of bilinear forms for all $\lambda \in \Lambda$. We have the result*

$$\int_0^T \|u - u_\lambda\|_H dt \le \frac{1}{m_\lambda^2} \int_0^T \left(\sup_{\|v\|=1} \left\langle \frac{d}{dt}u, v \right\rangle - \psi(v) - a_\lambda(u, v) \right)^2 dt,$$

where u_λ is the solution of (8.40) s.t. $u_\lambda(0) = u(0)$ and $u_\lambda(T) = u(T)$.

Proof. Computing, we have

$$m_\lambda \|u - (u_\lambda)\|_H^2 \le a(u - u_\lambda, u - u_\lambda)$$
$$= a(u, u - u_\lambda) - \left\langle \frac{d}{dt}(u_\lambda - u), u - u_\lambda \right\rangle$$
$$+ \psi(u - u_\lambda) - \left\langle \frac{d}{dt}u, u - u_\lambda \right\rangle$$

and, by simple calculations, we get

$$m_\lambda \|u - u_\lambda\|_H^2 - \frac{1}{2}\frac{d}{dt}\|u - u_\lambda\|_H^2$$
$$\le a(u, u - u_\lambda) + \psi(u - u_\lambda) - \left\langle \frac{d}{dt}u, u - u_\lambda \right\rangle.$$

Integrating both sides with respect to t and recalling that $u(0) = u_\lambda(0)$ and $u(T) = u_\lambda(T)$, we have

$$m_\lambda \int_0^T \|u - u_\lambda\|_H^2 \, dt$$

$$\leq \int_0^T \|u - u_\lambda\|_H \left\{ \sup_{\|v\|=1} a(u, v) + \psi(v) - \left\langle \frac{d}{dt} u, v \right\rangle \right\} dt$$

$$\leq \left(\int_0^T \|u - u_\lambda\|_H^2 dt \right)^{\frac{1}{2}} \left(\int_0^T \left(\sup_{\|v\|=1} a(u, v) + \psi(v) - \left\langle \frac{d}{dt} u, v \right\rangle \right)^2 dt \right)^{\frac{1}{2}},$$

and now the result follows.   □

Whenever $\inf_{\lambda \in \Lambda} m_\lambda \geq m > 0$, then the previous result states that in order to solve the inverse problem for the parabolic equation (8.39) one can minimize the functional

$$\int_0^T \left(\sup_{\|v\|=1} \langle u, v \rangle - \psi(v) - a_\lambda(u, v) \right)^2 dt \qquad (8.41)$$

over all $\lambda \in \Lambda$.

Example 8.9. Let us consider the equation

$$u_t = (k(x)u_x)_x + g(x, t), \ 0 < x < 1,$$
$$u_0 = 0,$$
$$u_1 = 0,$$

where $g(x, t) = tx(1 - x)$, subject to $u(x, 0) = 10\sin(\pi x)$ and $u(0, t) = u(1, t) = 0$. We set $k(x) = 1 + 3x + 2x^2$, solve for $u(x, t)$, and sample the solution at N^2 uniformly positioned grid points for $(x, t) \in [0, 1]^2$ to generate a collection of targets. Given these data and $g(x, t)$, we then seek an estimation of $k(x)$ in the form $k(x) = k_0 + k_1 x + k_2 x^2$. The results we obtain through the generalized collage method are summarized in Table 8.5.

8.1.3 Hyperbolic PDEs

Let us now consider the weakly formulated hyperbolic equation

Table 8.5: Collage coding results for the parabolic equation in Example 8.9.

ϵ	N	k_0	k_1	k_2
0	10	0.87168	2.90700	0.21353
0	20	0.93457	2.97239	1.49201
0	30	0.94479	2.98304	1.76421
0	40	0.94347	2.97346	1.85572
0.01	10	0.87573	2.82810	0.33923
0.01	20	0.92931	2.91536	1.32864
0.01	30	0.92895	2.84553	0.59199
0.10	10	0.90537	1.97162	0.59043
0.10	20	0.77752	0.92051	-0.77746
0.10	30	0.60504	-0.12677	-0.14565

$$\begin{cases} \left\langle \frac{d^2}{dt^2}u, v \right\rangle = \psi(v) + a(u,v), \\ u(0) = f, \\ \frac{d}{dt}u(0) = g, \end{cases} \tag{8.42}$$

where $\psi : H \to \mathbb{R}$ is a linear functional, $a : H \times H \to \mathbb{R}$ is a bilinear form, and $f, g \in H$ are the initial conditions. As in previous sections, the aim of the inverse problem for the system of equations above, consists of reconstructing the coefficients starting from a sample of observations of a target $u \in H$. We consider a family of bilinear functionals a_λ and let u_λ be the solution to

$$\begin{cases} \left\langle \frac{d}{dt}u_\lambda, v \right\rangle = \psi(v) + a_\lambda(u_\lambda, v), \\ u_0 = f, \\ \frac{d}{dt}u(0) = g, \end{cases} \tag{8.43}$$

We would like to determine if there exists a value of the parameter λ such that $u_\lambda = u$ or, more realistically, such that $\|u_\lambda - u\|$ is small enough. Theorem 8.10 states that the distance between the target solution u and the solution u_λ of (8.43) can be reduced by minimizing a functional that depends on parameters.

Theorem 8.10. *Let* $u : [0, T] \to L^2(D)$ *be the target solution that satisfies the initial condition in (8.42), and suppose that* $\frac{d^2}{dt^2}u$ *exists and belongs to* H. *Suppose that there exists a family of* $m_\lambda > 0$ *such that* $a_\lambda(v, v) \geq m_\lambda \|v\|^2$ *for all* $v \in H$. *We have the result*

$$\int_0^T \|u - u_\lambda\|^2 dt \leq \frac{1}{m_\lambda^2} \int_0^T \left(\sup_{\|v\|=1} \left\langle \frac{d^2}{dt^2}u, v \right\rangle - \psi(v) - a(u,v) \right)^2 dt,$$

where u_λ is the solution of (8.43) s.t. $u(0) = u_\lambda(0)$ and $u(T) = u_\lambda(T)$.

The proof of the theorem follows the same path as that of Theorem 8.8.

Proof. Computing, we have

$$m_\lambda \|u - u_\lambda\|^2 \le a(u - u_\lambda, u - u_\lambda)$$

$$= a(u, u - u_\lambda) - \left\langle \frac{d^2}{dt^2}(u - u_\lambda), u - u_\lambda \right\rangle$$

$$+ \psi(u - u_\lambda) - \left\langle \frac{d^2}{dt^2}u, u - u_\lambda \right\rangle.$$

Continuing, we get

$$m_\lambda \|u - u_\lambda\|^2 - \left\langle \frac{d^2}{dt^2}(u_\lambda - u), u - u_\lambda \right\rangle$$

$$\le a(u, u - u_\lambda) + \psi(u - u_\lambda) - \left\langle \frac{d^2}{dt^2}u, u - u_\lambda \right\rangle$$

$$\le \|u - u_\lambda\| \left(\sup_{\|v\|=1} a(u, v) + \psi(v) - \left\langle \frac{d^2}{dt^2}u, v \right\rangle \right).$$

Integrating both sides with respect to t, using integration by parts, and recalling that $u(0) = u_\lambda(0)$ and $u(T) = u_\lambda(T)$, we have

$$m_\lambda \int_0^T \|u - u_\lambda\|^2 ds + \int_0^T \left\| \frac{d}{ds}(u - u_\lambda) \right\|^2 ds$$

$$\le \left(\int_0^T \|u - u_\lambda\|^2 \, dt \right)^{\frac{1}{2}} \left(\int_0^T \left(\sup_{\|v\|=1} a(u, v) + \psi(v) - \left\langle \frac{d^2}{dt^2}u, v \right\rangle \right)^2 dt \right)^{\frac{1}{2}}$$

and the result follows.   □

Example 8.11. We adjust Example 8.9, considering

$$u_{tt} - (k(x)u_x)_x = g(x, t), \tag{8.44}$$

where $g(x, t) = tx(1 - x)$, subject to $u(x, 0) = \sin(\pi x)$ and $u_t(x, 0) = 0$ and $u(0, t) = u(1, t) = 0$. We set $k(x) = 1 + 3x + 2x^2$ and construct target data as in Example 8.9 and then seek to recover $k(x) = k_0 + k_1 x + k_2 x^2$ given these data and $g(x, t)$. The results we obtain from the generalized collage method are summarized in Table 8.6.

Table 8.6: Collage coding results for the hyperbolic equation in Example 8.11

ϵ	N	k_0	k_1	k_2
0	10	0.87168	2.90700	0.21353
0	20	0.93457	2.97239	1.49201
0	30	0.94479	2.98304	1.76421
0	40	0.94347	2.97346	1.85572
0.01	10	0.87573	2.82810	0.33923
0.01	20	0.92931	2.91536	1.32864
0.01	30	0.92895	2.84553	0.59199
0.10	10	0.90537	1.97162	0.59043
0.10	20	0.77752	0.92051	-0.77746
0.10	30	0.60504	-0.12677	-0.14565

8.1.4 An application: A vibrating string driven by a stochastic process

Before stating and solving two inverse problems, we begin by giving the details of and motivations for the specific model we are interested in studying. We consider the following system of coupled differential equations. The first one is a stochastic differential equation and the second one is a hyperbolic partial differential equation. On a domain $D \subset \mathbb{R}^d$, we have the equations

$$\begin{cases} dX_t = \left[\int_D g(u(t,y))d\mu_t(y) \right] X_t dt + X_t dB_t, \\ X_{t=0} = X_0, \end{cases} \qquad (8.45)$$

$$\begin{cases} \frac{d^2}{dt^2}u(t,y) + \nabla_y(\kappa_1(y)\nabla_y u(t,y)) = \kappa_2(y)\delta_{X_t}(y), \\ \qquad\qquad\qquad\qquad\qquad (t,y) \in [0,T] \times D, \\ u(0,y) = \phi_1(y), \\ \frac{\partial u}{\partial n}(t,y) = \phi_2(t,y), \qquad\qquad (t,y) \in [0,T] \times \partial D, \end{cases} \qquad (8.46)$$

where μ_t is the law of X_t and δ_{X_t} is the Dirac delta "function" at the point X_t.

For instance, imagine we have a flexible string directed along the x-axis, with the string kept stretched by a constant horizontal tension and forced to vibrate perpendicularly to the x-axis under random force $F(x,t)$. If u is the displacement of a point x at time t, it is well-known that u satisfies the equation

$$\frac{\partial}{\partial t^2}u - \frac{\partial}{\partial x^2}u = F(x,t), \qquad (8.47)$$

where $x \in D$ and $t > 0$. In our model, we suppose that $F(y, t) = \kappa_2(y)\delta_{X_t}(y)$, where X_t is a stochastic process that is a solution of the stochastic differential equation (8.45). In other words, the hyperbolic equation has a forcing term that is driven by this stochastic process. The random vibration on an infinite string has received recent attention (see [33] and [138]). Figure 8.7 presents some snapshots of the displacement for the related finite string problem. In the following sections, we present a method for solving two different parameter identification problems for this system of coupled differential equations: one for κ_1 and one for g. Both of these methods are based on the numerical schemes that were presented in previous sections.

Before we begin the analysis, a few words about (8.46) are in order since this equation contains the generalized function δ_{X_t}. That is, $\delta_{X_t}(y)$ has a meaning only when it is integrated with respect to a test function $\theta(y)$. Thus, the meaning of (8.46) is that for each $\theta \in H^1(D)$ we have

$$\int_D \theta(y) \left(\frac{d^2}{dt^2} u(t, y) + \nabla_y(\kappa_1(y)\nabla_y u(t, y)) \right) dy$$
$$= \int_D \theta(y)\kappa_2(y)\delta_{X_t}(y)dy = \theta(X_t)\kappa_2(X_t).$$

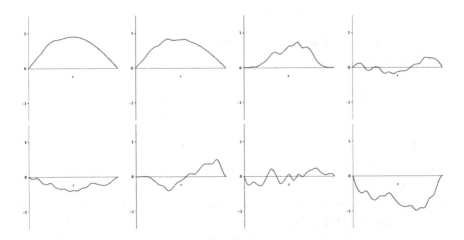

Fig. 8.7: Snapshots of a randomly forced vibrating string, with time increasing from left to right and top to bottom.

8.1.4.1 A parameter identification model for κ_1

For this parameter identification problem, we seek to estimate κ_1 given κ_2, g, and the observations of X_t. From the data, we recover the density f_{X_t} of the process X_t. Averaging (8.46) allows one to get a simpler model for the parameter identification problem (see [38]). This replaces the quantity δ_{X_t} by its expectation $\mathbb{E}(\delta_{X_t}(y))$. Since X_t is absolutely continuous then it is known that $\mathbb{E}(\delta_{X_t}(y)) = f_{X_t}(y)$, the density of the distribution of X_t, the previous model can be rewritten in an averaged form as

$$\begin{cases} dX_t = \left[\int_D g(\tilde{u}(t,y)) f_{X_t}(y) dy \right] X_t dt + X_t dB_t, \\ X_{t=0} = X_0, \end{cases} \tag{8.48}$$

coupled with the deterministic PDE

$$\begin{cases} \frac{d^2}{dt^2}\tilde{u}(t,y) + \nabla(\kappa_1(y)\nabla_y\tilde{u}(t,y)) = \kappa_2(y)f_{X_t}(y), \\ \qquad\qquad\qquad\qquad\qquad\qquad (t,y) \in [0,T] \times D, \\ \tilde{u}(0,y) = \phi_1(y), \\ \frac{\partial\tilde{u}}{\partial n}(t,y) = \phi_2(t,y), \qquad\qquad (t,y) \in [0,T] \times \partial D. \end{cases} \tag{8.49}$$

Note that the averaged equation (8.49) has solutions in the usual sense. For this particular case, we have that $\mathbb{E}(u(t,y)) = \tilde{u}(t,y)$. To see this, just notice that the operator $\frac{d^2}{dt^2} - \Delta_y$ is linear, so

$$\mathbb{E}\left(\frac{d^2}{dt^2}u + \nabla_y(\kappa_1(y)\nabla_y u)\right) = \frac{d^2}{dt^2}\mathbb{E}(u) + \nabla_y(\kappa_1(y)\nabla_y\mathbb{E}u).$$

Furthermore,

$$\mathbb{E}\left(\int_D \theta(y)\kappa_2(y)\delta_{X_t}(y)\,dy\right) = \int_D \theta(y)\kappa_2(y)f_{X_t}(y)\,dy$$

since κ_1 and κ_2 are deterministic functions. Thus, $\mathbb{E}(u)$ is the solution to the deterministic PDE

$$\begin{cases} \frac{d^2}{dt^2}\mathbb{E}(u(t,y)) - \kappa_1(y)\Delta_y\mathbb{E}(u(t,y)) = \kappa_2(y)f_{X_t}(y), (t,y) \in [0,T] \times D, \\ \mathbb{E}(u(0,y)) = \phi_1(y), \\ \frac{\partial\mathbb{E}(u)}{\partial n}(t,y) = \phi_2(t,y), \qquad\qquad\qquad (t,y) \in [0,T] \times \partial D. \end{cases}$$

However, clearly this is the same PDE for which $\tilde{u}$ is the solution. Thus we must have that $\mathbb{E}(u) = \tilde{u}$. This inverse problem can be solved using the techniques illustrated in section 8.1.3 on hyperbolic differential equations.

8.1.4.2 A parameter identification problem for g

For this parameter identification problem, we assume that κ_1 and κ_2 are known. The aim of the inverse problem consists of recovering the functional form of g and the density f_{X_t} of the process X_t starting from a sample of observations of the random process $u(t, y, \omega)$, say $(u(t, y, \omega_1), \ldots, u(t, y, \omega_n))$. Let $\tilde{u} = \mathbb{E}(u(t, y, \cdot))$. We do this in several steps. The first step is to use $u(t, y)$ and (8.46) to obtain an estimate of the distribution of X_t. Let us consider the averaged equation

$$\begin{cases} dX_t = \left[\int_D g(\tilde{u}(t,y)) f_{X_t}(y) dy \right] X_t dt + X_t dB_t \\ X_{t=0} = X_0, \end{cases} \tag{8.50}$$

coupled with the deterministic PDE

$$\begin{cases} \frac{d^2}{dt^2} \tilde{u}(t,y) - \kappa_1(y) \Delta_y \tilde{u}(t,y) = \kappa_2(y) f_{X_t}(y), \\ \hspace{4cm} (t,y) \in [0,T] \times D, \\ \tilde{u}(0,y) = \phi_1(y), \\ \frac{\partial \tilde{u}}{\partial n}(t,y) = \phi_2(t,y), \hspace{2cm} (t,y) \in [0,T] \times \partial D. \end{cases} \tag{8.51}$$

So now, putting $\tilde{u}$ in the previous hyperbolic equation and dividing by κ_2, we get f_{X_t}. Let us go back to the stochastic differential equation, and let us take the expectation of both sides, getting

$$\begin{cases} d\mathbb{E}(X_t) = \left[\int_D g(\tilde{u}(t,y)) f_{X_t}(y) dy \right] \mathbb{E}(X_t) dt, \\ \mathbb{E}(X_{t=0}) = \mathbb{E}(X_0). \end{cases} \tag{8.52}$$

We remember that the following relationship holds between the expectation of a random variable and its density:

$$\mathbb{E}(X_t) = \int_{\mathbb{R}^d} y f_{X_t}(y) dy. \tag{8.53}$$

Thus, we can obtain $\mathbb{E}(X_t)$, which now can be used for solving the inverse problem as we did in Sect. 8.1.4.1. We recover the function

$$\Phi_t = \int_D g(\tilde{u}(t,y)) f_{X_t}(y) dy. \tag{8.54}$$

The last step involves the analysis of (8.54). The only unknown in this model is g. Taking the L^2 expansions of Φ and g with respect to the same L^2 orthonormal basis $\{\phi_i\}$, we then get

$$\sum b_i \phi_i = \sum a_i \int_D \phi_i(\tilde{u}(t,y)) f_{X_t}(y) dy \tag{8.55}$$

and then

$$b_j = \sum a_i \int_{\mathbb{R}} \left[\int_D \phi(\tilde{u}(s,y)) f_{X_s}(y) dy \right] \phi_j(s) ds \qquad (8.56)$$

for $j = 0, \ldots,$ a linear system in a_i, the solution of which is our final step.

8.2 Self-similar objects in cone metric spaces

In the classical case of iterated function systems (IFSs), the existence of self-similar objects relies on a straightforward application of *Banach's theorem* on contractions. Fundamental ingredients of this theory are the use of complete metric spaces and the notion of contractivity, both of which strictly depend on the definition of distance. In the last several years, a great deal of attention has been devoted to the extension of the notion of metric space and, in particular, to the definition of cone metric space. In this context, the distance between two elements of the space is no longer a positive number but an element of a Banach space that has been equipped with an *ordering cone*. Many times it's a bit too simplistic to describe the distance between two objects using a positive number, as there are situations in which several aspects of the problem, that cannot be combined in a unique index have to be considered. Using a cone metric allows a better description of the complexity of the problem; of course, this implies that many results of classical theory of metric spaces need to be adapted. One relevant application of cone metric spaces can be found in the theory of image processing, in particular when studying the structural similarity of images. In this context, the difference between two images is calculated using several different criteria, which leads in a natural way to considering vector-valued distances.

8.2.1 Cone metric space

In the following, we will use $\mathbb{B}$ to denote a Banach space and $P \subset \mathbb{B}$ will be a *pointed cone* in $\mathbb{B}$. That is, P satisfies

1. $0 \in P$,
2. $\alpha, \beta \in \mathbb{R}$ with $\alpha, \beta \geq 0$ and $x, y \in P$ implies $\alpha x + \beta y \in P$,
3. $P \cap -P = \{0\}$.

The cone P induces an order in $\mathbb{B}$ in the usual way; that is, $x \le y$ if $y - x \in P$ or, said another way, there is some $p \in P$ such that $x + p = y$. The elements of P are said to be *positive* and the elements of the interior of P are *strictly positive*. We assume that P is closed and will also usually assume that $\text{int}(P) \ne \emptyset$. Notice that $p + \text{int}(P) \subseteq \text{int}(P)$ for every $p \in P$. We say that $x \ll y$ if $y - x \in \text{int}(P)$, so $0 \ll x$ means $x \in \text{int}(P)$. A *pointed wedge* satisfies properties 1 and 2, so every cone is a wedge but not conversely. A *cone metric* on a set $\mathbb{X}$ is a function $d : \mathbb{X} \times \mathbb{X} \to P$ such that

1. $d(x, y) = 0$ iff $x = y$,
2. $d(x, y) = d(y, x)$, and
3. $d(x, y) \le d(x, z) + d(z, y)$ for all $x, y, z \in \mathbb{X}$.

The idea for defining a cone metric seems to have first appeared in [152] and again, independently, in [115]. The recent interest in this concept was triggered by the paper [117], which also investigated fixed-point results in these spaces. Since [117] there have been a large number of other papers on cone metric spaces, extending various standard results and fixed-point theorems to cone metric spaces. As we show in the next section (and as others have also shown), the topology of a cone metric space is given by a regular metric and thus the real novelty of cone metric spaces is not in the convergence structure. For us, the novelty is more in using cone metric spaces as a framework for thinking about multiobjective optimization problems.

We say that a sequence (x_n) in $\mathbb{X}$ *converges to x in the cone metric* d if for any $c \in \text{int}(P)$ there is some $N \in \mathbb{N}$ such that for any $n \ge N$ we have $d(x_n, x) \ll c$. Notice the special use of points of the interior of P in this definition. However, by Proposition 8.12, it is enough to ask that $d(x_n, x) \le c$ rather than $d(x_n, x) \ll c$.

Proposition 8.12. *The sequence $x_n \to x$ in the cone metric d iff for every $c \in \text{int}(P)$ there is some N such that $n \ge N$ implies that $d(x_n, x) \le c$.*

Proof. One direction is obvious as $a \ll b$ implies $a \le b$.

For the other direction, we first notice that if $a \ll b \le c$ then $b - a \in \text{int}(P)$ and $c - b \in P$ so $c - a = (c - b) + (b - a) \in \text{int}(P)$ and thus $a \ll c$. So take $c \in \text{int}(P)$. Then we choose some $c' \in \text{int}(P)$ with $c' \ll c$ (simply select $c' \in B_\delta(c) \subset P$ such that $c - c' \in \text{int}(P)$). However, then by assumption there is some N such that $n \ge N$ implies $d(x_n, x) \le c' \ll c$ and thus $d(x_n, x) \ll c$. $\qquad\square$

8.2.2 Scalarizations of cone metrics

In this section, we investigate some properties of *scalarizations* of a cone metric. In particular, we show that each scalarization is a pseudometric (in the usual sense) and that the topology (notion of convergence) generated by a cone metric is equivalent to that of a standard metric. This vastly simplifies the considerations of topological questions in cone metric space. In particular, it also shows that, from the stand point of topology, the class of cone metric spaces is not new. However, this does not mean that there is no reason to consider cone metric spaces, as we discuss in Sect. 8.2.4.

Our results were discovered independently, but several others have established similar results [49, 4, 37, 7]. In particular, the results in [4] use the same approach as we take leading up to Corollary 8.21. The paper [49] has a very nice approach to the problem in a general Hausdorff locally convex topological vector space that uses the "nonlinear scalarization function" associated with the cone, defined as

$$\xi_e(y) = \inf\{r : y \in re - P\},$$

where $e \in \text{int}(P)$. This has the benefit of using the cone directly in the "scalarization" instead of indirectly (via the dual cone) as we do. The paper [37] can be viewed as a nice follow-up paper to [49].

Let $(\mathbb{X}, d)$ be a cone metric space that takes values in the cone $P \subset \mathbb{B}$. Recall that the *dual wedge* P^* is the set of all $p^* \in \mathbb{B}^*$ such that $p^*(q) \geq 0$ for all $q \in P$. Notice that by this definition P^* is always weak* closed as a subset of $\mathbb{B}^*$. It is possible for P^* not to be a cone, depending on P, but it is always a wedge. Let $S = \{p^* \in P^* : \|p^*\| = 1\}$ denote all those elements of P^* of norm 1. We note that S is a weak* compact *base* for P^* in that $P^* = \{tp^* : t \geq 0, p^* \in S\}$. For each $p^* \in S$, define d^p by

$$d^p(x, y) = p^*(d(x, y)).$$

Proposition 8.13. *For each $p^* \in S$, we have that d^p is a pseudometric on $\mathbb{X}$ and a metric on $\mathbb{X}$ if $p^* \in S \cap \text{int}(P^*)$.*

Proof. It is immediate from the definition that $d^p(x, y) \geq 0$ and $d^p(x, y) = d^p(y, x)$. Furthermore, we have that if $x, y, z \in \mathbb{X}$ then $d(x, y) \leq d(x, z) + d(y, z)$, which means that there is some $q \in P$ such that $d(x, y) + q = d(x, z) + d(z, y)$ so that $p^*(d(x, y)) + p^*(q) = p^*(d(x, z)) + p^*(d(z, y))$, but since $p^*(q) \geq 0$ this means that

$$p^*(d(x, y)) \leq p^*(d(x, z)) + p^*(d(z, y)) \quad \Rightarrow \quad d^p(x, y) \leq d^p(x, z) + d^p(z, y).$$

Thus d^p is a pseudometric for every $p^* \in S$. Note that the same argument works for any $p^* \in P^*$, but it is sufficient to consider $p^* \in S$ since $(\lambda p)^*(d(x,y)) = \lambda p^*(d(x,y))$ for any $\lambda \in \mathbb{R}^+$.

Take $p^* \in \text{int}(P^*)$ (assuming $\text{int}(P^*) \neq \emptyset$). Then we claim that $p^*(q) > 0$ for all nonzero $q \in P$. If not, then $p^*(q) = 0$ for some nonzero $q \in P$. However, then there is some $r^* \in \mathbb{B}^*$ with $r^*(q) = -1$ and $\|r^*\| = 1$. Now, as $p^* \in \text{int}(P^*)$, there is some $\delta > 0$, $\delta \in \mathbb{R}$ with $B_\delta(p^*) \subset P^*$. But then $p^* + \delta/2\, r^* \in P^*$. However, $(p^* + \delta/2\, r^*)(q) = -\delta/2$, which is a contradiction. Thus for all $p^* \in \text{int}(P^*)$ we have $p^*(q) > 0$ for all nonzero $q \in P$.   □

Example 8.14. On $\mathbb{R}^n$, the most natural example of a cone metric is given by

$$d(x,y) = (|x_1 - y_1|, |x_2 - y_2|, \ldots, |x_n - y_n|),$$

so the cone P is the positive orthant in $\mathbb{R}^n$. In this case, the dual wedge P^* is naturally identified with P. It might be natural to think of d^p as the distance between points in the direction p^*, but this is not correct. One way to see this is that if it were correct, then it would be impossible for d^p to be a metric for any p, as any two points whose projections onto the line $\mathbb{R}p^* = \{rp^* : r \in \mathbb{R}\}$ were the same would have zero d^p distance. In fact, for $p^* \in P^*$, we have $p^* = (p_1, p_2, \ldots, p_n)$ with $p_i \geq 0$ and thus

$$d^p(x,y) = p_1|x_1 - y_1| + p_2|x_2 - y_2| + \cdots + p_n|x_n - y_n|. \qquad (8.57)$$

Thus d^p is more correctly thought of as a weighted ℓ^1 norm. Notice that if all the weights are strictly positive (that is, each $p_i > 0$), then (8.57) defines a metric.

Lemma 8.15. *Suppose that $q \notin P$. Then there is some $p^* \in P^*$ such that $p^*(q) < 0$.*

Proof. Since $q \notin P$, by the Hahn-Banach theorem there is some $r^* \in \mathbb{B}^*$ and $c \in \mathbb{R}$ with $r^*(q) < c$ and $r^*(p) \geq c$ for all $p \in P$. Now, $0 \in P$, which implies that $0 \geq c$, so in fact $r^* \in P^*$.   □

Notice that Lemma 8.15 implies that if $p^*(q) \geq 0$ for all $p^* \in P^*$, then we must have $q \in P$. This is a dual to the fact that $p^*(q) \geq 0$ for all $q \in P$ implies that $p^* \in P^*$.

Lemma 8.16. *If $x, y \in P$ with $p^*(x) \leq p^*(y)$ for all $p^* \in P^*$, then $x \leq y$.*

Proof. Let $c = y - x$. Then $p^*(c) \geq 0$ for all $p^* \in P^*$, so in fact $c \in P$ and so $x + c = y$ or $x \leq y$. $\qquad\square$

Lemma 8.17. *If $p^* \in P^*$ and $\epsilon > 0$ is such that $B_\epsilon(p^*) \subset P^*$, then $p^*(q) \geq \epsilon$ for all $q \in P$.*

Lemma 8.18. *Suppose that $c \in \mathrm{int}(P)$. Then for any $p \in P$ there is some $\epsilon > 0$ such that $\epsilon p \leq c$. In fact, we can even arrange that $\epsilon p \ll c$.*

Proof. Since $c \in \mathrm{int}(P)$, there is some $\epsilon > 0$ such that $B_{2\epsilon}(c) \subset P$. But then $c - \epsilon p \in B_{2\epsilon}(c) \subset P$, and thus there is some $p' \in P$ with $\epsilon p + p' = c$ or $\epsilon p \leq c$. In fact, $\epsilon p \ll c$. $\qquad\square$

Definition 8.19. We say that $x_n \to x$ in d^p *uniformly over* $p \in S$ if for all $\epsilon > 0$ there is some $N \in \mathbb{N}$ with $n \geq N$ implying $d^p(x_n, x) < \epsilon$ for all $p \in S$.

Proposition 8.20. *The sequence $x_n \to x$ in the cone metric d iff $x_n \to x$ in d^p uniformly in p.*

Proof. Let $\epsilon > 0$ be chosen. Choose $c \in \mathrm{int}(P)$ with $\|c\| < \epsilon$. Then there is some N such that for all $n \geq N$ we have $d(x_n, x) \leq c$, which implies that $d^p(x_n, x) \leq p^*(c) \leq \|c\| < \epsilon$ for all $p \in S$.

Conversely, take $c \in \mathrm{int}(P)$. We know that $p^*(c) > 0$ for all $p^* \in P^*$ (by the same argument as in the proof of Proposition 8.13) in fact $\lambda = \inf_{p^* \in S} p^*(c) > 0$ since S is weak* compact. By the assumption of uniformity, there is some $N \in \mathbb{N}$ such that $n \geq N$ implies that $d^p(x_n, x) < \lambda/2 < p^*(c)$ for all $p \in S$, which implies that $d(x_n, x) \leq c$ by Lemma 8.16. $\qquad\square$

If we define $\rho(x, y) = \sup_{p \in S} d^p(x, y)$, then we have that ρ is a metric and convergence in ρ is equivalent to convergence in the cone metric d. Notice that for $p \in P$ we have $\sup_{p^* \in S} p^*(p) \leq \|p\|$ and thus

$$\rho(x, y) \leq \|d(x, y)\|.$$

In general, the inequality can be strict. While it is certainly true that $\sup_{\|q^*\|=1} q^*(p) = \|p\|$, it might not be the case that the maximizing q^* is in P. The metric ρ is the same metric as defined in [4].

Corollary 8.21. *For a sequence (x_n) in $(\mathbb{X}, d)$, we have $x_n \to x$ in the cone metric d iff $x_n \to x$ in the metric ρ.*

The notion of a Cauchy sequence is defined in the obvious way in $(\mathbb{X}, d)$. It is easy to see that if (x_n) is Cauchy in $(\mathbb{X}, d)$, then for all $p^* \in S$ we have that (x_n) is also Cauchy in the pseudometric d^p. We say that $(\mathbb{X}, d)$ is *complete* if every d-Cauchy sequence converges in the cone metric d.

Proposition 8.22. *Let $p \in \text{int}(P)$. Then $x_n \to x$ in the cone metric d iff for all $\epsilon > 0$ there is some $N \in \mathbb{N}$ such that $n \geq N$ implies that $d(x_n, x) \leq \epsilon p$.*

Proof. As $p \in \text{int}(P)$ and $\epsilon > 0$, we know that $\epsilon p \in \text{int}(P)$ as well. So, if $x_n \to x$ in d and $\epsilon > 0$ is given, then there must be some N such that $n \geq N$ implies that $d(x_n, x) \leq \epsilon p$. For the converse, let $c \in \text{int}(P)$ be given. Then by Lemma 8.18 there is some $\epsilon > 0$ such that $\epsilon p \leq c$. Now, by the assumption there is some $N \in \mathbb{N}$ such that $n \geq N$ implies that $d(x_n, x) \leq \epsilon p \leq c$ and thus $x_n \to x$ in d.   □

Definition 8.23. We say that the cone P satisfies *property* **B** if there is some $p^* \in P^*$ such that the set $(p^*)^{-1}([0, 1]) \cap P$ is a norm-bounded subset of P.

Notice that if property **B** is satisfied, then $(p^*)^{-1}([0, \lambda]) \cap P$ is norm bounded for any $\lambda \geq 0$ by the linearity of p^* and the fact that P is a cone. That is,

$$\lambda(p^*)^{-1}([0, 1]) = (p^*)^{-1}([0, \lambda])$$

for any $\lambda \geq 0$. The "ice cream cones" (see [3]) all satisfy property **B**, as do any finite-dimensional closed pointed cones.

Lemma 8.24. *Suppose that property* **B** *is satisfied for P and $c \in \text{int}(P)$. Then there is some $\epsilon > 0$ such that for all $q \in P$ with $p^*(q) < \epsilon$ we have $q \leq c$.*

Proof. Since $c \in \text{int}(P)$, there is some $\eta > 0$ such that $B_{2\eta}(c) \subset P$. By property **B**, the set $(p^*)^{-1}([0, 1])$ is bounded, say by $M > 0$. Then $(p^*)^{-1}([0, \epsilon])$ is bounded by ϵM. Choose ϵ such that $\epsilon M < \eta$. Then any $q \in (p^*)^{-1}([0, \epsilon])$ satisfies $\|q\| < s\eta$ and so $c - q \in P$ so $q \leq c$.   □

Proposition 8.25. *Suppose that P satisfies property* **B**. *Then $x_n \to x$ in the cone metric d iff $d^p(x_n, x) \to 0$.*

Proof. If $x_n \to x$ in the cone metric d, then $d^p(x_n, x) \to 0$, as seen above.

Conversely, suppose that $d^p(x_n, x) \to 0$. That is, for any $\epsilon > 0$ there is some $N \in \mathbb{N}$ such that $n \geq N$ implies that $p^*(d(x_n, x)) \leq \epsilon$. Let $c \in \text{int}(P)$ be given. Then, by Lemma 8.24, there is some $\epsilon > 0$ such that all $q \in P$ with $p^*(q) < \epsilon$ satisfy $q \leq c$. However, then by assumption there is some $N \in \mathbb{N}$ such that $n \geq N$ implies that $p^*(d(x_n, x)) \leq \epsilon$, which implies that $d(x_n, x) \leq c$. Thus $x_n \to x$ in the cone metric d.   □

What this means is that under the condition **B** on the cone P, convergence in the cone metric reduces to convergence in the particular metric d^p, so the cone metric topology is just a usual metric topology as given by d^p. This is a little different from the situation in Corollary 8.21, as there the metric is a bit more complicated, involving the supremum of d^p for all $p \in S$, while Proposition 8.25 involves only one (albeit special) d^p.

We mention that condition **B** is similar to the conditions discussed in the paper [157], in particular property II, which requires the set $(p^*)^{-1}([0,1]) \cap P$ to be relatively weakly compact. If the Banach space $\mathbb{B}$ is reflexive, then norm-bounded subsets are all relatively weakly compact. In fact, $\mathbb{B}$ is reflexive if and only if the unit ball of $\mathbb{B}$ is weakly compact. Thus, for reflexive spaces, condition **B** and property II are equivalent.

8.2.2.1 Completeness and contractivity

By arguments similar to those in the proofs of Corollary 8.21 and Proposition 8.25, we get the following result.

Proposition 8.26. *A sequence is Cauchy in* $(\mathbb{X}, d)$ *iff it is Cauchy in* $(\mathbb{X}, \rho)$. *Furthermore, if P satisfies property* **B**, *then (x_n) is Cauchy in* $(\mathbb{X}, d)$ *iff it is Cauchy in* $(\mathbb{X}, d^p)$. *Thus,* $(\mathbb{X}, d)$ *is complete iff* $(\mathbb{X}, \rho)$ *is complete, and if P satisfies property* **B**, *then* $(\mathbb{X}, d)$ *is complete iff* $(\mathbb{X}, d^p)$ *is complete.*

Definition 8.27. We say that $T : (\mathbb{X}, d) \to (\mathbb{X}, d)$ is *contractive* if there is some $k \in [0,1)$ such that for all $x, y \in X$ we have $d(T(x), T(y)) \le kd(x,y)$.

Proposition 8.28. *Suppose that T is contractive with contractivity k. Then* $\rho(T(x), T(y)) \le k\rho(x,y)$ *as well, so T is ρ-contractive. If P has property* **B** *with $p^* \in P^*$, then* $d^p(T(x), T(y)) \le kd^p(x,y)$, *so T is contractive in the metric d^p as well.*

Proof. The proof is simple and follows easily from the fact that for any $q^* \in P^*$ we have that if $u, v \in \mathbb{B}$ with $u \le v$, then $p^*(u) \le p^*(v)$. $\qquad\square$

In fact, more is true: T is contractive in the cone metric d with contractivity k if and only if it is contractive in ρ with contractivity k.

From the proposition, we easily get the following theorem (which is a slight strengthening of Theorem 1 in [117]). This result has also appeared in various forms in the literature (see [4] and the references therein).

Theorem 8.29. *Suppose that* $(\mathbb{X}, d)$ *is a complete cone metric space and* $T : (\mathbb{X}, d) \to (\mathbb{X}, d)$ *is a contraction. Then* T *has a unique fixed point.*

Proof. We simply see that T is also a contraction in the complete metric space $(\mathbb{X}, \rho)$ and thus has a unique fixed point.   □

A simple corollary to the proof of the contraction mapping theorem is the collage theorem, which is used in the theory of IFS fractal image compression (and other IFS fractal-based methods in analysis) and will feature heavily in our applications in Sect. 8.2.4.

Theorem 8.30. *(The collage theorem) Suppose that* $(\mathbb{X}, d)$ *is a complete cone metric space and* T *is a contraction on* $\mathbb{X}$ *with contractivity* k *and fixed point* $\bar{x}$. *Then, for any* $x \in X$, *we have*

$$\frac{1}{k+1}d(Tx, x) \leq d(\bar{x}, x) \leq \frac{1}{1-k}d(Tx, x).$$

Proof. We simply see that

$$d(\bar{x}, x) \leq d(\bar{x}, Tx) + d(Tx, x) =$$
$$d(T\bar{x}, Tx) + d(Tx, x) \leq kd(\bar{x}, x) + d(Tx, x),$$

which leads to the second inequality. The first inequality is obtained from

$$d(x, Tx) \leq d(x, \bar{x}) + d(\bar{x}, Tx) = d(x, \bar{x}) + d(T\bar{x}, Tx) \leq d(x, \bar{x}) + kd(x, \bar{x}).$$

□

The idea of using the collage theorem is that if we wish to find a contraction T whose fixed point $\bar{x}$ is "close" to some given x, then instead of minimizing $d(x, \bar{x})$ (which requires knowledge of $\bar{x}$) we can minimize $d(Tx, x)$, which is expressed entirely using the given data. Practically speaking, starting from a target element $x \in X$ and a family of operators T_λ, depending on a vector of parameters $\lambda \in \Lambda$, where $\Lambda \subset \mathbb{R}^s$ is a compact set, we wish to solve the program

$$\min_{\lambda \in \Lambda} \psi(\lambda) := \min_{\lambda \in \Lambda} d(T_\lambda x, x). \qquad (8.58)$$

Compared with the case of a real-valued distance, the objective function $\psi(\lambda)$ assigns values in the Banach space $\mathbb{B}$ (ordered by the pointed cone $P \subset \mathbb{B}$). As usual in vector optimization, a global solution to (8.58) is a vector $\lambda^* \in \Lambda$ such that $\psi(\lambda) \notin \psi(\lambda^*) - \text{int}(P)$ (see [121]).

There are two main approaches to dealing with multicriteria optimization programs, namely scalarization techniques and goal programming ([121, 149]). In this case, proceeding by scalarization, we get the estimates

$$\frac{1}{k+1}d^p(x,Tx) \leq d^p(x,\bar{x}) \leq \frac{1}{1-k}d^p(x,Tx)$$

for any $p^* \in S$ and thus

$$\frac{1}{k+1}\rho(x,Tx) \leq \rho(x,\bar{x}) \leq \frac{1}{1-k}\rho(x,Tx).$$

8.2.3 Cone with empty interior

In our application, we will be interested in considering cone metric spaces where perhaps the natural cone P has no interior points. This changes things considerably, as even the basic definition of convergence needs to be modified. So, suppose that P is a closed and pointed cone but with $\text{int}(P) = \emptyset$. As a motivating example, consider $\mathbb{B} = L^2[0,1]$ and $P = \{f : f(x) \geq 0 \text{ a.e. } x\}$. Then clearly $P = P^*$ and $\text{int}(P) = \emptyset$.

As there are no interior points, we must replace the condition on interior points with something else. We choose to use the *quasi-interior* points. For this purpose, we assume that $\mathbb{B}$ is separable and reflexive. Then the set of quasi-interior points of P is non empty and is characterized as (see [28], p. 17)

$$qi(P) = \{p \in P : q^*(p) > 0 \text{ for all } q^* \in P^* \setminus \{0\}\}.$$

We note that if $c \in qi(P)$ and $\epsilon > 0$, then $\epsilon c \in qi(P)$.

Thus we say that $x_n \to x$ in the cone metric d if for all $c \in qi(P)$ we have that eventually $d(x_n, x) \leq c$.

Proposition 8.13 is independent of whether or not $\text{int}(P) = \emptyset$, so again d^p is a pseudometric on $\mathbb{X}$ for any $p \in P^*$ and a metric if $p^* \in qi(P^*)$ (which is also always nonempty in this context). We again define $\rho(x,y) = \sup_{p^* \in S} d^p(x,y)$ and notice that it is a metric.

Proposition 8.31. *$x_n \to x$ in the cone metric if and only if $\rho(x_n, x) \to 0$.*

Proof. Suppose that $x_n \to x$ in the cone metric and let $\epsilon > 0$ be given. Choose some $c \in qi(P)$. Then $e = c(\epsilon/2\|c\|) \in qi(P)$ and so eventually $d(x_n, x) \leq e$, which implies that for all $p^* \in S$ we have

$$d^p(x_n, x) = p^*(d(x_n, x)) \leq p^*(e) = \epsilon/2 \quad \Rightarrow \quad \rho(x_n, x) < \epsilon.$$

Conversely, suppose that $\rho(x_n, x) \to 0$, and let $c \in qi(P)$. Then $p^*(c) > 0$ for all $p^* \in S$. In fact, since S is weak* compact, there is some $\lambda > 0$ such that $p^*(c) \geq \lambda > 0$ for all $p^* \in S$. Choose $\epsilon = \lambda/2$. Then eventually $\rho(x_n, x) < \epsilon$, which implies that

$$p^*(d(x_n, x)) \leq \rho(x_n, x) < \epsilon < p^*(c) \text{ for all } p^* \in S \quad \Rightarrow \quad d(x_n, x) \leq c.$$

$\square$

As usual, we say that $T : \mathbb{X} \to \mathbb{X}$ is a *contraction* if there is some $k \in [0, 1)$ with $d(Tx, Ty) \leq d(x, y)$ for all x, y. By the method of scalarization, it is easy to obtain the following.

Proposition 8.32. *Suppose that* $(\mathbb{X}, d)$ *is a complete cone metric space and T is a contraction on* $\mathbb{X}$*. Then T has a unique fixed point.*

Proof. The proof is a simple consequence of the fact that $\rho(Tx, Ty) \leq k\rho(x, y)$, and thus T is a contraction in the metric ρ. $\square$

Clearly we also obtain the collage theorem in this case, which we can express either in terms of the cone metric or in terms of ρ. As the ordering cone P in the Banach space $\mathbb{B}$ now has an empty interior, we need to modify the notion of a global solution to the minimization problem (8.58). In this case we have that a vector $\lambda^* \in \Lambda$ is a global solution if $\psi(\lambda) \notin \psi(\lambda^*) - qi(P)$ for all $\lambda \in \Lambda$. That is, we replace the interior with the quasi-interior. Again we can approach such vector optimization problems by means of scalarization and thus solve a family of scalar optimization problems.

8.2.4 Applications to image processing

8.2.4.1 Structural similarity index

Many times in image processing, people are interested in designing image quality indexes that better describe and measure visual quality and distortions between two images. The structural similarity (SSIM) index is an example of such an index, and in its original formulation it involves a product of three terms, each of which measures a particular aspect of two images (see [29, 30]). As was well-highlighted in [29], SSIM can be reformulated by considering vector-valued distances that are a particular case of cone metric distances in which the cone coincides with the positive orthant. Let's recall the basic assumptions of this theory. In [29], the authors define the following vector-valued distance: Given $x, y \in \mathbb{R}^N$,

$$d(x, y) = (d_1(x, y), d_2(x, y)) \in \mathbb{R}_+^2, \tag{8.59}$$

where d_1 describes the distance between the means of x and y, while d_2 measures the distortion between x and y. After assuming P equal to the positive orthant $\mathbb{R}_+^2$, it can be proved that d is a cone metric distance. For more details and numerical experiments, one can see [29]. However, it is worth listing some comments that could be useful to extend this index to more general contexts and to create better compression algorithms. First, the assumption that the ordering cone coincides with the positive orthant can be too restrictive; in fact, roughly speaking, this corresponds to assigning the same importance to vector components, while in some situations it could be convenient to assume a lexicographic ordering cone that assigns priorities to vector components or even more complicated cones. This could imply that d is no longer a cone metric and some modifications to its definition are required. Second, in order to use a cone-metric to analyze approximation problems, it could happen that one has to solve optimization programs that involve multiobjective functions. In the case of a positive orthant, this can be done by scalarization techniques (which lead to some sort of weighted combination of d_1 and d_2) or goal programming algorithms. When different cones are assumed, these numerical techniques need to be adapted by using elements of the dual cone (or its quasi-interior).

8.2.4.2 A Hausdorff cone metric

In this section, we will again let $(\mathbb{X}, d)$ be a cone metric space that takes values in the cone $P \subset \mathbb{B}$. We assume that $\text{int}(P^*) \neq \emptyset$.

Let $\mathbb{H}(\mathbb{X})$ denote the collection of all nonempty and compact subsets of $\mathbb{X}$. Our purpose in this section is to define on $\mathbb{H}(\mathbb{X})$ a cone metric analogue of the usual Hausdorff distance between sets. The particular choice we make is very natural and thus has some very nice properties. However, as we will see, the Banach space in which our new cone metric takes values is not $\mathbb{B}$.

Recalling that $S = \{p^* \in P^* : \|p^*\| = 1\}$, we define

$$\mathcal{F} = \{\text{all bounded } f : S \to \mathbb{R}\}$$

and note that $\mathcal{F}$ is a Banach space under the norm $\|f\| = \sup_x |f(x)|$. The cone $\mathbb{P} \subset \mathcal{F}$ is defined by $f \in \mathbb{P}$ whenever $f(p^*) \geq 0$ for all $p^* \in S$. The zero element of $\mathcal{F}$ is the zero function, which is included in $\mathbb{P}$.

Definition 8.33. Let $A, B \in \mathbb{H}(\mathbb{X})$. Define $d_H(A, B) \in \mathcal{F}$ by

$$d_H(A, B)(p^*) = d_H^p(A, B),$$

where d_H^p is the Hausdorff pseudometric on $\mathbb{H}(\mathbb{X})$ induced by d^p on $\mathbb{X}$.

Clearly $d_H(A, B) \in \mathbb{P}$ for any A, B. Most of the cone metric properties of d_H follow from the fact that each d^p is a pseudometric. For any $p^* \in \text{int}(P^*)$, we know that d^p is a metric which then means that d_H^p is actually a metric as well and thus $d_H^p(A, B) = 0$ if and only if $A = B$. Thus, if $A \neq B$ we have $d_H(A, B) \neq 0$, and thus d_H is a cone metric.

The following properties are all easy to verify:

1. $A_1, A_2, B_1, B_2 \in \mathbb{H}(\mathbb{X})$ implies that

$$d_H(A_1 \cup A_2, B_1 \cup B_2) \leq \max\{d_H(A_1, B_1), d_H(A_2, B_2)\},$$

   where the maximum is taken in a pointwise fashion.
2. If $(\mathbb{X}, d)$ is complete, then $(\mathbb{H}(\mathbb{X}), d_H)$ is also complete. If $\mathbb{X}$ is 000 compact, then so is $\mathbb{H}(\mathbb{X})$.
3. If $T : \mathbb{X} \to \mathbb{X}$ is a contraction with contractivity k, then $T : \mathbb{H}(\mathbb{X}) \to \mathbb{H}(\mathbb{X})$ is also contractive with contractivity k. This is easy to see as $d^p(Tx, Ty) \leq k d^p(x, y)$ for each p^*.
4. If $T_i : \mathbb{X} \to \mathbb{X}, i = 1, 2, \ldots, N$, are contractive with contractivity k, then so is $T : \mathbb{H}(\mathbb{X}) \to \mathbb{H}(\mathbb{X})$, defined by

$$T(A) = \bigcup_i T_i(A).$$

With this formalism it is possible to construct geometric fractals in cone metric spaces using the standard IFS theory.

8.2.4.3 IFSM on cone metric spaces and self-similarity

Many times in image processing, one has to deal with the notion of distance; a fundamental aspect in this field is to decide "how far" two images are from each other. The choice of a suitable definition of distance is not at all easy; what seems to be close in one metric can be very far in another one. This leads quite naturally to defining an environment in which many possible metrics can be considered simultaneously, and the notion of a cone metric lends itself naturally to this requirement. This is the case, for instance, in the analysis of the structural similarity index of images (see [29, 30]), which is used to improve the measure assessing visual distortions between two images, or in the case of fractal-based measure approximation with entropy maximization and sparsity constraints (see [109]). In both these contexts, the difference between two images is calculated using different criteria, which leads in a natural way to considering vector-valued distances.

Example 8.34. As an illustrative example, let us consider a possible application in the area of image compression through generalized fractal transforms and, in particular, the case of iterated function systems on mapping (IFSMs). The classical IFSM theory requires one to specify *a priori* the L^p space in which the image is embedded; an approach based on cone metrics allows an all-in-one environment that preserves the complexity of the problem. For simplicity, let $\mathbb{X} = [0,1]$ and consider the cone metric $d : L^\infty(\mathbb{X}) \times L^\infty(\mathbb{X}) \to L^\infty([1,+\infty])$, where $d(u,v)(p) = \|u - v\|_p$ for all $p \in [1,+\infty]$. Let $L^\infty_+([1,+\infty])$ denote the cone of all a.e. positive functions. It is easy to prove that $(L^\infty(\mathbb{X}), d)$ is a complete $L^\infty_+([1,+\infty])$ cone metric space. We can further define an IFSM on $L^\infty(\mathbb{X})$ in the usual manner,

$$Tu(x) = \sum_{i=1}^{n} \alpha_i u(w_i^{-1}(x)) + \beta_i, \qquad (8.60)$$

where $w_i : \mathbb{X} \to \mathbb{X}$ is a set of nonoverlapping maps with contractivity factors K_i. We have $T : L^\infty(\mathbb{X}) \to L^\infty(\mathbb{X})$, and a classical result (see [67]) shows that

$$\|Tu - Tv\|_p \leq K\|u - v\|_p$$

for all $p \in [1,+\infty]$, where $K = \max_i K_i$. This implies that

$$d(Tu, Tv) \leq Kd(u, v)$$

with respect to the order induced by $L^\infty_+([1,+\infty])$.

As a numerical example, let us compute the collage distance (the objective function ψ from (8.58)) for a simple illustrative case. Let $w_1(x) = \frac{1}{2}x$, $w_2(x) = \frac{1}{2}x + \frac{1}{2}$, $\phi_1(x) = \frac{1}{2}x + \frac{1}{4}$ and $\phi_2(x) = \frac{1}{2}x + \frac{1}{2}$. If we assume $u(x) = 1$ then $Tu(x) = \frac{3}{4}I_{[0,1/2]}(x) + I_{[1/2,1]}(x)$ where I is the indicator function. The collage distance $d(u, Tu)$ leads to the result

$$d(u, Tu)(p) = \|u - Tu\|_p = \frac{1}{4}\left(\frac{1}{2}\right)^{\frac{1}{p}}.$$

Example 8.35. We now present a different type of example to illustrate the flexibility of the cone metric framework for image analysis problems. Our cone metric in this case will measure the difference between two images in each "resolution" level separately. To this end let $\psi_{i,j}$ be a wavelet basis for our image space (for example, either $L^2(\mathbb{R}^2)$ or $L^2([0,1]^2)$ with periodic boundary conditions). The book [123] is a great source for information about wavelets (and many other beautiful topics).

For any image $f \in L^2$, we can expand f in the wavelet basis to obtain

$$f = \sum_{i,j} f_{i,j} \psi_{i,j}.$$

In this, the first subscript, i, represents the "scale" and the second subscript, j, represents the "location" at that scale. For two images f, g, we define our cone metric to be

$$d(f,g)(i) = \left(\sum_j (f_{i,j} - g_{i,j})^2 \right)^{1/2}.$$

Thus $d(f,g)(i)$ represents the ℓ^2 distance between the images f and g on the wavelets at the scale i. Formally, $d : L^2 \times L^2 \to \ell^2_+$, as $d(f,g)(i) \geq 0$ for all i. We see that the cone ℓ^2_+ has empty interior, so we must use the quasi-interior rather than the interior (as in Sect. 8.2.3).

Again we use an IFSM operator T, as in (8.60), for our image analysis. For appropriate maps w_k and ϕ_k, this induces a type of IFS operation on the wavelet coefficients (see [134]). If c_k is the Jacobian associated to w_k (which is strictly less than 1 as w_k is contractive) and K_k is the contractivity of the "grey-level map" ϕ_k, the one can show that (compare with Sect. 3.3 in [67])

$$d(Tf, Tg)(i) \leq \left(\sum_k |c_k| K_k^2 \right)^{1/2} d(f,g)(i-1),$$

which leads to conditions under which T is contractive in the cone metric d.

The benefit of using the cone metric framework is that we preserve the information about how well the approximation is at each distinct resolution level. Thus in an image recovery operation, we can either attempt to perform a vector optimization where we obtain a Pareto optimal point (non-dominated for all resolutions) or we can attempt the simpler problem of focusing on the visually important resolutions (usually the lower resolutions). It also allows one to truncate the expansion to any dimension.

Appendix A
Topological and Metric Spaces

This appendix will be devoted to the introduction of the basic properties of metric, topological, and normed spaces. A metric space is a set where a notion of distance (called a metric) between elements of the set is defined. Every metric space is a topological space in a natural manner, and therefore all definitions and theorems about topological spaces also apply to all metric spaces. A normed space is a vector space with a special type of metric and thus is also a metric space.

All of these spaces are generalizations of the standard Euclidean space, with an increasing degree of structure as we progress from topological spaces to metric spaces and then to normed spaces.

A.1 Sets

In this book, we use classical notation for the union and intersection of sets. $A \backslash B$ denotes the difference of A and B, and $A \triangle B$ is the symmetric difference. We use the notation $A \subset B$ for strict inclusion, while $A \subseteq B$ if $A = B$ is allowed. Furthermore, $A_n \uparrow A$ means that A_n is a non decreasing sequence of sets and $A = \cup A_n$, while $A_n \downarrow A$ is a non increasing sequence of sets and $A = \cap A_n$. If A and B are two sets, their *Cartesian product* $A \times B$ consists of all pairs (a, b), $a \in A$, and $b \in B$. If we have an infinite collection of sets, A_λ for $\lambda \in \Lambda$, then the Cartesian product of this collection is defined as:

$$\prod_\lambda A_\lambda = \left\{ f : \Lambda \to \bigcup_\lambda A_\lambda : f(\lambda) \in \Lambda \right\}.$$

The *Axiom of Choice* states that this product is nonempty as long as each A_λ is nonempty and $\Lambda \neq \emptyset$.

A.2 Topological spaces

The basic idea of a topological space is to try to find a very general setting in which the notions of convergence and continuity can be defined in some meaningful way. The key idea is to think of "open" sets as somehow measuring closeness. In this section, we give many of the definitions of the basic concepts in topology. As such, this section is primarily a list of definitions.

A.2.1 Basic definitions

Definition A.1. (Topology and topological space) A set $\mathbb{X}$ is a *topological space* if a topology $\mathcal{T}$ on $\mathbb{X}$ has been chosen. A *topology on* $\mathbb{X}$, $\mathcal{T}$, is a family of subsets of $\mathbb{X}$, called the *open sets*, that satisfies the following properties:

- $\emptyset, \mathbb{X} \in \mathcal{T}$.
- $A_i \in \mathcal{T}$ for $i = 1, 2, \ldots, n$ implies that $\bigcap_i A_i \in \mathcal{T}$.
- $A_\lambda \in \mathcal{T}$ for $\lambda \in \Lambda$ implies $\bigcup_\lambda A_\lambda \in \mathcal{T}$.

In words, a topology on $\mathbb{X}$ is a family of subsets that is closed under arbitrary unions and finite intersections. We call the complement of an open set a *closed* set.

Sometimes it is more convenient to specify a collection smaller than the collection of all possible open sets. This is the idea behind the following definitions.

Definition A.2. (Base and Subbase) A subfamily $\mathcal{T}_0 \subseteq \mathcal{T}$ is called a *base of* $\mathcal{T}$ if each open set $A \in \mathcal{T}$ can be written as a union of sets of $\mathcal{T}_0$. We also say that $\mathcal{T}_0$ *generates the topology* $\mathcal{T}$. A *subbase* of a topology is a family of sets $\mathcal{B}$ such that their finite intersections form a base for the topology $\mathcal{T}$; that is, such that the collection

$$\left\{ \bigcap_{i=1}^n A_i : A_i \in \mathcal{B} \right\}$$

is a base for $\mathcal{T}$.

If $A \subseteq \mathbb{X}$, the *induced* or *subspace* topology on A is given by intersections $A \cap O$, for all $O \in \mathcal{T}$. It is easy to verify that this collection satisfies the properties required for it to be a topology on A.

If $\mathbb{X}$ and $\mathbb{Y}$ are two topological spaces with topologies $\mathcal{T}_1$ and $\mathcal{T}_2$, respectively, then their Cartesian product $\mathbb{X} \times \mathbb{Y}$ is a topological space with the topology $\mathcal{T}$ generated by the base $\{A \times B, A \in \mathcal{T}_1, A \in \mathcal{T}_2\}$. This topology is referred to as the *product topology* on $\mathbb{X} \times \mathbb{Y}$. The product topology can also be defined for infinite products via the weak topology (see below).

Definition A.3. (Closure, interior, boundary) Let $E \subseteq \mathbb{X}$.

- The set $\bigcap \{C \subseteq \mathbb{X} : E \subseteq C, C \text{ is closed}\}$ is called the *closure* of E and is denoted by $\operatorname{cl} E$ or $\overline{E}$.
- An open set A such that $E \subseteq A$ is said to be a *neighbourhood* of E.
- A point $x \in E$ is said to be an *interior* point of E if there exists a neighbourhood $A \subseteq E$ of x.
- The set of all interior points of E is called the *interior* of E and denoted by $\operatorname{int} E$. The interior of E is also equal to $\operatorname{int} E = \bigcup \{A \subseteq \mathbb{X} : A \subseteq E, A \text{ is open}\}$.
- The *boundary* of E, ∂E, is the set $\operatorname{cl} E \setminus \operatorname{int} E$.

If we have two topologies $\mathcal{T}_1$ and $\mathcal{T}_2$ on the same set $\mathbb{X}$, we say that $\mathcal{T}_1$ is *weaker* than $\mathcal{T}_2$ (or $\mathcal{T}_2$ is *stronger* or *finer* than $\mathcal{T}_1$) if $\mathcal{T}_1 \subseteq \mathcal{T}_2$. It is easy to see that the intersection of any family of topologies is a topology. Furthermore, for any collection of topologies, there is a unique finest topology that is stronger than all of them.

The finest topology of all is the *discrete topology*, in which every set is open. The coarsest topology is the *trivial* or *indiscreet* topology, in which the only open sets are the empty set and the whole space.

A.2.2 Convergence, countability, and separation axioms

Definition A.4. (Convergence) A sequence of points $x_n \in \mathbb{X}$, $n \geq 1$, converges to $x \in \mathbb{X}$ as $n \to +\infty$ if every neighbourhood of x contains all x_n with $n \geq n_0$ for some n_0.

A partially ordered set $(\Lambda, \preceq)$ is called a *directed system* if for every $\alpha, \beta \in \Lambda$ there is a $\gamma \in \Lambda$ with $\alpha \preceq \gamma$ and $\beta \preceq \gamma$. A *net* in $\mathbb{X}$ is a function $x : \Lambda \to \mathbb{X}$ from a directed system to $\mathbb{X}$. The net x_λ *converges* to $x \in \mathbb{X}$ if for every neighbourhood N of x there is some $\gamma \in \Lambda$, so if $\lambda \in \Lambda$ with $\gamma \preceq \lambda$, then $x_\lambda \in N$.

All sequences are nets, but not all nets are sequences. It is possible to show that a set C is closed if and only if it contains the limit for every convergent net of its points. In general, sequential convergence is not sufficient to define a given topology. However, if the space is first countable, then sequential convergence is sufficient to determine the topology (since each point has countably many sets to determine convergence and thus a net can be reduced to a sequence).

A point $x \in E$ is said to be *isolated* if $\{x\}$ is an open set. One consequence of this is that there is no sequence $x_n \to x$ with $x_n \neq x$ for all n.

Definition A.5. (First and second countable, separable)

- A *neighbourhood base* at a point $x \in X$ is a collection of sets $\mathcal{N}$ such that that for each neighbourhood U of x there is some $N \in \mathcal{N}$ with $x \in N \subseteq U$.
- A topological space X is *first countable* if each point has a countable neighbourhood base.
- A topological space X is *second countable* if there is a countable base for the topology.
- A topological space X is *separable* if there is a countable set $\{x_n\}$ such that that for any point $x \in X$ and any neighbourhood U of x there is some x_n with $x_n \in U$. In particular, this means that $\mathrm{cl}\{x_n : n \in \mathbb{N}\} = X$.

As an example, $\mathbb{R}^n$ with the usual notion of convergence (topology) is second countable. Every second countable space is first countable. Most of the "usual" spaces one encounters are first countable, or even second countable. Thus, for most situations sequential convergence is sufficient. Every second countable space is also separable. Conversely, any separable and first countable space is also second countable.

Definition A.6. (Separation conditions)

- A space is said to be T_0 if for all distinct $x, y \in X$ there is some open set U such that either $x \in U$ and $y \notin U$ or $x \notin U$ and $y \in U$.
- A space is said to be T_1 if $\{x\}$ is a closed set for all $x \in X$.
- A space is said to be *Hausdorff* or T_2 if for all distinct $x, y \in X$ there are disjoint open sets U, V with $x \in U$ and $y \in V$.

These conditions are listed in order of increasing strength. That is, $(Hausdorff)\, T_2 \Rightarrow T_1 \Rightarrow T_0$.

A.2.3 Compactness

The concept of *compactness* is central in many areas of topology. Roughly, it is a type of finiteness condition.

Definition A.7. (Compactness)

- A set $K \subseteq \mathbb{X}$, is *compact* if each open covering of K admits a finite subcovering (i.e., whenever $K \subseteq \bigcup_\lambda A_\lambda$ for open sets A_λ, there is some finite subset $\{\lambda_1, \lambda_2, \lambda_3, \dots, \lambda_n\} \subseteq \Lambda$ such that $K \subseteq \bigcup_j A_{\lambda_j}$).
- If $\mathbb{X}$ itself is a compact set, then $\mathbb{X}$ is called a *compact space*.
- A set $E \subseteq \mathbb{X}$ is said to be *relatively compact* if cl B is a compact set.
- $\mathbb{X}$ is called *locally compact* if each point $x \in \mathbb{X}$ has a neighbourhood with compact closure.
- If $\mathbb{X}$ can be represented as a countable union of compact sets, then it is said to be *σ-compact*.

The set of all nonempty compact subsets of $\mathbb{X}$ is denoted by $\mathcal{K}(\mathbb{X})$.

There is a standard construction that adds one point, located at "infinity", to a locally compact space to obtain a compact space. The resulting space is called the *one-point* or *Aleksandrov compactification*. The corresponding open sets are open sets of $\mathbb{X}$, and the added point ∞ has neighbourhoods that are complements to compact sets.

If the topology $\mathcal{T}$ has a countable base, then the compactness property of a set K is equivalent to the fact that every sequence $x_n \in K$ admits a convergent subsequence with its limit in K. In general, K is compact if and only if every net $x_\lambda \in K$ has a subnet with its limit in K.

A nonempty closed set E is said to be *perfect* if E does not have *isolated points*. IFS fractals are typically compact and perfect sets.

In Hausdorff space, each compact set K is also closed. (This is not true in general if the space is not Hausdorff.)

A.2.4 Continuity and connectedness

After defining the spaces, the natural next step is to define functions between these spaces. For the category of topological spaces, the natural functions are the continuous functions.

Definition A.8. (Continuity) A function $f : \mathbb{X} \to \mathbb{Y}$ that maps the topological space $\mathbb{X}$ into another topological space $\mathbb{Y}$ is *continuous* if for each open set $A \subseteq \mathbb{Y}$ the inverse image $f^{-1}(A) = \{x \in \mathbb{X} : f(x) \in A\}$ is an open set in $\mathbb{X}$.

If both f and f^{-1} are continuous, then we say that f is a *homeomorphism*.

If $f : \mathbb{X} \to \mathbb{Y}$ is a homeomorphism, then we consider $\mathbb{X}$ and $\mathbb{Y}$ topologically indistinguishable. This is similar to the situation in linear algebra where if $L : \mathbb{V} \to \mathbb{W}$ is a bijective linear map, then $\mathbb{V}$ and $\mathbb{W}$ are basically the same as linear spaces.

Definition A.9. (Connectedness) A set $A \subseteq \mathbb{X}$ is *connected* if A cannot be written as the union of two disjoint open sets.

Theorem A.10. *(Properties of continuous functions) Suppose that $f : \mathbb{X} \to \mathbb{Y}$ is continuous.*

- *If $g : \mathbb{Y} \to \mathbb{Z}$ is continuous, then so is $g \circ f : \mathbb{X} \to \mathbb{Z}$.*
- *If $K \subseteq \mathbb{X}$ is compact, then $f(K) \subseteq \mathbb{Y}$ is also compact,*
- *If $K \subseteq \mathbb{X}$ is compact and connected and $f : \mathbb{X} \to \mathbb{R}$ is continuous, then $f(K) \subset \mathbb{R}$ is a closed and bounded interval.*
- *If f is a bijection and $\mathbb{X}$ is compact, then f is a homeomorphism.*

Notice that the finer the topology on $\mathbb{X}$ the easier it is for $f : \mathbb{X} \to \mathbb{Y}$ to be continuous. In particular, if we place the discrete topology on $\mathbb{X}$, then for any topology on $\mathbb{Y}$, $f : \mathbb{X} \to \mathbb{Y}$ will be continuous. Similarly, if we place the indiscreet topology on $\mathbb{Y}$, then any function $f : \mathbb{X} \to \mathbb{Y}$ will be continuous.

Definition A.11. (Weak topology) Let $\mathbb{X}$ be a set, $\mathbb{Y}_\lambda$ be a collection of topological spaces, and $f_\lambda : \mathbb{X} \to \mathbb{Y}_\lambda$ be a collection of functions. The *weak topology on $\mathbb{X}$ by $\{f_\lambda\}$* is the coarsest topology on $\mathbb{X}$ for which each f_λ is continuous.

We can use this definition in defining the product topology. There are always natural projections $\pi_\mathbb{X} : \mathbb{X} \times \mathbb{Y} \to \mathbb{X}$ and $\pi_\mathbb{Y} : \mathbb{X} \times \mathbb{Y} \to \mathbb{Y}$ for any sets $\mathbb{X}, \mathbb{Y}$. If $\mathbb{X}$ and $\mathbb{Y}$ are also topological spaces, then we can place the weak topology on $\mathbb{X} \times \mathbb{Y}$ by these projections and this weak topology coincides with the product topology.

Thus, for an arbitrary collection of topological spaces $\mathbb{X}_\lambda$ we define the product topology on $\prod \mathbb{X}_\lambda$ as the weak topology by the collection of projections $\pi_\lambda : \prod_\gamma \mathbb{X}_\gamma \to \mathbb{X}_\lambda$.

Weak topologies are fundamental in functional analysis, where the "weak" topology on a Banach space is the weak topology by the collection of bounded linear functionals and the "weak-*" topology is a topology on a dual space generated by the collection of all point evaluations and thus is a topology of pointwise convergence.

A.3 Metric spaces

A metric space is a very special type of topological space with nice "geometric" properties. This is because the topology is given by a distance function, the metric.

Definition A.12. A metric space is an ordered pair $(\mathbb{X}, d)$ where $\mathbb{X}$ is a set and $d : \mathbb{X} \times \mathbb{X} \to \mathbb{R}_+$ is a metric on $\mathbb{X}$; that is, a function such that for any x, y, and z in $\mathbb{X}$ the following properties hold:

- $d(x, y) = 0$ if and only if $x = y$ (identity of indiscernibles).
- $d(x, y) = d(y, x)$ (symmetry).
- $d(x, z) \leq d(x, y) + d(y, z)$ (triangle inequality).

The function d is also called a *distance function* or simply a *distance*.

The discrete topology is given by the *discrete metric* defined as $d(x, x) = 0$ and $d(x, y) = 1$ if $x \neq y$. In a sense, this is the simplest type of metric to place on a set.

The metric is used to generate a subbase for a topology, the *metric topology*. From this, all the usual objects in a topology are easily defined. Given x in a metric space $\mathbb{X}$, we define the open ball of radius $\epsilon > 0$ at x as the set

$$B_\epsilon(x) = \{y \in \mathbb{X} : d(x, y) < \epsilon\}.$$

A subset $O \subseteq \mathbb{X}$ is called *open* if for every x in O there exists an $\epsilon > 0$ such that $B_\epsilon(x) \subseteq O$. As usual, the complement of an open set is called *closed*. A neighbourhood of the point x is any subset of $\mathbb{X}$ that contains an open ball at x as a subset. The concepts of connectedness and separability are as in general topological spaces.

A.3.1 Sequences in metric spaces

The set of balls $B_{1/n}(x)$ forms a countable neighbourhood basis at x, and thus any metric space is first countable. This means that sequential convergence is sufficient to determine the metric topology.

A sequence x_n in a metric space $\mathbb{X}$ is said to *converge* to the limit $x \in \mathbb{X}$ iff for every $\epsilon > 0$ there exists a natural number N such that $d(x_n, x) < \epsilon$ for all $n > N$; this is the same as saying that x_n is eventually in every $B_\epsilon(x)$. A subset $C \subseteq \mathbb{X}$ is closed if and only if every sequence in C that converges to a limit in $\mathbb{X}$ has its limit in C.

Definition A.13. (Cauchy sequence, completeness, Polish space)

- A sequence $x_n \in \mathbb{X}$ is said to be a *Cauchy sequence* if for all $\epsilon > 0$ there exists a natural number N such that for all $n, m \geq N$ we have $d(x_n, x_m) \leq \epsilon$.
- A metric space $\mathbb{X}$ is said to be *complete* if every Cauchy sequence converges in $\mathbb{X}$.
- Complete separable metric spaces are called *Polish* spaces.

If M is a complete subset of the metric space $\mathbb{X}$, then M is closed in $\mathbb{X}$.

A.3.2 Bounded, totally bounded, and compact sets

A set $A \subseteq \mathbb{X}$ is called *bounded* if there exists some number $M > 0$ such that $d(x, y) \leq M$ for all $x, y \in A$. The smallest possible such M is called the *diameter* of A. The set $A \subseteq \mathbb{X}$ is called *precompact* or *totally bounded* if for every $\epsilon > 0$ there exist a finite number of open balls of radius ϵ whose union covers A. Not every bounded A is totally bounded. As a simple example the discrete metric is bounded with diameter 1 but not totally bounded, unless the set A is finite.

Theorem A.14. *(Equivalence of notions of compactness) Let $\mathbb{X}$ be a metric space and $A \subseteq \mathbb{X}$. Then the following are equivalent:*

- *Every sequence $x_n \in A$ has a convergent subsequence with its limit in A.*
- *Every open cover of A has a finite subcover.*
- *A is complete and totally bounded.*

A metric space that satisfies any one of these conditions is said to be *compact*. The equivalence of these three concepts is not true in a general topological space. Since every metric space is Hausdorff, every compact subset is also closed.

The Heine-Borel theorem gives the equivalence of the third condition with the others. That is, it states that a subset of a metric space is compact if and only if it is complete and totally bounded.

A metric space is *locally compact* if every point has a compact neighbourhood. A space is *proper* if every closed ball $\{y : d(x, y) \leq \epsilon\}$ is compact. Proper spaces are locally compact, but the converse is not true in general.

A.3.3 Continuity

Definition A.15. (Continuity) Suppose $(\mathbb{X}_1, d_1)$ and $(\mathbb{X}_2, d_2)$ are two metric spaces. The map $f : \mathbb{X}_1 \to \mathbb{X}_2$ is *continuous* if it has one (and therefore all) of the following equivalent properties:

1. (general topological continuity) For every open set $A \subseteq \mathbb{X}_2$, the preimage $f^{-1}(A)$ is open in $\mathbb{X}_1$.
2. (sequential continuity or Heine continuity) If x_n is a sequence in $\mathbb{X}_1$ that converges to $x \in \mathbb{X}_1$, then the sequence $f(x_n)$ converges to $f(x) \in \mathbb{X}_2$.
3. (ϵ-δ continuity or Cauchy continuity) For every $x \in \mathbb{X}_1$ and every $\epsilon > 0$ there exists $\delta := \delta(x, \epsilon) > 0$ such that for all $y \in B_\delta(x)$ we have $f(y) \in B_\epsilon(f(x))$.

As in a general topological space, the image of every compact set under a continuous function is compact, and the image of every connected set under a continuous function is connected. Given a point $x \in \mathbb{X}$ and a compact set $K \subseteq \mathbb{X}$, we know that the function $d(x, y)$ has at least one minimum point $\bar{y}$ when $y \in K$. So we have $d(x, \bar{y}) \leq d(x, y)$ for all $y \in K$. We call $\bar{y}$ the *projection* of the point x on the set K and denote it as $\bar{y} = \pi_x K$. The point $\bar{y}$ is not unique, so we choose one of the minima.

Definition A.16. (Uniform continuity) The map $f : \mathbb{X}_1 \to \mathbb{X}_2$ is *uniformly continuous* if for every $\epsilon > 0$ there exists $\delta := \delta(\epsilon) > 0$ such that for all $y \in B_\delta(x)$ we have $f(y) \in B_\epsilon(f(x))$.

Clearly every uniformly continuous map f is continuous. The converse is true if $\mathbb{X}_1$ is compact (HeineCantor theorem). Uniformly continuous maps turn Cauchy sequences in $\mathbb{X}_1$ into Cauchy sequences in $\mathbb{X}_2$.

Definition A.17. (Lipschitz function) Given a number $K > 0$, the map $f : \mathbb{X}_1 \to \mathbb{X}_2$ is *K-Lipschitz continuous* if $d(f(x), f(y)) \leq K d(x, y)$ for all $x, y \in \mathbb{X}_1$.

Every Lipschitz continuous map is uniformly continuous, but the converse is not true in general.

Definition A.18. (Contraction) A 1-Lipschitz continuous function is called a *contraction*.

Theorem A.19. *(Banach fixed-point theorem or contraction mapping theorem) Suppose that $f : \mathbb{X} \to \mathbb{X}$ is a contraction and $\mathbb{X}$ is a complete metric space. Then f admits a unique fixed point $\bar{x}$ and, for any $x \in \mathbb{X}$, $d(f^n(x), \bar{x}) \to 0$ as $n \to \infty$.*

If $\mathbb{X}$ is compact, the condition can be weakened: $f : \mathbb{X} \to \mathbb{X}$ admits a unique fixed point if $d(f(x), f(y)) < d(x, y)$ for all $x, y \in \mathbb{X}$. For this, compactness is essential. As a simple example, consider the function $f : [1, \infty) \to [1, \infty)$ given by $f(x) = x + 1/x$. Clearly f has no fixed point, but it is easy to show that $|f(x) - f(y)| < |x - y|$.

A.3.4 Spaces of compact subsets

Let $(\mathbb{X}, d)$ be a complete metric space and $\mathbb{H}(\mathbb{X})$ denote the space of all nonempty closed and bounded subsets of $\mathbb{X}$. The *Hausdorff distance* between $A, B \in \mathbb{H}(\mathbb{X})$ is defined as

$$d_{\mathbb{H}}(A, B) = \max \left\{ \sup_{x \in A} d(x, B), \sup_{x \in B} d(x, A) \right\}.$$

Here, $d(x, A)$ denotes the usual distance between the point x and the set A,

$$d(x, A) = \inf_{y \in A} d(x, y).$$

We shall define $d(A, B) := \max_{x \in A} d(x, B)$ such that

$$d_{\mathbb{H}}(A, B) = \max\{d(A, B), d(B, A)\}.$$

Note that in general $d(A, B) \neq d(B, A)$.

For certain considerations, the following alternative definition of Hausdorff metric may be more appealing. For $E \subseteq \mathbb{X}$ and $\epsilon \geq 0$, the set

$$E_\epsilon := \{x \in \mathbb{X} \ : \ d(x, E) \leq \epsilon\} = \{x \in \mathbb{X} : d(x, y) < \epsilon \text{ for some } y \in E\}$$

is called the ϵ–*neighbourhood* or the ϵ-*dilation* of the set E.

Then, for any $A, B \in \mathbb{H}(\mathbb{X})$, it holds that

$$d_{\mathbb{H}}(A, B) < \varepsilon \quad \Longleftrightarrow \quad A \subseteq B_\varepsilon \text{ and } B \subseteq A_\varepsilon.$$

Thus we have

$$d_{\mathbb{H}}(A, B) = \inf\{\epsilon \geq 0 : A \subseteq B_\epsilon \text{ and } B \subseteq A_\epsilon\}.$$

Recalling the notion of Minkowski addition, we can also think of the ϵ-dilation as the Minkowski addition $E \oplus B_\epsilon(0)$ (i.e. as the dilation of E by a ball with radius ϵ).

Theorem A.20. *(Completeness and compactness of $\mathbb{H}(\mathbb{X})$) The space $(\mathbb{H}(\mathbb{X}), d_H)$ is a complete (compact) metric space if $(\mathbb{X}, d)$ is complete (compact).*

Let $(\mathbb{X}, d)$ be a metric space. The following are some useful properties of the Hausdorff distance:

1. For all $x, y \in \mathbb{X}$, $C \subset \mathbb{X}$, we have $d(x, C) \le d(x, y) + d(y, C)$.
2. If $A \subseteq B$, then $d(C, A) \ge d(C, B)$ and $d(A, C) \le d(B, C)$ for all $C \subset \mathbb{X}$.
3. For all $x, y \in \mathbb{X}$ and $A, B \subset \mathbb{X}$, we have $d(x, A) \le d(x, y) + d(y, B) + d(B, A)$.
4. For all $x \in \mathbb{X}$ and $A, B \subset \mathbb{X}$, we have $d(x, A) \le d(x, B) + d(B, A)$.
5. Suppose that $(\mathbb{X}, \|\|)$ is a real normed space and $E \subset \mathbb{X}$ a convex subset of $\mathbb{X}$. Let $A_i, B_i \subseteq E$ and $\lambda_i \in [0, 1]$ for $i = 1, 2, \ldots, N$ and $\sum_i \lambda_i = 1$. Then

$$d_\mathbb{H}\left(\sum_i \lambda_i A_i, \sum_i \lambda_i B_i\right) \le \sum_i \lambda_i d_\mathbb{H}(A_i, B_i).$$

6. Let $A, B_i \in \mathbb{H}(E)$, $\lambda_i \in [0, 1]$ for $i = 1, 2, \ldots, N$ and such that $\sum \lambda_i = 1$. Suppose that A is convex. Then

$$d_H\left(A, \sum_i \lambda_i B_i\right) \le \sum_i \lambda_i d_H(A, B_i).$$

7. Let $A, B \in \mathbb{H}(\mathbb{X})$ and $b \in B$. Then for all $\epsilon > 0$, there is an element $a_\epsilon \in A$ such that
$$d(a_\epsilon, b) \le d_H(A, B) + \epsilon.$$

8. Let $A, B, I \subset \mathbb{R}^n$. Then $d_h(A + I, B + I) \le d_h(A, B)$.

Appendix B
Basic Measure Theory

In this appendix, we briefly review the basic definitions and results from measure theory. This is certainly not intended as a course in measure theory but is only a reminder of that part of measure theory that is needed for this book. For a very good introduction to measure theory, see any of the books [150, 50, 8].

B.1 Introduction

The prototypical example of a measure is the Lebesgue measure on $\mathbb{R}$, a generalization of measuring the length of an interval. For notational purposes, if $I \subset \mathbb{R}$ is an interval, we denote by $\text{length}(I)$ the length of I.

For a set S that is a finite union of intervals, the "length" of S should be the sum of the lengths of the constituent intervals. Similarly, if S is a disjoint union of countably many intervals, then the total "length" should just be the sum of the individual lengths. For a more general set S, we define

$$\lambda(S) = \inf \left\{ \sum_i \text{length}(I_i) : S \subseteq \bigcup_i I_i \right\}, \tag{B.1}$$

where the infimum is taken over all countable covers of S by intervals I_i. It is not so hard to see that $\lambda(I) = \text{length}(I)$ in the case where I is an interval and that (B.1) also gives the natural measure of a set that is either a finite or countable union of disjoint intervals. With this definition, λ has some very nice properties:

1. $\lambda(\emptyset) = 0$.

2. $A \subseteq B \Rightarrow \lambda(A) \leq \lambda(B)$.
3. $\lambda(\bigcup_n A_n) \leq \sum_n \lambda(A_n)$.
4. $\lambda(S) = \lambda(S + a)$ for any $a \in \mathbb{R}$, where $S + a = \{s + a : s \in S\}$.
5. $\lambda(cS) = |c|\lambda(S)$ for any $c \in \mathbb{R}$, where $cS = \{c\,s : s \in S\}$.

It is possible to get equality in property 3 for disjoint A_n but not for arbitrary subsets A_n, only for sets from a special subclass of subsets, the *measurable sets*. A set $S \subset \mathbb{R}$ is *Lebesgue measurable* if for any $A \subset \mathbb{R}$ we have

$$\lambda(A) = \lambda(A \cap S) + \lambda(A \setminus S). \tag{B.2}$$

Somehow a measurable set is nice enough to split an arbitrary set in such a way that the measures of the parts are additive. Strangely enough, not all subsets of $\mathbb{R}$ are Lebesgue measurable (however, the existence of nonmeasurable sets is a consequence of the Axiom of Choice, so nonmeasurable sets are necessarily strange and hard to define).

B.2 Measurable spaces and measures

We now abstract the definition of Lebesgue measure to more general spaces. Let Ω be a nonempty set. We first set the collection of subsets of Ω that we will "measure."

Definition B.1. A *σ-algebra in Ω* is a collection $\mathcal{A}$ of subsets of Ω with the following properties:

1. $\emptyset, \Omega \in \mathcal{A}$.
2. $A \in \mathcal{A} \Rightarrow \Omega \setminus A \in \mathcal{A}$.
3. $A_n \in \mathcal{A}$ for all $n \in \mathbb{N}$ $\Rightarrow \bigcup_n A_n \in \mathcal{A}$.

The pair $(\Omega, \mathcal{A})$ is called a *measurable space*.

It is easy to see that if $\mathcal{A}_\alpha$ is a collection of σ-algebras, then $\cap_\alpha \mathcal{A}_\alpha$ is also a σ-algebra. Thus, if $\mathcal{C}$ is a collection of subsets of Ω, then there is a unique smallest σ-algebra $\mathcal{A}$ that contains $\mathcal{C}$. We call this σ-algebra the *σ-algebra generated by $\mathcal{C}$*.

Definition B.2. Let $(\Omega, \mathcal{A})$ be a measurable space. A nonnegative function μ defined on $\mathcal{A}$ is a *measure* if it satisfies

1. $\mu(\emptyset) = 0$,
2. $0 \leq \mu(A) \leq +\infty$ for all $A \in \mathcal{A}$,
3. $A \subseteq B \Rightarrow \mu(A) \leq \mu(B)$, and

4. for all pairwise disjoint sequences $\{A_n\}$ in $\mathcal{A}$, $\mu(\bigcup_n A_n) = \sum_n \mu(A_n)$.

The triple $(\Omega, \mathcal{A}, \mu)$ is called a *measure space*.

We include property 2 just to emphasize that it is possible for a measure μ to take the value $+\infty$ on some set. An example of this is that $\lambda(\mathbb{R}) = +\infty$, where λ is the Lebesgue measure. If $\mu(\Omega) < \infty$, then we say that μ is a *finite measure*. Furthermore, even though property 3 is a consequence of properties 2 and 4, we include it just to emphasize the monotonicity property of μ. Finally, if $\Omega = \bigcup_n E_n$ with $E_n \in \mathcal{A}$ and $\mu(A_n) < \infty$, we say that $(\Omega, \mathcal{A}, \mu)$ is a *σ-finite measure space*.

Some basic properties of measures are given in the next proposition.

Proposition B.3. *Let $(\Omega, \mathcal{A}, \mu)$ be a measure space. Then:*

1. *If $E_1 \subseteq E_2 \subseteq \cdots$ is an increasing sequence of measurable sets (i.e., each $E_n \in \mathcal{A}$), then $\mu(\bigcup_n E_n) = \lim_n \mu(E_n) = \sup_n \mu(E_n)$.*
2. *If $E_1 \supseteq E_2 \supseteq \cdots$ is a decreasing sequence of measurable sets and $\mu(E_k) < \infty$ for some k, then $\mu(\bigcap_n E_n) = \lim_n \mu(E_n) = \inf_n \mu(E_n)$.*

Definition B.4. We say that a property occurs for μ *almost all* $x \in \Omega$ if there is some set $A \subset \Omega$ such that the property occurs for all $x \in A$ and $\mu(\Omega \setminus A) = 0$.

For instance, f_n converges pointwise to f for μ almost all points if there is some set A, so for all $x \in A$ we have $f_n(x) \to f(x)$ and $\mu(\Omega \setminus A) = 0$.

Definition B.5. Two measures μ and ν on the same measurable space $(\Omega, \mathcal{A})$ are said to be *mutually singular*, written $\mu \perp \nu$, if there are two measurable sets A, B with $\mu(B) = 0 = \nu(A)$ and $\Omega = A \cup B$.

If μ is a measure with $\mu(\Omega) = 1$, then μ is called a *probability measure*. The collection of all probability measures on Ω is denoted by $\mathcal{P}(\Omega)$.

B.2.1 Construction of measures

We briefly indicate how measures can be constructed. A *premeasure* on Ω is a set function ν defined on some class $\mathcal{C}$ of subsets of Ω such that

1. $\emptyset \in \mathcal{C}$,

2. $0 \leq \nu(C) \leq +\infty$ for all $C \in \mathcal{C}$,
3. $\nu(\emptyset) = 0$.

As an example, the class $\mathcal{C}$ could be equal to $\{[a, b) : a, b \in \mathbb{R}\}$ and $\nu([a, b)) = b - a$.

An *outer measure* on the set Ω is a set function μ^* defined on all subsets of Ω that satisfies

1. $\mu^*(\emptyset) = 0$,
2. $A \subseteq B \Rightarrow \mu^*(A) \leq \mu^*(B)$, and
3. $E \subseteq \bigcup_n E_n \Rightarrow \mu^*(E) \leq \sum_n \mu^*(E_n)$.

We define a set $E \subset \Omega$ to be *measurable with respect to* μ^* if for all $A \subseteq \Omega$ we have

$$\mu^*(A) = \mu^*(A \cap E) + \mu^*(A \setminus E). \tag{B.3}$$

It turns out that the set of all μ^*-measurable sets is a σ-algebra and that μ^* restricted to this σ-algebra is countably additive.

The relationship between premeasures and outer measures is simple. Given a premeasure ν, we define

$$\mu^*(S) = \inf\left\{\sum_n \nu(C_n) : S \subset \bigcup_n C_n\right\}, \tag{B.4}$$

where the infimum is taken over all countable covers of S by elements of $\mathcal{C}$. If there are no such covers, we use the convention that $\inf \emptyset = +\infty$. The set function μ^* defined in (B.4) is an outer measure. Restricting μ^* to the σ-algebra of μ^*-measurable sets results in a measure.

As another example, we briefly discuss Lebesgue-Stieltjes measures. Let $g : \mathbb{R} \to \mathbb{R}$ be a nondecreasing and right-continuous function. Define the premeasure $\nu((a, b]) = g(b) - g(a)$. The measure generated by this premeasure is similar to the Lebesgue measure but has "weight function" g. That is, the "measure" of an interval is given not by its length but by the function g.

B.2.1.1 Product measures

If $(\mathbb{X}, \mathcal{X}, \mu)$ and $(\mathbb{Y}, \mathcal{Y}, \nu)$ are two measure spaces, there is a natural way to construct both a σ-algebra and a measure on $\mathbb{X} \times \mathbb{Y}$. A *rectangle* in $\mathbb{X} \times \mathbb{Y}$ is a set of the form $A \times B$ with $A \in \mathcal{X}$ and $B \in \mathcal{Y}$. The σ-algebra generated by the collection of all rectangles is called the *product σ-algebra* and is denoted by $\mathcal{X} \times \mathcal{Y}$.

For a rectangle $A \times B$, we define $(\mu \times \nu)(A \times B) = \mu(A) \cdot \nu(B)$ (where here we take $0 \cdot \infty = 0$). This defines a countably additive set function on the collection of rectangles. The extension to $\mathcal{X} \times \mathcal{Y}$ is a measure and is called the *product measure*.

As an example of this, the "usual" measure on $\mathbb{R}^n$ is the n-fold product of the Lebesgue measure on $\mathbb{R}$.

It is also possible to consider infinite product spaces and infinite product measures. We will only need this for the case of countable products of probability measures, so we restrict our attention to this case. Let $(\mathbb{X}_n, \mathcal{X}_n, \mu_n)$ be probability measure spaces for $n \in \mathbb{N}$. We will define a natural σ-algebra and measure on $\mathbb{X} = \prod_\lambda \mathbb{X}_n$ (we defined an arbitrary product in Appendix A). For the σ-algebra $\mathcal{X}$, we use the σ-algebra generated by all finite-dimensional measurable rectangles. These are sets of the form

$$B(i_1, i_2, \ldots, i_n) = \left\{ x \in \prod_n \mathbb{X}_n : x_{i_j} \in B_{i_j} \right\},$$

where $i_1, i_2, \ldots, i_n \in \mathbb{N}$ and $B_{i_j} \in \mathcal{X}_{i_j}$ for each j. That is, we only specify finitely many of the coordinates of the points in $B(i_1, i_2, \ldots, i_n)$, and these coordinates have to be specified as $\mathcal{X}_i$ measurable sets. Finally, for the measure μ, it is defined on the measurable rectangles as

$$\mu(B(i_1, i_2, \ldots, i_n)) = \mu_{i_1}(B_{i_1}) \cdot \mu_{i_2}(B_{i_2}) \cdots \mu_{i_n}(B_{i_n}).$$

Another way to think of this is to define B_n for all n as $B_n = B_{i_j}$ if $n = i_j$ and otherwise let $B_n = \mathbb{X}_n$. Then

$$B(i_1, i_2, \ldots, i_n) = \prod_{n \in \mathbb{N}} B_n$$

and so

$$\mu(B(i_1, i_2, \ldots, i_n)) = \prod_{n \in \mathbb{N}} \mu_n(B_n) = \mu_{i_1}(B_{i_1}) \cdot \mu_{i_2}(B_{i_2}) \cdots \mu_{i_n}(B_{i_n})$$

as before since $\mu_n(\mathbb{X}_n) = 1$.

B.2.1.2 Push-forward measures

If $(\mathbb{X}, \mathcal{X}, \mu)$ is a measure space and $f : \mathbb{X} \to \mathbb{Y}$ is a function, we can use f to construct a *push-forward* measure on $\mathbb{Y}$. For this, we use the largest σ-algebra for which f is measurable. That is,

$$\mathcal{Y} = \{B \subseteq \mathbb{Y} : f^{-1}(B) \in \mathcal{X}\}.$$

It is straightforward to show that this defines a σ-algebra on $\mathbb{Y}$. The measure $f_\#(\mu)$ is equally simple to describe:

$$f_\#(\mu)(B) = \mu(f^{-1}(B)) \quad \text{for all } B \in \mathcal{Y}.$$

B.2.2 Measures on metric spaces and Hausdorff measures

If the space Ω is also a metric space, it is possible for the metric and any measures defined on Ω to interact and thus exhibit a richer structure.

The *Borel sets* or *Borel σ-algebra*, denoted by $\mathcal{B}$, is defined to be the smallest σ-algebra that contains all the open sets in Ω. A measure μ defined on the class of Borel sets is called a *Borel measure*.

The *support* of a measure μ is the closed set

$$\text{supp}(\mu) = \Omega \setminus \bigcup \{U : U \text{ is an open set with } \mu(U) = 0\}$$

and can be characterized (in the separable case) as the closed set C such that $\mu(\Omega \setminus C) = 0$ and $\mu(C \cap O) > 0$ for all open sets with $O \cap C \neq \emptyset$.

A measure μ on Ω is said to be *tight* if for all $\epsilon > 0$ there is some compact set $K_\epsilon \subset \Omega$ with $\mu(\Omega \setminus K_\epsilon) < \epsilon$.

Proposition B.6. *Let Ω be a complete metric space and μ be a Borel measure on Ω. Then, for any two $A, B \in \mathcal{B}$ with $\inf\{d(a,b) : a \in A, b \in B\} > 0$, we have $\mu(A \cup B) = \mu(A) + \mu(B)$. Furthermore, if μ is a finite measure, then for all $B \in \mathcal{B}$ we have*

$$\mu(B) = \inf\{\mu(U) : B \subseteq U, U \text{ open }\} = \sup\{\mu(C) : C \subset B, C \text{ closed}\}.$$

Finally, if Ω is also separable and μ is finite, then μ is tight.

An important class of measures on a metric space are the *Hausdorff measures*. These measures are nice generalizations of the Lebesgue measure on $\mathbb{R}$ to a general metric space. For a subset $S \subseteq \Omega$, we define the *diameter of S* to be

$$|S| = \sup\{d(x,y) : x, y \in S\}. \tag{B.5}$$

For $\delta > 0$ and $S \subset \Omega$, a δ-*cover* of S is a countable collection of sets E_n with $|E_n| < \delta$ for each n and $S \subseteq \bigcup_n E_n$.

Let $d \geq 0$ and $\delta > 0$ be fixed. We define

$$\mathcal{H}_\delta^d(S) = \inf \left\{ \sum_i |E_n|^d : \{E_n\} \text{ is a } \delta\text{-cover of } S \right\}.$$

It is not hard to see that $\mathcal{H}_\delta^d(S)$ is increasing as δ decreases (since the collection of allowable covers decreases), and thus

$$\mathcal{H}^d(S) = \sup_{\delta > 0} \mathcal{H}_\delta^d(S) = \lim_{\delta \to 0} \mathcal{H}_\delta^d(S) \qquad (B.6)$$

exists. This limit is defined as the *Hausdorff d-dimensional measure* on Ω. It is a bit of an effort to show that $\mathcal{H}^d$ is a Borel measure on Ω. Clearly $\mathcal{H}^d$ is finite only if Ω is bounded, but it might be unbounded even if Ω is compact. For example,

$$\mathcal{H}^0(S) = \begin{cases} +\infty & \text{if } S \text{ is an infinite set,} \\ n & \text{if } S \text{ contains } n \text{ elements,} \end{cases}$$

so that $\mathcal{H}^0$ is a *counting measure* on Ω. We mention that it is possible to greatly generalize the construction of $\mathcal{H}^d$ by introducing a *gauge function*. A function $g : [0, \infty) \to [0, \infty)$ is a gauge function if g is nondecreasing and continuous and $g(x) \to 0$ as $x \to 0$. Then we replace $|E_n|^d$ by $g(|E_n|)$ in the definition of $\mathcal{H}_\delta^d$ and we get a measure $\mathcal{H}^g$. See [126, 147] for more on Hausdorff measures.

B.3 Measurable functions and integration

After defining measurable spaces, the next natural objects to study are mappings between measurable spaces.

Definition B.7. Let $(\mathbb{X}, \mathcal{X})$ and $(\mathbb{Y}, \mathcal{Y})$ be two measurable spaces. A function $f : \mathbb{X} \to \mathbb{Y}$ is said to be *measurable* if $f^{-1}(S) \in \mathcal{X}$ for all $S \in \mathcal{Y}$.

The special case where $Y = \mathbb{R}$ and $\mathcal{Y} = \mathcal{B}$ (the Borel sets) is of particular interest. In many cases, by measurable function we will mean this special case.

Proposition B.8. *If c is a constant and f, g are measurable, then so are $f + g, f \cdot g, cf, \max\{f, g\}, \min\{f, g\}$. Furthermore, if f_n is a sequence of measurable functions, then the functions $\sup_n f_n, \inf_n f_n, \limsup f_n,$ and $\liminf f_n$ are all measurable.*

For a set $S \subset \Omega$, we use χ_S to denote the *characteristic function of* S defined by

$$\chi_S(t) = \begin{cases} 1 & \text{if } x \in S, \\ 0 & \text{if } x \notin S. \end{cases}$$

Notice that S is a measurable set if and only if χ_S is a measurable function.

Definition B.9. Let $(\Omega, \mathcal{A})$ be a measurable space. A *simple function* on Ω is any function of the form

$$\varphi = \sum_i \alpha_i \chi_{S_i},$$

where $\alpha_1, \alpha_2, \ldots, \alpha_n \in \mathbb{R}$ and $S_i \in \mathcal{A}$.

That is, a simple function is a finite linear combination of characteristic functions of measurable sets.

Proposition B.10. *Let f be a nonnegative measurable function on a measurable space $(\Omega, \mathcal{A})$. Then there is an increasing sequence of simple functions $\varphi_n \leq \varphi_{n+1}$ with $\varphi_n(x) \to f(x)$ for each $x \in \Omega$.*

B.3.1 The integral

In this section assume that we have a measure space $(\Omega, \mathcal{A}, \mu)$.

The definition of the integral of a simple function is completely straightforward. That is, if

$$\varphi = \sum_i \alpha_i \chi_{E_i},$$

then we define

$$\int_\Omega \phi(x) \, d\mu(x) = \sum_i \alpha_i \mu(E_i). \tag{B.7}$$

The motivation is clear; the integral is the "area under the curve," (that is, the "height" α_i multiplied by the "length" $\mu(E_i)$).

For a positive measurable function $f : \Omega \to [0, \infty)$, we define

$$\int_\Omega f \, d\mu = \sup \left\{ \int_\Omega \varphi \, d\mu : 0 \leq \varphi \leq f, \varphi \text{ simple} \right\}. \tag{B.8}$$

Furthermore, for $A \subset \Omega$ a measurable set, we define

$$\int_A f(x) \, d\mu(x) = \int_\Omega f(x) \chi_A(x) \, d\mu(x).$$

Proposition B.11. *Let f, g be nonnegative measurable functions and c be a nonnegative constant. Then*

$$\int_\Omega f + g \, d\mu = \int_\Omega f \, d\mu + \int_\Omega g \, d\mu$$

and

$$\int_\Omega cf \, d\mu = c \int_\Omega f \, d\mu.$$

If $f(x) \le g(x)$ for μ almost all x, then

$$\int_\Omega f \, d\mu \le \int_\Omega g \, d\mu.$$

Finally, if A, B are disjoint measurable sets, then

$$\int_{A \cup B} f \, d\mu = \int_A f \, d\mu + \int_B f \, d\mu.$$

Theorem B.12. *(Fatou's lemma) Let f_n be a sequence of nonnegative measurable functions. Then*

$$\int_\Omega \liminf f_n \, d\mu \le \liminf \int_\Omega f_n \, d\mu.$$

Theorem B.13. *(monotone convergence theorem) Let f_n be a sequence of nonnegative measurable functions that converge μ almost everywhere to a function f. Suppose further that $f_n \le f$ for all n. Then*

$$\int_\Omega f \, d\mu = \lim_n \int_\Omega f_n \, d\mu.$$

Corollary B.14. *Let f_n be a sequence of nonnegative measurable functions. Then*

$$\int_\Omega \sum_n f_n \, d\mu = \sum_n \int_\Omega f_n \, d\mu.$$

We now define the integral of a (somewhat) arbitrary measurable function f. Notice that $f^+ = \max\{f, 0\}$ and $f^- = -\min\{f, 0\}$ are both nonnegative measurable functions. We define

$$\int_\Omega f \, d\mu = \int_\Omega f^+ \, d\mu - \int_\Omega f^- \, d\mu,$$

on the condition that only one of the terms in the sum is infinite.

Definition B.15. We say that the measurable function f is *integrable* if

$$\int_\Omega |f| \, d\mu = \int_\Omega f^+ \, d\mu + \int_\Omega f^- \, d\mu < \infty.$$

Proposition B.16. *Let f, g be integrable functions and α, β be constants. Then*

1. $\int_\Omega \alpha f + \beta g \, d\mu = \alpha \int_\Omega f \, d\mu + \beta \int_\Omega g \, d\mu,$
2. *if $|h| \leq |f|$ for μ almost all x and h is measurable, then h is integrable, and*
3. *if $f \geq g$ for μ almost all x, then $\int_\Omega f \, d\mu \geq \int_\Omega g \, d\mu.$*

Theorem B.17. *(Lebesgue dominated convergence theorem) Let g be integrable, and suppose that f_n is a sequence of measurable functions that satisfy $|f_n| \leq |g(x)|$ for μ almost all x. Suppose further that $f_n(x) \to f(x)$ for μ almost all x. Then f is integrable and*

$$\lim_n \int_\Omega f_n \, d\mu = \int_\Omega f \, d\mu.$$

We can also include a sequence of measures in the convergence theorems.

Theorem B.18. *Suppose that μ_n is a sequence of measures on the measurable space $(\Omega, \mathcal{A})$ with $\mu_n(A) \to \mu(A)$ for all $A \in \mathcal{A}$. Suppose further that f_n is a sequence of nonnegative measurable functions that converge pointwise μ almost everywhere to the function f. Then*

$$\int_\Omega f \, d\mu \leq \liminf \int_\Omega f_n \, d\mu_n.$$

Finally, we give a theorem that deals with integration in product spaces and gives some sufficient conditions under which the integral in the product space can be computed by computing integrals in each factor space.

Theorem B.19. *(Fubini's theorem) Let $(\mathbb{X}, \mathcal{X}, \mu)$ and $(\mathbb{Y}, \mathcal{Y}, \nu)$ be two σ-finite measure spaces. Let $\pi = \mu \times \nu$ on $\mathbb{X} \times \mathbb{Y}$ and $\mathcal{A} = \mathcal{X} \times \mathcal{Y}$ be the product measure and σ-algebra, respectively. Suppose that $F : \mathbb{X} \times \mathbb{Y} \to \mathbb{R}$ is integrable with respect to π. Then the functions*

$$f(x) = \int_\mathbb{Y} F(x, y) \, d\nu(y) \quad and \quad g(y) = \int_\mathbb{X} F(x, y) \, d\mu(x)$$

are defined μ (respectively ν) almost everywhere and are $\mathcal{X}$ (respectively $\mathcal{Y}$) measurable. Furthermore,

$$\int F(x,y) \, d\pi(x,y) = \int_X \int_Y F(x,y) \, d\nu(y) \, d\mu(x) = \int_X f(x) \, d\mu(x)$$
$$= \int_Y \int_X F(x,y) \, d\mu(x) \, d\nu(y) = \int_Y g(y) \, d\nu(y).$$

B.4 Signed measures

A *signed measure* on the measurable space $(\Omega, \mathcal{A})$ is a set-function μ defined on $\mathcal{A}$ that satisfies the following properties:

1. ν assumes at most one of the values $-\infty$ or $+\infty$
2. $\nu(\emptyset) = 0$.
3. $\nu(\bigcup_n E_n) = \sum_n \nu(E_n)$ for any sequence $\{E_n\}$ of measurable sets.

We sometimes denote the collection of finite signed measures on Ω by $\mathcal{M}(\Omega)$ or $\mathcal{M}(\Omega, \mathbb{R})$ to emphasize the fact that the measures take values in $\mathbb{R}$.

Given a signed measure ν, we say that $P \in \mathcal{A}$ is a *positive set* if $\nu(P) \geq 0$ and $\nu(S) \geq 0$ for all $S \subset P$ with $S \in \mathcal{A}$. A similar definition applies for a *negative set*.

Theorem B.20. *(Hahn decomposition theorem) Let ν be a signed measure on $(\Omega, \mathcal{A})$. Then there is a positive set P and a negative set N such that $\Omega = P \cup N$ and $N \cap P = \emptyset$.*

Theorem B.21. *(Jordan decomposition theorem) Let ν be a signed measure on $(\Omega, \mathcal{A})$. Then there are two measures ν^+ and ν^- that are mutually singular and such that $\nu = \nu^+ - \nu^-$. Furthermore, this decomposition is unique.*

For a given signed measure ν, the positive measure $|\nu| = \nu^+ + \nu^-$ is called the *absolute value* or *total variation* of ν. If $|\nu|$ is a finite measure, then we say that ν is a finite signed measure or has *finite variation*. The space of finite signed measures on Ω is a normed linear space under the norm

$$\|\mu\| = |\mu|(\Omega) = \mu^+(\Omega) + \mu^-(\Omega). \tag{B.9}$$

This norm is called the *total variation norm*.

Definition B.22. Given two measures μ and ν, we say that ν *is absolutely continuous with respect to* μ, denoted by $\nu \ll \mu$, if $\nu(A) = 0$ whenever $\mu(A) = 0$.

For two signed measures, we say $\nu \ll \mu$ if $|\nu| \ll |\mu|$.

Finally, a measurable function f on Ω is *integrable* with respect to the signed measure μ if $|f|$ is integrable with respect to $|\mu|$. In this case, we define

$$\int_\Omega f \, d\mu = \int_\Omega f^+ \, d\mu^+ + \int_\Omega f^- \, d\mu^- - \int_\Omega f^+ \, d\mu^- - \int_\Omega f^- \, d\mu + .$$

In general, as long as this sum does not include terms of the form $+\infty - \infty$, then one can have a well-defined notion of an integral yielding the value $+\infty$ or $-\infty$.

Theorem B.23. *(Radon-Nikodym theorem)* Let $(\Omega, \mathcal{A}, \mu)$ be a σ-finite measure space and ν be a signed measure defined on $\mathcal{A}$. Suppose that $\nu \ll \mu$. Then there is a measurable function f such that for each $E \in \mathcal{A}$ we have

$$\nu(E) = \int_E f \, d\mu.$$

Furthermore, the function f is uniquely defined μ almost everywhere.

The function f from the Radon-Nikodym theorem is called the *Radon-Nikodym derivative of ν with respect to μ.*

B.5 Weak convergence of measures

There are several different ways in which a sequence of measures can converge. One way we have already discussed is if $\mu_n(A) \to \mu(A)$ for all $A \in \mathcal{A}$. For this section, we assume that Ω is a metric space and that we use the Borel sets $\mathcal{B}$ and that all measures are Borel measures.

Definition B.24. Let μ_n be a sequence of Borel measures and μ be a Borel measure. We say that μ_n *converges weakly to* μ, denoted by $\mu_n \Rightarrow \mu$, if for all continuous and bounded $f : \Omega \to \mathbb{R}$ we have

$$\int_\Omega f \, d\mu_n \to \int_\Omega f \, d\mu.$$

Since constant functions are continuous and bounded, if $\mu_n \Rightarrow \mu$ we know that $\mu_n(\Omega) \to \mu(\Omega)$. Clearly the converse is not true. However, if $\mu_n(A) \to \mu(A)$ for every $A \in \mathcal{B}$, then $\mu_n \Rightarrow \mu$.

Proposition B.25. *Let μ_n and μ be finite measures. Then the following conditions are equivalent:*

1. $\mu_n \Rightarrow \mu$.
2. $\int f \, d\mu_n \to \int f \, d\mu$ *for all bounded and uniformly continuous f.*
3. $\limsup \mu_n(F) \leq \mu(F)$ *for all closed sets F and $\mu_n(\Omega) \to \mu(\Omega)$.*
4. $\liminf \mu_n(G) \geq \mu(G)$ *for all open sets G and $\mu_n(\Omega) \to \mu(\Omega)$.*
5. $\mu_n(S) \to \mu(S)$ *for all Borel sets S with $\mu(\partial S) = 0$.*

The reason that this convergence is called weak convergence is made a little more clear in the next theorem. First we recall some notions. Given a topological space $\mathbb{X}$, $C_c(\mathbb{X})$ denotes the collection of all continuous $f : \mathbb{X} \to \mathbb{R}$ such that there is some compact $K \subset \mathbb{X}$ with $f(\mathbb{X} \setminus K) = 0$. That is, each function in $C_c(\mathbb{X})$ has *compact support*. Clearly $C_c(\mathbb{X})$ is a linear space. We make $C_c(\mathbb{X})$ into a normed linear space by defining $\|f\| = \sup_{x \in \mathbb{X}} |f(x)|$. The completion of $C_c(\mathbb{X})$ in this norm is

$$C_0(\mathbb{X}) = \{f : f \text{ continuous}, \forall \epsilon > 0, \exists \text{ compact } K, f(\mathbb{X} \setminus K) \subset (-\epsilon, \epsilon)\}$$

which is a Banach space. A linear functional on $C_c(\mathbb{X})$ is a linear function $\Phi : C_c(\mathbb{X}) \to \mathbb{R}$, and it is *bounded* if there is some $C > 0$ such that $|\Phi(f)| \leq C\|f\|$.

For simplicity and convenience, we give a slightly less general version of the following theorem. More general versions are available in the references.

Theorem B.26. *(Riesz representation theorem)* *Let Ω be a locally compact separable metric space. Then, to each bounded linear functional Φ on $C_c(\Omega)$ there corresponds a unique finite signed Borel measure μ such that*

$$\Phi(f) = \int_\Omega f \, d\mu.$$

If Ω is compact, then $\mathbb{C}_c(\Omega) = C(\Omega) = C^*(\Omega)$ and thus Theorem B.26 gives a representation of bounded linear functionals on the space of all real-valued continuous functions on Ω.

This theorem indicates that the topology of weak convergence of measures is the weak* topology on the space of measures induced by this representation. In fact, it says that the space of finite signed Borel measures normed by the total variation is the dual space (in the sense of a Banach space) to the space $C_0(\Omega)$. From the abstract Banach-Alaoglu theorem (see [170]), we get the following fact.

Proposition B.27. *Let Ω be a locally compact separable metric space and let $c > 0$ be a fixed constant. Then the set of finite signed Borel measures μ with $|\mu| \leq c$ is weakly compact.*

B.5.1 Monge-Kantorovich metric

In certain nice situations, the weak* topology on the space of probability measures can be metrized. One particularly simple case is when Ω is a complete separable metric space. While it is certainly possible to obtain more general results, we will content ourselves with this simple case, as it is usually sufficient. The Monge-Kantorovich metric arises out of considerations in the area of mass transportation problems. For more general results and a historical discussion, see [72, 73, 88, 164].

Recall (see Appendix A) that we use $\mathrm{Lip}(\Omega)$ to denote the collection of Lipschitz functions $f : \Omega \to \mathbb{R}$ and $\mathrm{Lip}_1(\Omega)$ denotes the space of Lipschitz functions with Lipschitz constant bounded by 1.

Definition B.28. (Monge-Kantorovich metric) For two probability measures $\mu, \nu \in \mathcal{P}(\Omega)$, we define

$$d_{MK}(\mu, \nu) = \sup \left\{ \int_\Omega f \, d(\mu - \nu) : f \in \mathrm{Lip}_1(\Omega) \right\}. \qquad (B.10)$$

The definition of d_{MK} makes it clear that there is a relationship between convergence in this metric and weak convergence. In fact, in the case of a compact metric space, convergence of probability measures in the Monge-Kantorovich metric and weak convergence are equivalent, as is seen in the next proposition.

Proposition B.29. *Let Ω be a compact metric space. Then d_{MK} metrizes the topology of weak convergence on $\mathcal{P}(\Omega)$. Furthermore, $\mathcal{P}(\Omega)$ is compact under d_{MK}.*

For the purposes of IFS fractal constructions, the Monge-Kantorovich metric has the nice property that the distance between two probability measures μ and ν is linked with the underlying distance function in Ω. For instance, for two point masses, $d_{MK}(\delta_x, \delta_y) = d(x, y)$. It is also possible to get bounds on $d_{MK}(\mu, \nu)$ based on the Hausdorff distance between the supports of μ and ν. Of course, this connection between the distance on Ω and that on $\mathcal{P}(\Omega)$ is not surprising given that the Monge-Kantorovich metric came out of considerations involving optimal transport problems.

It is also possible to consider the case where Ω is not compact but is only assumed to be complete and separable. In this case, it is not possible to use (B.10) on all of $\mathcal{P}(\Omega)$, as it can be unbounded. To see this, take $\Omega = \mathbb{R}$, $\mu = \delta_0$, and

$$\nu = \sum_{n=1}^{\infty} 2^{-n} \delta_{2^n}.$$

Then

$$\int_{\mathbb{R}} x \, d(\nu - \mu) = \sum_{n} 2^{-n} 2^n = \infty.$$

The problem here is that $\int_{\mathbb{R}} x \, d\nu = \infty$. The solution is to assume some sort of uniform tightness on all the measures.

Proposition B.30. *Let Ω be a complete separable metric space and $a \in \Omega$. Define the set of probability measures*

$$\mathcal{P}_1(\Omega) = \left\{ \mu \in \mathcal{P}(\Omega) : \int_{\Omega} d(x, a) \, d\mu(x) < \infty \right\}.$$

Then $\mathcal{P}_1(\Omega)$ is complete under the Monge-Kantorovich metric d_{MK}.

Other related topologies and metrics can be placed on the space $\mathcal{P}(\Omega)$ (or even $\mathcal{M}(\Omega, \mathbb{R})$) by choosing some class of "test functions" as integrators. The relations between these various topologies can be quite complicated and depend in an intricate way on the precise properties of the space Ω. There is some discussion of this in [50].

B.6 L^p spaces

Throughout this section, we take $(\Omega, \mathcal{A}, \mu)$ to be a fixed measure space.

Definition B.31. (L^p *spaces*) Let $p \geq 1$. Define

$$L^p(\Omega, \mu) = \left\{ f : \Omega \to \mathbb{R} : \int_{\Omega} |f|^p \, d\mu < \infty \right\}.$$

Furthermore, for $f \in L^p(\Omega, \mu)$, define

$$\|f\|_p = \left(\int_{\Omega} |f|^p \, d\mu \right)^{1/p}.$$

From the definition, it is immediate that, for any constant α, we have $|\alpha| \|f\|_p = \|\alpha f\|_p$. Furthermore, $\|f\|_p = 0$ if and only if $f(x) = 0$ for μ almost all x.

For $p \geq 1$, the *conjugate exponent* q is the $q \geq 1$ such that $1/p + 1/q = 1$. In the case $p = 1$, we set $q = \infty$.

Proposition B.32.

1. *(Minkowski's Inequality)*   Let $f, g \in L^p(\Omega, \mu)$. Then $f + g \in L^p(\omega, \mu)$ and $\|f + g\|_p \leq \|f\|_p + \|g\|_p$.
2. *(Hölder's Inequality)*   Let $f \in L^p(\Omega, \mu)$ and $g \in L^q(\Omega, \mu)$, where p and q are conjugate exponents. Then $fg \in L^1(\Omega, \mu)$ and $\|fg\|_1 \leq \|f\|_p \|g\|_q$.
3. *(Cauchy-Bunyakovskii-Schwarz inequality)*   If $f, g \in L^2(\Omega, \mu)$, then $fg \in L^1(\Omega, \mu)$ and

$$\left| \int fg \, d\mu \right| \leq \|f\|_2 \|g\|_2.$$

Theorem B.33. $L^p(\Omega, \mu)$ *is a complete metric space (and thus is a Banach space).*

Proposition B.34.

1. *The class of all simple functions* φ *with* $\mu(\phi^{-1}(\mathbb{R} \setminus \{0\})) < \infty$ *is dense in* $L^p(\Omega, \mu)$.
2. *If* Ω *is a locally compact metric space, then* $C_c(\Omega)$ *is dense in* $L^p(\Omega, \mu)$.

In the special situation of $L^2(\Omega, \mu)$, we have more structure. In addition to the norm $\|f\|_2$, we can define an *inner product*

$$\langle f, g \rangle = \int_\Omega f g \, d\mu. \tag{B.11}$$

By the Cauchy-Bunyakovskii-Schwarz inequality, we know that $\langle f, g \rangle \in \mathbb{R}$ if $f, g \in L^2(\Omega, \mu)$, so this is well-defined. We have the following properties.

Proposition B.35. *The inner product satisfies the following properties for all* $f, g, h \in L^2(\Omega, \mu)$ *and constants* α:

1. $\langle f, g \rangle = \langle g, f \rangle$.
2. $\langle \alpha f, g \rangle = \alpha \langle f, g \rangle$.
3. $\langle f + g, h \rangle = \langle f, h \rangle + \langle g, h \rangle$.
4. $\langle f, f \rangle \geq 0$ *and* $\langle f, f \rangle = 0$ *if and only if* $f(x) = 0$ *for* μ *almost all* x.

Appendix C
Basic Notions from Set-Valued Analysis

This appendix will be devoted to introducing of the basic properties and results of set-valued analysis. For a more complete introduction, see [9].

C.1 Basic definitions

Given two metric spaces $\mathbb{X}$ and $\mathbb{Y}$, a *multifunction* is a set-valued mapping (i.e. a map $F : \mathbb{X} \rightrightarrows \mathbb{Y}$ from the space $\mathbb{X}$ to the power set $2^{\mathbb{Y}}$). We recall that the graph of F is the following subset of $\mathbb{X} \times \mathbb{Y}$:

$$\text{graph } F = \{(x, y) \in \mathbb{X} \times \mathbb{Y} : y \in F(x)\}.$$

If $F(x)$ is a closed, compact or convex set for each $x \in \mathbb{X}$, we say that F is *closed, compact, or convex valued*, respectively. There are two ways to define the inverse image under a multifunction F of a subset S:

1. $F^{-1}(S) = \{x \in \mathbb{X} : F(x) \cap S \neq \emptyset\}$.
2. $F^{+1}(S) = \{x \in \mathbb{X} : F(x) \subseteq S\}$.

The subset $F^{-1}(S)$ is called the *inverse image* of S under F and F^{+1} is called the *core* of S under F.

Let $(\mathbb{X}, \mathcal{B}, \mu)$ be a finite measure space; a multifunction $F : \mathbb{X} \rightrightarrows \mathbb{Y}$ is said to be *measurable* if for each open $O \subset \mathbb{Y}$ we have

$$F^{-1}(O) = \{x \in \mathbb{X} : F(x) \cap O \neq \emptyset\} \in \mathcal{B}.$$

A function $f : \mathbb{X} \to \mathbb{Y}$ is a *selection* or *selector* of F if $f(x) \in F(x)$, $\forall x \in \mathbb{X}$.

A multifunction $F : \mathbb{X} \rightrightarrows \mathbb{X}$ is said to be *Lipschitzian* if there exists a $K > 0$ such that $d_{\mathbb{H}}(F(x), F(y)) \leq Kd(x, y)$ for all $x, y \in \mathbb{X}$, where $d_{\mathbb{H}}$ denotes the Hausdorff metric.

If F is Lipschitz and $\mathbb{Y}$ is an inner product space, then the functions $x \to \mathrm{supp}\,(p, F(x))$, where

$$\mathrm{supp}\,(p, F(x)) = \sup_{s \in F(x)} p \cdot s,$$

are also Lipschitz with the same constant K for any $p \in S^{d-1}$, where

$$S^{d-1} = \{y \in \mathbb{Y} : \|y\| = 1\}.$$

C.1.1 Contractive multifunctions, fixed points, and collage theorems

A Lipschitz multifunction with $K \in [0, 1)$ is said to be a *contraction*. For a contractive multifunction, there exists a (not necessarily unique) *fixed point* $\bar{x} \in \mathbb{X}$ that satisfies the *inclusion relation* $\bar{x} \in F(\bar{x})$. Note that the fixed point $\bar{x}$ is not necessarily unique. As a simple example, $\mathbb{X} = \mathbb{Y} = \mathbb{R}$ and $F(x) = [0, 1]$ has fixed point $\bar{x}$ for any $\bar{x} \in [0, 1]$.

Theorem C.1. *Let $(\mathbb{X}, d)$ be a complete metric space and $F : \mathbb{X} \to \mathbb{H}(\mathbb{X})$ be a contractive multifunction such that $d_{\mathbb{H}}(F(x), F(y)) \leq cd(x, y)$ for all $x, y \in \mathbb{X}$ with $c \in [0, 1)$. Then:*

1. *For all $x_0 \in \mathbb{X}$, there exists a point $\bar{x} \in \mathbb{X}$ such that $x_{n+1} = P(x_n) \to \bar{x}$ when $n \to +\infty$, where $P(x) = \pi_x(F(x))$ is the projection of x onto $F(x)$.*
2. *$\bar{x}$ is a fixed point; that is, $\bar{x} \in F(\bar{x})$.*

Theorem C.2. *Given a contractive multifunction $F : \mathbb{X} \rightrightarrows \mathbb{X}$, let*

$$\mathbb{X}_F = \{x \in \mathbb{X} : x \in F(x)\}$$

be the set of all fixed points of F. If $(\mathbb{X}, d)$ is a complete (compact) metric space, then $\mathbb{X}_F$ is complete (compact).

Theorem C.3. *(Collage and anti-collage) Let $(\mathbb{X}, d)$ be a complete metric space and $F : \mathbb{X} \to \mathbb{H}(\mathbb{X})$ a contraction multifunction with contractivity factor $c \in [0, 1)$. The following properties hold:*

1. *(Generalized collage theorem) For all $x \in \mathbb{X}$, there exists a fixed point $\bar{x}$ such that*

$$d(x_0, \bar{x}) \leq \frac{d(x, F(x))}{1-c},$$

so that

$$d(x, \mathbb{X}_F) \leq \frac{d(x, F(x))}{1-c}.$$

2. *(Generalized anti-collage theorem) $d(x, F(x)) \leq (1+c)d(x, \mathbb{X}_F)$.*

If our multifunctions have compact values, a natural metric between two multifunctions is to use a supremum norm. Specifically, let $(\mathbb{X}, d)$ be a complete metric space and $F_1, F_2 : \mathbb{X} \rightrightarrows \mathbb{X}$ be two multifunctions on $\mathbb{X}$. We define the following distance between F_1 and F_2:

$$d_\infty(F_1, F_2) = \sup_{x \in \mathbb{X}} d_{\mathbb{H}}(F_1(x), F_2(x)).$$

Theorem C.4. *(Continuity of fixed-point sets) Suppose that $\mathbb{X}_{F_1}$ and $\mathbb{X}_{F_2}$ are compact sets and F_1 and F_2 are compact-valued contractive multifunctions with contractivities c_1 and c_2. Then*

$$d_{\mathbb{H}}(\mathbb{X}_{F_1}, \mathbb{X}_{F_2}) \leq \frac{d_\infty(F_1, F_2)}{1 - \min\{c_1, c_2\}}.$$

Let us suppose that $(\mathbb{X}, d)$ is a compact metric space. Starting from a multifunction $F : \mathbb{X} \to \mathbb{H}(\mathbb{X})$, it is quite natural to define an associated multifunction map F^* on $\mathbb{H}(\mathbb{X})$ as follows:

$$F^*(A) := \bigcup_{a \in A} F(a), \quad \forall A \in \mathbb{H}(\mathbb{X}). \tag{C.1}$$

Then $F^* : \mathbb{H}(\mathbb{X}) \to \mathbb{H}(\mathbb{X})$ and $F^* : \mathbb{H}(\mathbb{X}) \to \mathbb{H}(\mathbb{X})$ is contractive in the $d_{\mathbb{H}}$ metric if F is a contraction. From Banach's theorem, there exists a unique fixed point $A \in \mathbb{H}(\mathbb{X})$ such that $F^*(A) = A$. What is the relation between the solution of the inclusion $a \in F(a)$ and the equation $A = F^*(A)$? This relation is simple. Suppose that $a \in F(a)$ and $A = F^*(A)$. Then $a \in A$. Thus the fixed set of F^* contains all the fixed points of the multifunction F.

C.1.2 Convexity and multifunctions

Let us now suppose that $\mathbb{X}$ and $\mathbb{Y}$ are two Banach spaces and that $\mathbb{Y}$ is ordered by a pointed convex cone C (which means that we consider $y_1 \leq y_2$ if $y_2 - y_1 \in C$). A multifunction $F : \mathbb{X} \rightrightarrows \mathbb{Y}$ is said to be C-convex if

$$tF(x) + (1 - t)F(y) \subseteq F(tx + (1 - t)y) + C$$

whenever $x, y \in \mathbb{X}$ and $t \in (0, 1)$. When $F : \mathbb{X} \to \mathbb{Y}$ is a point-to-point mapping then this definition is the classical notion of C-convexity.

A point (u_0, f_0) is a *local weak minimum* for the multifunction F if $f_0 \in F(u_0)$ and there exists a neighbourhood U of u_0 such that

$$F(x) \subset f_0 + (-\operatorname{int} C)^c$$

for all $x \in U$. Let $\mathbb{X}$ be a convex set and $F : \mathbb{X} \rightrightarrows \mathbb{Y}$ be a C-convex multifunction. It is immediate to prove that if (u_0, f_0) with $f_0 \in F(u_0)$ is a local weak minimum of F, then it is a global weak minimum.

C.2 Integral of multifunctions

We now recall the definition of the integral of multifunctions. Let $(\mathbb{X}, \mathcal{B}, \mu)$ be a measure space and $\mathbb{Y}$ be a normed space. We suppose that $F : \mathbb{X} \rightrightarrows \mathbb{Y}$ is a multifunction that takes nonempty, convex, and compact subsets of $\mathbb{Y}$ as values. We denote by $\mathcal{F}(X, Y)_\mu$ the space of all integrable selections of F; that is

$$\mathcal{F}(X, Y)_\mu = \{f \in L^1(\mathbb{X}, \mathbb{Y}) : f(x) \in F(x), \text{ a.e. } x \in \mathbb{X}\}.$$

A multifunction F is said to be *integrably bounded* if there exists a nonnegative function $k \in L^1(\mathbb{X})$ such that

$$F(x) \subset k(x)B_1(0)$$

for a.e. $x \in \mathbb{X}$. Clearly, in this case all measurable selections are integrable. Notice that if F is Lipschitz and either F or μ has bounded support, then F is integrably bounded as well.

We give the definition of the integral of a multifunction F in the Aumann sense (see Section 8.6 in [9]). The integral of F on $\mathbb{X}$ is the set of integrals of integrable selections of F; that is,

$$\int_{\mathbb{X}} F \, d\mu = \left\{ \int_{\mathbb{X}} f \, d\mu : f \in \mathcal{F}(X,Y)_{\mu} \right\}.$$

The following result states some properties of integrals of multifunctions.

Theorem C.5. *(Properties of multifunction integration)*

- *Let us consider measurable, integrably bounded set-valued maps F_i : $\mathbb{X} \rightrightarrows \mathbb{Y}$ and set $G(x) = \overline{F_1(x) + F_2(x)}$. Then*

  - $\overline{\int_{\mathbb{X}} G d\mu} = \overline{\int_{\mathbb{X}} F_1 \, d\mu + \int_{\mathbb{X}} F_2 \, d\mu}$ *and*
  - *for all $p \in \mathbb{Y}^*$, supp $\left(\int_{\mathbb{X}} F \, d\mu, p \right) = \int_{\mathbb{X}} \text{supp}\,(F(x), p) \, d\mu(x)$.*

- *Suppose F is an integrably bounded multifunction. Then $\int_{\mathbb{X}} F(x) \, d\mu(x)$ is a bounded set.*
- *We have $d_{\mathbb{H}}(\int_{\mathbb{X}} F \, d\mu, \int_{\mathbb{X}} G \, d\mu) \leq \int_{\mathbb{X}} d_{\mathbb{H}}(F(x), G(x)) \, d\mu(x)$.*
- *Suppose μ_n converges to μ in the Monge-Kantorovich metric and F is Lipschitz with constant K. Then, for all $\epsilon > 0$, there exists N such that $\forall n \geq N$,*

$$\int_{\mathbb{X}} \text{supp}\,(p, F(x)) d(\mu_n(x) - \mu(x)) < K\epsilon,$$

  for all $p \in S^{d-1}$.
- *Suppose $F : \mathbb{X} \rightrightarrows \mathbb{Y}$ is a Lipschitz multifunction and $\mu_n \to \mu$ in the Monge-Kantorovich metric on $\mathbb{X}$. Then we also have that supp $(p, \int_{\mathbb{X}} F \, d\mu_n) \to$ supp $(p, \int_{\mathbb{X}} F \, d\mu)$, and this convergence is uniform with respect to $p \in S^{d-1}$.*

References

[1] S.K. Alexander. *Multiscale approaches in image modelling and image processing*. PhD thesis, Department of Applied Mathematics, University of Waterloo, 2005.

[2] S.K. Alexander, E.R. Vrscay, and S. Tsurumi. A simple, general model for the affine self-similarity of images. In A. Campilho and M. Kamel, editors, *Image Analysis and Recognition*, volume 5112 of Lecture Notes in Computer Science, pages 192–203, Springer-Verlag, Berlin, 2008.

[3] C.D. Aliprantis and R. Tourky. *Cones and Duality*, volume 84 of Graduate Studies in Mathematics. AMS, Providence, 2007.

[4] A. Amini-Harandi and M. Fakhar. Fixed point theory in cone metric spaces obtained via the scalarization method. *Comput. Math. Appl.*, 59:3529–3534, 2010.

[5] L. Andersson. Inverse eigenvalue problems for a Sturm-Liouville equation in impedance form. *Inverse Probl.*, 4:927–971, 1988.

[6] Z. Artstein. Set-valued measures. *Trans. Amer. Math. Soc.*, 165:103–125, 1972.

[7] M. Asadi, S.M. Vaezpour, and H. Soleimani. Metrizability of cone metric spaces. Technical report, arXiv:1102.2353v1 [math.FA], 2011.

[8] R.B. Ash. *Probability and Measure Theory*, 2nd edition. Harcourt, San Diego, 2000.

[9] J.P. Aubin and H. Frankowska. *Set-Valued Analysis*. Birkhauser, Boston, 1990.

[10] M. Bajraktarevic. Sur une équation fonctionelle. *Glas. Mat. Fiz. Astron.*, 12:201–205, 1957.

[11] S. Banach. Sur les opérations dans les ensembles abstraits et leurs applications aux équations intégrales. *Fundam. Math.*, 3:133–

181, 1922.

[12] M. Barnsley, M. Berger, and H. Soner. Mixing Markov chains and their images. *Prob. Eng. Inf. Sci.*, 2:387–414, 1988.

[13] M. Barnsley, S. Demko, J. Elton, and J. Geronimo. Invariant measures for Markov processes arising from iterated function systems with place-dependent probabilities. *Ann. Inst. Henri Poincare*, 24(3):367–394, 1988.

[14] M. Barnsley, J. Elton, and D. Hardin. Recurrent iterated function systems. *Constr. Approx.*, 5:3–31, 1989.

[15] M.F. Barnsley. Fractal functions and interpolation. *Constr. Approx.*, 2:303–329, 1986.

[16] M.F. Barnsley. *Fractals Everywhere*. Academic Press, New York, 1988.

[17] M.F. Barnsley and S. Demko. Iterated function systems and the global construction of fractals. *Proc. Roy. Soc. London Ser. A*, 399:243–275, 1985.

[18] M.F. Barnsley, V. Ervin, D. Hardin, and J. Lancaster. Solution of an inverse problem for fractals and other sets. *Proc. Natl. Acad. Sci.*, 83:1975–1977, 1985.

[19] M.F. Barnsley and L.P. Hurd. *Fractal Image Compression*. AK Peters, Ltd., Wellesley, MA., 1992.

[20] K. Baron and A. Lasota. Markov operators on the space of vector measures: coloured fractals. *Ann. Polon. Math.*, 69(3):217–234, 1998.

[21] M. Berger. Wavelets as attractors of random dynamical systems. In E. Mayer-Wolf, Ely Merzbach, and Adam Shwartz, editors, *Stochastic Analysis: Liber Amicorum for Moshe Zakai*, pages 75–90. Academic Press, Boston, 1991.

[22] M. Berger. Random affine iterated function systems: Curve generation and wavelets. *SIAM Rev.*, 34(3):361–385, 1992.

[23] M. Berger. Random affine iterated function systems: Mixing and encoding. In *Diffusion Processes and Related Problems in Analysis, Vol II: Stochastic Flows*, number 27 in Progress in Probability, pages 315–346. Birkhauser, Boston, 1992.

[24] M. Berger and Y. Wang. Multidimensional two-scale dilation equations. In C.K. Chui, editor, *Wavelets – A Tutorial in Theory and Applications*, pages 295–323. Academic Press, Boston, 1992.

[25] A. Bielecki. Une remarque sur la méthode de Banach-Caccioppoli-Tikhonov dans la théorie des équations differentielles ordinaires. *Bull. Acad. Polon Sci. Cl. III*, 4:261–264, 1956.

[26] H.G. Bock. Recent advances in parameter identification for ordinary differential equations. In P. Deuflhard and E. Hairer, edi-

tors, *Progress in Scientific Computing*, pages 95–121, Birkhäuser, Boston, 1983.

[27] E. Bolker. A class of convex bodies. *Trans. Amer. Math. Soc.*, 145:312–319, 1969.

[28] R.I. Bot, S. Grad, and G. Wanka. *Duality in Vector Optimization*. Springer, Berlin, 2009.

[29] D. Brunet, E.R. Vrscay, and Z. Wang. A class of image metrics based on the structural similarity quality index. In M. Kamel and A.C. Campilho, editors, *International Conference on Image Analysis and Recognition (ICIAR '11), Burnaby, BC, Canada*, volume 6753 of Lecture Notes in Computer Science, page 100–110, Springer, Berlin, June 22–24 2011.

[30] D. Brunet, E.R. Vrscay, and Z. Wang. Structural similarity-based affine approximation and self-similarity of images revisited. In M. Kamel and A.C. Campilho, editors, *International Conference on Image Analysis and Recognition (ICIAR '11), Burnaby, BC, Canada*, volume 6753 of Lecture Notes in Computer Science, page 264–275, Springer, Berlin, June 22–24 2011.

[31] A. Buades, B. Coll, and J. Morel. Image denoising methods: A new nonlocal principle. *SIAM Rev.*, 52(1):113–147, 2005. This is an updated version of the authors' original paper, "A review of image denoising algorithms, with a new one," *Multiscale Model. Simul.*, 4, 490–530, 2005.

[32] A. Bucci and D. La Torre. Population and economic growth with human and physical capital investments. *Int. Rev. Econ.*, 56(1):17–27, 2009.

[33] E.M. Cabaña. On the vibrating string forced by white noise. *Z. Wahrscheinlichkeitstheorie verw. Geb.*, 15:111–130, 1970.

[34] C. Cabrelli, C. Heil, and U. Molter. Accuracy of lattice translates of several multidimensional refinable functions. *J. Approx. Theory*, 95(1):5–52, 1998.

[35] C. Cabrelli, C. Heil, and U. Molter. Accuracy of several multidimensional refinable distributions. *J. Fourier Anal. Appl.*, 6(5):483–502, 2000.

[36] C.A. Cabrelli, B. Forte, U.M. Molter, and E.R. Vrscay. Iterated fuzzy set systems: A new approach to the inverse problem for fractals and other sets. *J. Math. Anal. Appl.*, 171:79–100, 1992.

[37] H. Cakalli, A. Sonmez, and C. Genc. Metrizability of topological vector space valued cone metric spaces. Technical report, arXiv:1007.3123v2 [math.GN], 2010.

[38] V. Capasso, H.W. Engl, and S. Kindermann. Parameter identification in a random environment exemplified by a multi-

scale model for crystal growth. *SIAM Multiscale Model. Simul.*, 7(2):814–841, 2008.

[39] P. Centore and E.R. Vrscay. Continuity of attractors and invariant measures of iterated function systems. *Can. Math. Bull.*, 37:315–329, 1994.

[40] E.A. Coddington and N. Levinson. *Theory of Ordinary Differential Equations*. McGraw-Hill, New York, 1955.

[41] C. Colapinto and D. La Torre. *Mathematical and Statistical Methods in Insurance and Finance*, pages 83–90. Springer, New York, 2006.

[42] J. Conway. *A Course in Functional Analysis*, volume 96 of Graduate Texts in Mathematics. Springer-Verlag, New York, 1990.

[43] E.T. Copson. *Metric Spaces*. Cambridge University Press, Cambridge, 1968.

[44] I. Daubechies. *Ten Lectures on Wavelets*. CBMS-NSF Regional Conference Series in Applied Mathematics. Society for Industrial and Applied Mathematics, Philadelphia, 1992.

[45] I. Daubechies and J. Lagarias. Two-scale difference equations II: Local regularity, infinite products of matrices and fractals. *SIAM J. Math. Anal.*, 23:1031–1079, 1992.

[46] G. Davis. Self-quantized wavelet subtrees: A wavelet-based theory for fractal image compression. In H. Szu, editor, *1995 Data Compression Conference Proceedings*, volume 2491, page 232–241, SPIE, Orlando, 1995.

[47] P. Diaconis and D. Freedman. Iterated random functions. *SIAM Rev.*, 41:41–76, 1999.

[48] J. Diestel and J.J. Uhl Jr. *Vector Measures*. Number 15 in Mathematical Surveys. AMS, Providence, 1977.

[49] W.S. Du. A note on cone metric fixed point theory and its equivalence. *Nonlin. Anal.*, 72:2259–2261, 2010.

[50] R.M. Dudley. *Real Analysis and Probability*, 2nd edition. Cambridge University Press, Cambridge, 2002.

[51] N. Dunford and J.T. Schwartz. *Linear Operators Part I*. Interscience, New York, 1967.

[52] G. Edgar. *Measure, Topology, and Fractal Geometry*, 2nd edition. Undergraduate Texts in Mathematics. Springer, New York, 2008.

[53] J. Elton. An ergodic theorem for iterated maps. *J. Erg. Theory Dyn. Sys.*, 7:481–488, 1987.

[54] J. Elton. A multiplicative ergodic theorem for Lipschitz maps. *Stochastic Process. Appl.*, 34(1):39–47, 1990.

[55] K. Falconer. *The Geometry of Fractal Sets.* Cambridge University Press, Cambridge, 1985.

[56] K. Falconer. *Fractal Geometry: Mathematical Foundations and Applications.* Wiley, New York, 1990.

[57] K. Falconer. *Techniques in Fractal Geometry.* Wiley, New York, 1997.

[58] K. Falconer and T. O'Neil. Vector-valued multifractal measures. *Proc. Roy. Soc. London Ser. A*, 452(1949):1433–1457, 1996.

[59] R.S. Falk. Error estimate for the numerical identification of a variable coefficient. *Math. Comput.*, 40:537–546, 1983.

[60] Y. Fisher. *Fractal Image Compression, Theory and Applications.* Springer-Verlag, New York, 1995.

[61] B. Forte and F. Mendivil. A classical ergodic property for IFS: A simple proof. *J. Erg. Theory Dyn. Sys.*, 18(3):609–611, 1998.

[62] B. Forte, F. Mendivil, and E.R. Vrscay. 'Chaos games' for iterated function systems with grey level maps. *SIAM J. Math. Anal.*, 29(4):878–890, 1998.

[63] B. Forte, F. Mendivil, and E.R. Vrscay. IFS operators on integral transforms. In M. Dekking, J. Levy-Vehel, E. Lutton, and C. Tricot, editors, *Fractals: Theory and Applications in Engineering*, Springer-Verlag, New York, 1999.

[64] B. Forte and E.R. Vrscay. Solving the inverse problem for function and image approximation using iterated function systems. *Dyn. Cont. Impul. Sys.*, 1(2):177–231, 1995.

[65] B. Forte and E.R. Vrscay. Solving the inverse problem for measures using iterated function systems: A new approach. *Adv. Appl. Prob.*, 27:800–820, 1995.

[66] B. Forte and E.R. Vrscay. Inverse problem methods for generalized fractal transforms. In Y. Fisher, editor, *Fractal Image Coding and Analysis*, volume 159 of NATO ASI Series F, Springer-Verlag, Berlin, 1998.

[67] B. Forte and E.R. Vrscay. Theory of generalized fractal transforms. In Y. Fisher, editor, *Fractal Image Coding and Analysis*, volume 159 of NATO ASI Series F, Springer-Verlag, Berlin, 1998.

[68] A. Friedman and F. Reitich. Analysis of a mathematical model for the growth of tumors. *J. Math. Biol.*, 38:262–284, 1999.

[69] E.O. Frind and G.F. Pinder. Galerkin solution of the inverse problem for aquifer transmissivity. *Water Res.*, 9:1297–1410, 1973.

[70] L. Fumanelli, M. Iannelli, H.A. Janjua, and O. Jousson. Mathematical modeling of bacterial virulence and host-pathogen inter-

actions in the Dictyostelium/Pseudomonas system. *J. of Theor. Biol.*, 270:19–24, 2011.

[71] P.R. Halmos. *Lectures on Ergodic Theory*. Chelsea, New York, 1956.

[72] L.G. Hanin. Kantorovich-Rubinstein norm and its application in the theory of Lipschitz spaces. *Proc. Amer. Math. Soc.*, 115(2):345–352, 1992.

[73] L.G. Hanin. An extension of the Kantorovich norm. In L.A. Caffarelli and M. Milman, editors, *Monge-Ampere Equation: Applications to Geometry and Optimization*, volume 226 of *Contemporary Mathematics*, pages 113–130, American Mathematical Society, Providence, RI, 1999.

[74] F. Hiai. Radon-Nikodym theorems for set-valued measures. *J. Multivariate Anal.*, 8:96–118, 1978.

[75] J. Hutchinson. Fractals and self-similarity. *Indiana Univ. Math. J.*, 30:713–747, 1981.

[76] J. Hutchinson, D. La Torre, and F. Mendivil. IFS Markov operators on multimeasures. in progress.

[77] K. Ito and K. Kunisch. Augmented Lagrangian SQP-methods, Hilbert spaces and application to control in the coefficients. *SIAM J. Optim.*, 6:96–125, 1996.

[78] K. Ito, K. Kunisch, and Z. Li. Level-set function approach to an inverse interface problem. *Inverse Probl.*, 17(5):1225–1242, 2001.

[79] K. Ito and J. Zou. Identification of some source densities of the distribution type. *J. Comput. Appl. Math.*, 132:295–308, 2001.

[80] A. Jacquin. Image coding based on a fractal theory of iterated contractive image transformations. *IEEE Trans. Image Proc.*, 1:18–30, 1992.

[81] A.E. Jacquin. Fractal image coding based on a theory of iterated contractive image transformations. *Proc. SPIE*, 1360:227-239, 2005.

[82] S. Jaffard. Multifractal formalism for functions, i. *SIAM J. Math. Anal.*, 28:944–970, 1997.

[83] S. Karlin. Some random walks arising in learning models, i. *Pacific J. Math.*, 3:725–756, 1953.

[84] H. Keller. *Numerical Methods for Two-Point Boundary-Value Problems*. Dover, New York, 1992.

[85] Y.L. Keung and J. Zou. An efficient linear solver for nonlinear parameter identification problems. *SIAM J. Sci. Comput.*, 22:1511–1526, 2000.

[86] M. Kisielewicz. *Differential Inclusions and Optimal Control*. Kluwer Academic, Boston, 1990.

[87] G. Koshevoy. The Lorenz zonotope and multivariate majorizations. *Soc. Choice Welf.*, 15(1):1–14, 1998.

[88] A.S. Kravchenko. Completeness of the space of separable measures in the Kantorovich-Rubinshtein metric. *Sib. Math. J.*, 47(1):68–76, 2006.

[89] H. Kunze and S. Gomes. Solving an inverse problem for Urison-type integral equations using Banach's fixed point theorem. *Inverse Probl.*, 19:411–418, 2003.

[90] H. Kunze and K. Heidler. The collage coding method and its application to an inverse problem for the Lorenz system. *J. Appl. Math. Comput.*, 186(1):124–129, 2007.

[91] H. Kunze, J. Hicken, and E.R. Vrscay. Inverse problems for ODEs using contraction maps: suboptimality of the "collage method". *Inverse Probl.*, 20:977–991, 2004.

[92] H. Kunze and S. Murdock. Solving inverse two-point boundary value problems using collage coding. *Inverse Probl.*, 22:1179–1190, 2006.

[93] H. Kunze and D. La Torre. Solving inverse problems for differential equations by the collage method and applications to economics. *Int. J. Optim. Theory, Methods, Appl.*, 1(4):26–35, 2009.

[94] H. Kunze, D. La Torre, and K. Levere. An inverse problem for the Hammerstein integral equation with application to imaging. In *10th European Congress of Stereology and Image Analysis Proceedings*, 2009.

[95] H. Kunze, D. La Torre, K. Levere, and E.R. Vrscay. Solving inverse problems for the Hammerstein integral equation and its random analog using the 'collage method' for fixed points. *Int. J. Pure Appl. Math.*, 60(4):393–408, 2010.

[96] H. Kunze, D. La Torre, and E.R. Vrscay. Contractive multifunctions, fixed point inclusions and iterated multifunction systems. *J. Math. Anal. Appl.*, 330(1):159–173, 2007.

[97] H. Kunze, D. La Torre, and E.R. Vrscay. Random fixed point equations and inverse problems using the collage method for contraction mappings. *J. Math. Anal. Appl.*, 334:1116–1129, 2007.

[98] H. Kunze, D. La Torre, and E.R. Vrscay. Solving inverse problems for random equations and applications. In G. Dulikravich, editor, *Inverse Problems and Dynamic Optimization Proceedings*, 2007.

[99] H. Kunze, D. La Torre, and E.R. Vrscay. From iterated function systems to iterated multifunction systems. *Commun. Appl. Nonlinear Anal.*, 15(4):1–13, 2008.

[100] H. Kunze and S. Vasiliadis. Using the collage method to solve ODEs inverse problems with multiple datasets. *Nonlinear Anal. Theory, Methods and Appl.*, 71:1298–1306, 2009.

[101] H. Kunze and E.R. Vrscay. Solving inverse problems for ordinary differential equations using the Picard contraction mapping. *Inverse Probl.*, 15:745–770, 1999.

[102] H. Kunze and E.R. Vrscay. Using the Picard contraction mapping to solve inverse problems for ordinary differential equations. In M.F. Barnsley, D. Saupe, and E.R. Vrscay, editors, *Proceedings of the Fractals in Multimedia Symposium*, volume 132 of The IMA Volumes in Mathematics and its Applications, pages 157–174, New York, Springer, 2001.

[103] M. Kwieciński. A note on continuity of fixed points. *Univ. Iagel. Acta Math.*, 29:19–24, 1992.

[104] D. La Torre and S. Marsiglio. Endogenous technological progress in a multi-sector growth model. *Econ. Model.*, 27(5):1017–1028, 2010.

[105] D. La Torre and F. Mendivil. Iterated function systems on multifunctions and inverse problems. *J. Math. Anal. Appl.*, 340(2):1469–1479, 2008.

[106] D. La Torre and F. Mendivil. Union-additive multimeasures and self-similarity. *Commun. Math. Anal.*, 7(2):51–61, 2009.

[107] D. La Torre, F. Mendivil, and E.R. Vrscay. Iterated function systems on multifunctions. In G.Aletti et al, editor, *Math Everywhere: Deterministic and Stochastic Modeling in Biomedicine, Economics and Industry*, pages 125–138. Springer, New York, 2006.

[108] D. La Torre and E.R. Vrscay. A generalized fractal transform for measure-valued images. *Nonlinear Analysis: Theory, Methods and Applications*, 71(12):e1598–e1607, 2009.

[109] D. La Torre and E.R. Vrscay. Fractal-based measure approximation with entropy maximization and sparsity constraints. In *MaxEnt 2011, Waterloo, Canada*, July 10–15, 2011.

[110] D. La Torre and E.R. Vrscay. Fractal transforms and self-similarity: Recent results and applications. *Image Anal. Stereol.*, 18(2):63–79, 2011.

[111] D. La Torre and E.R. Vrscay. Random measure-valued image functions, fractal transforms and self-similarity. *Appl. Math. Lett.*, 24(8):1405–1410, 2011.

[112] D. La Torre, E.R. Vrscay, M. Ebrahimi, and M. Barnsley. Measure-valued images, associated fractal transforms and the

affine self-similarity of images. *SIAM J. Imaging Sci.*, 2(2):470–507, 2009.

[113] K. Leeds. *Dilation Equations with Matrix Dilations.* PhD thesis, Georgia Institute of Technology, 1997.

[114] G. Lewellen. Self-similarity. *Rocky Mountain J. Math.*, 23(3):1023–1040, 1993.

[115] S.D. Lin. A common fixed point theorem in abstract spaces. *Ind. J. Pure Appl. Math.*, 18(8):685–690, 1987.

[116] *LNLA 2008, The 2008 International Workshop on Local and Non-Local Approximation in Image Processing*, Lausanne, Switzerland, August 23-24 2008. http://www.eurasip.org/Proceedings/Ext/LNLA2008/contents.html.

[117] H. Long-Guang and Z. Xian. Cone metric spaces and fixed point theorems of contractive mappings. *J. Math. Anal. Appl.*, 33:1468–1476, 2007.

[118] E. Lorenz. Deterministic nonperiodic flow. *J. Atmos. Sci.*, 20:130–141, 1963.

[119] A.J. Lotka. *Elements of Physical Biology.* Williams and Wilkins, Baltimore, 1925.

[120] N. Lu. *Fractal Imaging.* Academic Press, New York, 1997.

[121] D.T. Luc. *Theory of Vector Optimization.* Springer-Verlag, Berlin, 1989.

[122] M. MacClure. Generating self-affine tiles and their boundaries. *Mathematica J.*, 11(1):4–31, 2008.

[123] S. Mallat. *A Wavelet Tour of Signal Processing,* 2nd edition. Academic Press, Burlington, MA, 2009.

[124] B. Mandelbrot. *The fractal geometry of nature.* WH Freeman, 1982.

[125] S. Marsili-Libelli. Parameter estimation of ecological models. *Ecol. Model.*, 62:233–258, 1992.

[126] P. Mattila. *Geometry of Sets and Measures on Euclidean Spaces.* Cambridge University Press, Cambridge, 1995.

[127] R.D. Mauldin. Infinite iterated function systems: Theory and applications. *Prog. Prob.*, 37:91–110, 1995.

[128] R.D. Mauldin and S.C. Williams. Hausdorff dimension in graph directed constructions. *Trans. Amer. Math. Soc.*, 309(2):811–829, 1988.

[129] G.S. Mayer. *Resolution enhancement in magnetic resonance imaging by frequency extrapolation.* PhD thesis, Department of Applied Mathematics, University of Waterloo, 2008.

[130] G.S. Mayer and E.R. Vrscay. Iterated fourier transform systems: A method for frequency extrapolation. In M. Kamel

and A. Campilho, editors, *Image Analysis and Recognition*, volume 4633 of Lecture Notes in Computer Science, pages 728–739, Springer, Berlin, 2007.

[131] G.S. Mayer, E.R. Vrscay, M.L. Lauzon, B.G. Goodyear, and J.R. Mitchell. Self-similarity of images in the fourier domain, with applications to mri. In M. Kamel and A. Campilho, editors, *Image Analysis and Recognition*, volume 5112 of *Lecture Notes in Computer Science*, pages 43–52, Springer, Berlin, 2008.

[132] F. Mendivil. A generalization of ifs with probabilities to infinitely many maps. *Rocky Mountain J. Math.*, 28(3):1043–1051, 1998.

[133] F. Mendivil and J. Silver. Chaos games for wavelet analysis and wavelet synthesis. *Comput. Math. Appl.*, 46(1):45–59, 2003.

[134] F. Mendivil and E.R. Vrscay. Correspondence between fractal-wavelet transforms and iterated function systems with grey-level maps. In J. Levy-Vehel, E. Lutton, and C. Tricot, editors, *Fractals in Engineering: From Theory to Industrial Applications*, pages 54–64, Springer Verlag, London, 1997.

[135] F. Mendivil and E.R. Vrscay. Fractal vector measures and vector calculus on planar fractal domains. *Chaos, Solitons Fractals*, 14:1239–1254, 2002.

[136] C.A. Micchelli and H. Prautzsch. Uniform refinement of curves. *Linear Algebra Appl.*, 114/115:841–870, 1989.

[137] S.B. Nadler. Multi-valued contraction mappings. *Pacific J. Math.*, 30(2):475–487, 1969.

[138] E. Orsinger. Randomly forced vibrations of a string. *Ann. Inst. Henri Poincaré (B) Prob. Stat.*, 18(4):367–394, 1982.

[139] N. Papageorgiou. Radon-nikodym theorems for multimeasures and transition multimeasures. *Proc. Amer. Math. Soc.*, 111(2):465–474, 1991.

[140] A. Perazzolo. IDFS, un nuovo approccio alla compressione e decompressione dei dati immagine. Tesi di laurea in scienze dell'informazione, Facoltà di Scienze, M.F.N., Università degli Studi di Verona, 1997.

[141] K. Petersen. *Ergodic Theory*. Cambridge University Press, Cambridge, 1983.

[142] M.P. Polis and R.E. Goodson. Parameter identification in distributed systems: A synthesizing overview. *Proc. IEEE*, 64:45–61, 1976.

[143] A.H. Read. The solution of a functional equation. *Proc. Roy. Soc. Edinburgh*, A, 63:336–345, 1951.

[144] G.R. Richter. An inverse problem for the steady state diffusion equation. *SIAM J. Appl. Math.*, 41:210–221, 1981.

[145] A. Robert. *A course in p-adic analysis*, volume 198 of Graduate Texts in Mathematics. Springer-Verlag, New York, 2000.

[146] R.T. Rockafellar and R.J-B. Wets. *Variational Analysis.* Springer-Verlag, New York, 1998.

[147] C.A. Rogers. *Hausdorff Measures.* Cambridge University Press, Cambridge, 1970.

[148] D. Romani. Transformata frattale IDFS su uno schema finito. Tesi di laurea in scienze dell'informazione, Facoltà di Scienze, M.F.N., Università degli Studi di Verona, 1997.

[149] C. Romero. *Handbook of Critical Issues in Goal Programming.* Pergamon Press, New York, 1991.

[150] H.L. Royden. *Real Analysis,* 3rd edition. Macmillan, New York, 1988.

[151] M. Ruhl and H. Hartenstein. Optimal fractal coding is NP-hard. In J. Storer and M. Cohn, editors, *Data Compression Conference'97,* pages 261–270, Snowbird, Utah, 1997.

[152] B. Rzepecki. On fixed point theorems of maia type. *Publications de Institut Mathematique,* 28(42):179–186, 1980.

[153] D. Saupe, R. Hamzaoui, and H. Hartenstein. Fractal image compression – an introductory overview. Technical report, Albert-Ludwigs University at Freiburg, 1997.

[154] A. Sitz, U. Schwarz, J. Kurths, and H.U. Voss. Estimation of parameters and unobserved components for nonlinear systems from noisy time series. *Phys. Rev. E,* 66(016210):1–9, 2002.

[155] D.R. Smart. *Fixed Point Theorems.* Cambridge University Press, London, 1974.

[156] M.W. Smiley and S.R. Proulx. Gene expression dynamics in randomly varying environments. *J. Math. Biol.,* 61:231–251, 2010.

[157] A. Sterna-Karwat. Remarks on convex cones. *J. Optim. Theory Appl.,* 59:335–340, 1988.

[158] G. Strang and T. Nguyen. *Wavelets and Filter Banks.* Wellesley-Cambridge Press, Cambridge, 1996.

[159] R.S. Strichartz and Y. Wang. Geometry of self-affine tiles. I. *Indiana Univ. Math. J.,* 48(1):1–23, 1999.

[160] K. Stromberg. *Probability for Analysts.* Chapman and Hall, New York, 1994.

[161] A.S. Sznitman. *Topics in Propagation of Chaos,* volume 1464 of Lecture Notes in Mathematics. Springer-Verlag, Berlin, 1991.

[162] C. Tricot and R. Riedi. Attracteurs, orbites et ergodicité. *Ann. Math. Blaise Pascal,* 6(1):55–72, 1999.

[163] A. van de Walle. Relating fractal compression to transform methods. Master of mathematics, Department of Applied Mathematics, University of Waterloo, 1995.

[164] A.M. Vershik. Kantorovich metric: Initial history and little-known applications. *J. Math. Sci.*, 133(4):1410–1417, 2006.

[165] C.R. Vogel. *Computational Methods for Inverse Problems.* SIAM, New York, 2002.

[166] E.R. Vrscay. A generalized class of fractal-wavelet transforms for image representation and compression. *Can. J. Elect. Comput. Eng.*, 23:69–84, 1998.

[167] E.R. Vrscay and C. Roehrig. Iterated function systems and the inverse problem of fractal construction using moments. In E. Kaltofen and S.M. Watt, editors, *Computers and Mathematics*, pages 250–259. Springer-Verlag, New York, 1989.

[168] P. Walters. *An Introduction to Ergodic Theory.* Springer-Verlag, New York, 1982.

[169] Y. Wang. Self-affine tiles (Hong Kong 1997). In *Advances in Wavelets*, pages 261–282. Springer-Verlag, Singapore, 1999.

[170] A. Wilansky. *Functional Analysis.* Blaisdell Publishing, New York, 1964.

[171] N. Wilansky. *Lipschitz Algebras.* World Scientific, Singapore, 1999.

[172] R.F. Williams. Composition of contractions. *Bol. Soc. Brasil. Mat.*, 2:55–59, 1971.

[173] Y.S. Yoon and W-G Yeh. Parameter identification in an inhomogeneous medium with the finite-element method. *Soc. Pet. Eng. J.*, 16:217–226, 1976.

[174] K. Yosida. *Lectures on Differential and Integral Equations.* Interscience, New York, 1960.

[175] A. Zygmund. *Trigonometric Series*, volume 1. Cambridge University Press, Cambridge, 1964.

Index